Trading Caterpillar Fungus in Tibet

International Institute for Asian Studies

Publications

The International Institute for Asian Studies is a research and exchange platform based in Leiden, the Netherlands. Its objective is to encourage the interdisciplinary and comparative study of Asia and to promote (inter)national cooperation. IIAS focuses on the humanities and social sciences and on their interaction with other sciences. It stimulates scholarship on Asia and is instrumental in forging research networks among Asia Scholars. Its main research interest are reflected in the three book series published with Amsterdam University Press: Global Asia, Asian Heritages and Asian Cities.

IIAS acts as an international mediator, bringing together various parties in Asia and other parts of the world. The Institute works as a clearinghouse of knowledge and information. This entails activities such as providing information services, the construction and support of international networks and cooperative projects, and the organisation of seminars and conferences. In this way, IIAS functions as a window on Europe for non-European scholars and contributes to the cultural rapprochement between Europe and Asia.

IIAS Publications Officer: Paul van der Velde
IIAS Assistant Publications Officer: Mary Lynn van Dijk

Trading Caterpillar Fungus in Tibet

When Economic Boom Hits Rural Area

Emilia Roza Sulek

Routledge
Taylor & Francis Group
LONDON AND NEW YORK

International Institute for Asian Studies

Publications

First published in 2019 by Amsterdam University Press Ltd.

Published 2025 by Routledge
4 Park Square, Milton Park, Abingdon, Oxon OX14 4RN
605 Third Avenue, New York, NY 10158

Routledge is an imprint of the Taylor & Francis Group, an informa business

ISBN: 9789462985261 (hbk)
ISBN: 9781041189701 (pbk)
ISBN: 9781003708452 (ebk)
NUR 740

Cover image: Bargaining over caterpillar fungus, Gabde County, Golok
Source: Emilia Roza Sulek

Cover design: Coördesign, Leiden

DOI: 10.5117/9789462985261

Table of Contents

List of Illustrations, Maps and Tables

Illustrations

Maps

Tables

All photos except Figure 30 by Emilia Roza Sulek.
Figure 30 by Sonamjid.
Figure 14 by Francesca La Vigna.
Maps by Maya Schneeberger (Map 4 after Daniel Winkler).

Acknowledgements

It had long been my dream to conduct research in Golok. Many things had to happen before this dream could come true and many people helped me along the way. In Warsaw, where this story began, I owe my gratitude to Agata Bareja-Starzyńska, Jerzy Sławomir Wasilewski and Anna Malewska-Szałygin for inspiring my interest in Asia and in social anthropology. The training I received at the University of Warsaw helped me discover that anthropology is what I wanted to do and that Central Asia would be my main destination. I also thank Adam Kozieł, who has been both a critical reader and a passionate supporter of my work, and Jerzy Bayer who kindly translated Chinese documents for this book and helped organize my first field research in China.

In Germany, I worked at the Humboldt University of Berlin, together with Toni Huber and Mona Schrempf, who supported me during my doctorate writing years. I also had the honour collaborating with Melvyn Goldstein. I am grateful for his concrete feedback and support, so important to young scholars. I also thank Daniel Winkler, a well-known scholar of caterpillar fungus, who was generous enough to share his data and discuss my research results. I owe much to my colleagues and friends: David Holler, with whom I have spent long hours talking about Tibetan pastoralism; Diana Lange, who kept telling me of the fantastic projects awaiting me after my doctorate; and Bianca Horlemann, with whom I had many inspiring discussions about Golok, our common region of interest. I owe much to the late Andreas Gruschke, who shared both his knowledge and valuable contacts in China. I also thank Saverio Krätli and other colleagues from the Commission on Nomadic Peoples and, most importantly, Philip Salzman for his friendship and support. I am also greatly indebted to Tsering Thar, who assisted me in many ways during my research, and Per Kvaerne for his support in getting my first research grant – it had a symbolic meaning for me and has not been forgotten.

Arriving in Golok for the first time, I was unaware of the scale of my project and I was only able to carry it out with the help of many people. Firstly, I thank Alexandru Anton-Luca for his friendship and for being my 'godfather' in Qinghai. I thank Anne-Laure Cromphout and Holly Galley for a crash course on how to survive fieldwork in Golok, and Angus Cargill for his hospitality. In addition, I would never have succeeded in conducting my research without my Tibetan friends and collaborators. I owe the most to Sonamjid and Danpa Darje, who gave me a home, worked with me and stood by my side in times of trouble. My thanks also to Ache Tseko, Churtod and his family, and Pamatso-lolo whose home was my second home/refuge. I thank Chukorjab and Yulha and their families, who I got to know well over the years and who, hopefully, will remain my friends for many more. I thank Gesar and his family for helping me through many difficulties. A person I always think about with respect and gratitude is Cholo, a charismatic leader of Yunzhi *dewa*, who honoured me with his friendship. I thank Golok Dabhe and Lhako, my good companions and hosts. Finally, I am greatly indebted to Aku Suoba for his friendship and trust as well as his very professional help, which continued long after my fieldwork. This list could be filled with many more names and memories, but there are also those who I should allow to remain anonymous. I thank all my informants, those who were cooperative and those who were not, but who always brought something new to my work. The protagonists of this book are either unnamed or hide behind pseudonyms, but they will know that it is about them.

Writing this book was possible thanks to the generous support of the International Institute of Asian Studies (IIAS) in Leiden. I want to express my gratitude to the Institute, to Mary Lynn van Dijk and Paul van der Velde, as well as other members of IIAS for their hospitality and patience. Another important person in this story is Alex McKay, who has been my enthusiastic supporter and was the first person to suggest publishing this book with Amsterdam University Press. I also thank John Bray for the years of friendship and intellectual exchange and for his support during the writing process. I thank Tak-Wing Ngo, editor of the Global Asia Series, and the two anonymous reviewers for their support and constructive criticism, which helped me improve this book manuscript, as well as Saskia Gieling, Jaap Wagenaar and everyone at Amsterdam University Press for their professional care and advice. Thanks go also to David Hawkins for his sincere interest in my writing and his professional help, as well as Maya Schneeberger for creating maps for this book. I finished writing this book in Switzerland, during my work at the Department of Social Anthropology and Cultural Studies, University of Zürich; I thank my colleagues there for creating a working atmosphere that facilitated this. I thank both the Humboldt University and the University of Zürich for the funding that has allowed both research in China and the completion of this publication. Although technically, one can only have a single alma mater, it seems that I have several.

Many thoughts expressed in this book were discussed with my friends, both anthropologists and scholars of Tibet. I could always count on Lilian Iselin and Yusuke Bessho, who shared their immense experience and knowledge of Tibet, as well as Yangdon Dhondup, especially during my stay in the Netherlands, whose opinion mattered to me a great deal. In Berlin, I had the support of Paweł Lewicki: we were fellow travellers through the exciting but painful process of doctorate writing. I also thank Li Li, my advisor, translator and a dedicated scholar of Golok, Anna Napiontkówna for her faith in this project, Francesca La Vigna for her care and for giving me a home in Berlin. Finally, I thank Thomas Studer, who accompanied me during the final stretches of work on this book, for his support and for the new perspectives.

The most important people are always mentioned at the end. These are my parents, Róża and Antoni, whose unconditional support made both conducting the research and writing this book possible. I will never be able to repay what I owe them. This book is dedicated to them, because they also dedicated themselves to it.

Notes on Transliteration

Most of my research was conducted in the Tibetan language and in places where Tibetan is the main medium of communication. A logical consequence of this is that priority is given to Tibetan in this book. However, rendering Golok pronunciation is not easy, especially given the lack of a standard system for Tibetan phonetics. In this situation, I decided to use my own simple phonetic equivalents of local terms and names. The spellings – using the Wylie system of transliteration – are given in the glossary at the end of the book. Sometimes, providing a spelling was impossible: some local words do not have a written form or have many possible variants.

When writing about the local geography, I use Tibetan or Chinese toponyms, or both simultaneously, although the latter only in the case of key place names. All other Golok toponyms in the book appear in their Tibetan version, unless my informants preferred the Chinese name, i.e., when Tibetan names are translations of new administrative units but have not taken root in the local language. Tibet is part of the People's Republic of China, whether we like it or not, but replacing Tibetan toponyms with their Chinese equivalents would falsify the local people's perception of space. Many places where I worked do not have Chinese names anyway, and if one wanted to make a map of Golok using only Chinese language, it would require leaving large parts blank or filling them with ad hoc toponyms. In this situation, the best solution was to follow the local people's choices. Foreign words, given in italics in the text, are most often Tibetan. Whenever it is not clear from the context, I add abbreviations: 'Tib.' for Tibetan, 'Chin.' for Chinese, 'Mong.' for Mongolian and, finally, 'Skt.' and 'Lat.' for occasional Sanskrit and Latin words.

This book was born between several languages: the research was conducted in Tibetan and Chinese, with the occasional use of English. I wrote it in English, whilst living in a German-, Dutch- or Swiss-speaking country. However, English is not my mother (or father) tongue. The quotations 'from myself' that appear in this book, and which come from my fieldnotes and field diaries, were written in Polish. I edited and translated them for this publication. These quotations, unless their source is clear from the context, are marked with 'FN' for fieldnotes and 'FD' for field diaries.

The names of all the protagonists and my main informants are pseudonyms: they are all names common in Golok. As with many scholars working in Tibet, I struggled between a desire to portray this book's protagonists as lively and complex individuals and the need to protect their privacy. I know that in concealing my informants' identities I disappoint many of them, who would like to see their contribution to this book manifested in a more literal way. I hope they will forgive me for doing so.

List of Abbreviations

AC	Autonomous County
CCTV	China Central Television
FD	field diary
FN	fieldnotes
PLA	People's Liberation Army
PRC	People's Republic of China
RCC	Rural Credit Cooperative
SARS	Severe Acute Respiratory Syndrome
SNNR	Sanjiangyuan National Nature Reserve
SU	sheep unit
SSU	sheep stocking unit
TAP	Tibetan Autonomous Prefecture
SU	Tibet Autonomous Region
TQAP	Tibetan and Qiang Autonomous Prefecture

List of Units of Measurement and Currency

jin (Chin.)	unit of weight, half a kilogram
jama (Tib.)	unit of weight, half a kilogram
yuan (Chin.)	unit of currency, in the period of my research 1 euro was worth about 10 yuan
jiao (Chin.)	unit of currency, one-tenth of a yuan
fen (Chin.)	unit of currency, one-hundredth of a yuan
shang (Tib.)	unit of weight and formerly of currency, one-tenth of a *jin* or *jama,* fifty grams
karma (Tib.)	unit of currency, one-hundredth of a *shang* (equivalent of Chin. *fen*)
mu (Chin.)	unit of area, one-fifteenth of a hectare

Introduction

Crowds thronging the streets in Golok, a pastoral region on the Tibetan plateau, signal the arrival of spring. Everyone is feverishly buying and selling. Groups of people huddle over trays of a strange-looking product. Men in black suits sit under big umbrellas with telephones glued to their ears, discussing something in hushed voices. A group of women squat on the ground, cleaning something with toothbrushes. A few steps away there is a cardboard box in the sun; a young woman guards its contents against curious onlookers. It seems as if everyone is carrying at least some of this curious product. People in restaurants pull it from their pockets and show what looks like a small, brown, dry caterpillar to their neighbours.

Wherever you go, people talk about one thing. Instead of exchanging the usual greetings, they inquire 'do you have many *yartsa*?' 'What's the latest price?', passengers on a bus say as a conversation starter. The words *yartsa gumbu* are on everyone's lips.[1] *Yartsa gumbu* is the Tibetan name of a rather unusual organism, which looks like a larva with a horn growing from its head. It is a parasitic fungus (*Ophiocordyceps sinensis*) that feeds on caterpillars of certain species of moth that inhabit the Tibetan plateau. Advertised as a wonder drug, it commands high prices on the Chinese market and thousands of kilograms of *yartsa gumbu* are sold in China and abroad. But there is only one part of the world where it grows: this species of caterpillar fungus is endemic to the Tibetan plateau and the Himalayas.

Golok is one of the Tibetan regions that produce caterpillar fungus and where people have built their livelihoods upon it. During the several weeks a year when this fungus is dug from the ground, it becomes everyone's primary concern. It captures the attention of people from all walks of life and often appears in unexpected contexts.

In Dawu, Golok's biggest town, in the house of a Buddhist monk, people wait for divinations. In gratitude for the monk's help, they offer him tea bricks, money, and caterpillar fungus. In a photo studio, a trader waits for his photographs to be developed. He radiates with pride, showing the other people around: 'Look at how large these *yartsa* are!' A monk in the monastery reveals that he dreamt that the *yartsa* will soon be extinct. During a boring

1 The Tibetan name *yartsa gumbu* is shortened in Golok to *yartsa*, and I use it in this form in quotations from the interviews. The Chinese name *donchong xiacao* is also abbreviated to *chongcao*: Chinese speakers quoted in this book use this form.

class, a school pupil draws a gigantic specimen of caterpillar fungus in his notebook, along with its name in Chinese and a price: 500 yuan.

In the Golok highlands, in high-altitude valleys where pastoralists have their tents, houses, and livestock, the atmosphere is just as fervid as in the town. Columns of people walk into the mountains, all of them carrying metal hoes. Miniature silhouettes move across the pastures as if searching for something in the grass. One sees them on high mountain slopes, in areas where people hardly ever venture. In the afternoon, on dusty roads, there are surprising crowds of people in this sparsely populated land. You wonder: has there been a car accident, a fight, or a political protest? No, these are the impromptu markets where people sell the caterpillar fungus.

Golok is in a state similar to a gold rush. On the one hand, it seems like a festival of joy: traders and diggers visiting the town party into the early hours. On the other hand, there is an air of conspiracy: people exchange information about police checkpoints on the roads and whisper that the authorities have been confiscating caterpillar fungus found on illegal diggers. Here, at 4000 m, in a region seldom visited, the streets suddenly swarm with people. There are many exotic faces: Huis, Salars, and Han Chinese mingle with Tibetan farmers. Groups of beggars roam the streets. They come from neighbouring Tibetan regions as well as more distant parts of China: caterpillar fungus season makes people generous. The beggars disappear when the season is over – just like the prostitutes, who also come to Golok in spring: this is the best time to earn money.

Aim of this Book

The most obvious aspects of the seasonal phenomenon described above are hard to miss for anyone who visits Golok. But its inner mechanisms and consequences are more difficult to observe. How far and how deep does its influence reach?

This book is a detailed record of the caterpillar fungus boom as observed in the region of Golok during the first decade of the twenty-first century. Golok at that time was already a place of intense economic life. The digging and sale of caterpillar fungus was a lucrative business engaging thousands of pastoralists and drawing people, both diggers and traders, from other regions. Therefore, although the book tells the story of one place in particular, it also reveals much about other parts of the Tibetan plateau, where the same phenomenon was observed.

Several groups feature in our story. The main protagonists are pastoralists from one particular township. The second group, the pastoralists' trading partners, work in the township and town markets. The third group remains invisible at first: these are seasonal migrants, who dig for caterpillar fungus in Golok. They are the pastoralists' partners in another economic arrangement: they lease the pastoralists' land for digging. The last, fourth actor is the state, both as abstract legislative body and control apparatus represented by state officials of various levels. Through its legislative power, the state tries to regulate the actions of the other three groups and the officials, as the arm of the state, are responsible for controlling them. Thus, the relations between these actors can be described as follows: the pastoralists earn their income from caterpillar fungus with the help of traders and migrant diggers and do so either in spite of the regulations introduced by the state or in cooperation with state organs.

The two stages on which the events described in this book take place are the pastureland and the market. The pastureland is naturally the field of the pastoralists' everyday activities – they live in high-altitude valleys and concentrate on raising livestock. Yet, the pastureland is not only grazing space for their herds, it is also a supplier of this expensive fungus. By digging it out of the ground, the pastoralists *produce* it for the market. The market is where the caterpillar fungus trade is set in motion: it starts its journey through the commodity chain to reach customers in distant corners of China and abroad. Here, at the Golok marketplaces, it is also converted into cash. However, the pastoralists' income is derived not just from the market. Golok is the location for migrations of people who are neither traders, nor pastoralists, but who also want to benefit from the caterpillar fungus boom. These venturers pay the pastoralists to lease their land. Thus, the pastoralists' income also comes from selling temporary rights to dig the fungus. The land is thus a production stage on two levels: where the caterpillar fungus is produced as a commodity for the market, and where income from the land leases accrues. These land leases, in turn, enable the diggers to search for caterpillar fungus and so produce both the fungus for the market and income for themselves.

Caterpillar fungus is a magnet that draws people together in Golok. And so, the two stages of action described above are not only zones of production, but also of contact between different groups linked by different types of economic arrangements. The pastoralists leave their high mountain valleys to sell the fungus in the town. Sometimes, traders go to the highlands to buy the fungus there. The physical channel of contact among these groups is the road between Dawu, the prefectural capital, and the township. Yet, this is also the road that the protagonists of this book – pastoralists, traders, and diggers – must pass along on their symbolic road to wealth. It is also where the seasonal influx of people into the township is controlled during spring. If, in this simplified sketch, the pastureland is the habitat of the pastoralists and the marketplace of the traders, the road (or the part of it where controls are enforced) is the state's domain.

However, the goal of this ethnography is not only to analyse interactions between different actors or the financial side of this area of local production; this book also considers the consequences of the emergence of the whole caterpillar fungus economy and its impact on society. The caterpillar fungus economy denotes here a broad field of income-generating practices related to this resource: not only the digging and sale of the fungus, but a whole range of concomitant practices that have emerged with this boom. The book argues that this economy has not only brought about improvement of the material situation of the rural population in regions such as Golok but has also led to a complex transformation in the local society.

The caterpillar fungus boom has created a significant new source of income for the pastoralists. This would not be so special if not for the fact that Golok used to be one of the poorest parts of Qinghai Province, which itself is very low in terms of national economic statistics.[2] In the 1980s and 1990s, Golok had little infrastructure, many areas were accessible only on horseback, and many facilities were beyond the reach of its population (Horlemann 2002: 255). Some of these observations remained valid during the years of my research: the township roads were often unmetalled tracks, some areas remained unreachable with motor vehicles, and many services were difficult to reach for pastoralists in more remote valleys. And yet, Golok towns continued to grow in size. In Dawu, new buildings rose, including apartment blocks with central heating and other facilities. There is a football stadium, a television station, and a number of modern buildings hosting government units. Many of these developments were possible thanks to the state investment programme called 'Open up the West' (Chin. *Xibu da kaifa*). This programme was launched in 2001 to improve living standards in the less economically prosperous western regions of China, to foster their integration with the state, and increase state security (Goodman 2004: 67). The initiative did help 'lift' Golok up the ladder of infrastructure development, but it did not directly translate into 'economic happiness'

2 Golok occupied the second-to-last place in terms of GDP per capita income (2964 yuan; QPSY 2009). Other prefectures with similarly bad results were Yushu and Malho, also pastoral regions.

for its inhabitants, especially those living from the land. However, signs of economic improvement were observed in the lives of the pastoralists too. But these had less to do with state investments and more with the caterpillar fungus economy.

From my first days in Golok I heard remarks to the effect that pastoralists waste the money they earn from collecting caterpillar fungus: 'They go to town, get drunk, and spend everything on prostitutes and gambling'. Such comments came from people who were not directly engaged in the caterpillar fungus economy or for whom the pastoralists were 'others', people they did not identify with. These were members of the local intelligentsia, Buddhist monks, but also farmers and town dwellers. Although their words may hold a grain of truth, I perceived in them an unpleasant superior tone. They suggested that the pastoralists 'did not know' how to handle money. My own observations did not confirm this stereotype: I did not see the town being invaded by hordes of drunken pastoralists, even though I witnessed how town life changed during the caterpillar fungus season.

My goal became to ascertain what really happened with the pastoralists' income. When the harvests are over and money is in people's pockets, there is a time of spending. What does this money go on? With this question in my mind, I started investigating the pastoralists' money management decisions. The picture that emerged was indeed of the pastoralists having money and spending it, but not necessarily of them squandering it. There was no doubt that the pastoralists had more money than before the boom. They had *much* more money now, and part of it they spent on consumer goods. But a closer investigation showed that they also used it for other purposes and sometimes did not spend it at all, but saved it for the future. The money that the pastoralists spent fulfilled their needs, both those more immediate and those concerned with improving their status longer term, translating here into changes in their standard of living, measured by the amount and quality of material goods owned, technical equipment, or even food diversity. Other investments become visible only when one leaves the town; these pertain to complex and long-lasting changes in the pastoralists' world. They also translate into enhanced quality of life, but this time measured by the quality of infrastructure, living conditions, or transport facilities. These are the more *systemic* changes in the pastoral regions.

This book shows that money earned thanks to the caterpillar fungus boom is being used by the pastoralists to transform their region. In other words, it demonstrates that this economic boom has led to a profound transformation of the local society and life realities. This has been accomplished with the minds, hands, and money earned by the pastoralists from various economic practices related to the boom. It partly corresponds with 'development' as it is envisioned by development planners and the Chinese state, and partly diverges from it. Thus, the argument of this book can be paraphrased as follows: the money derived from caterpillar fungus is used by pastoralists to *develop* their region. This sentence can only be formulated this way if special precautions have been taken regarding the meanings vested in the term 'development'.

Development and Modernity

The concept of development is a problematic and contested one.[3] It is the product of a particular geopolitical situation in the post-war and postcolonial world. As an ideology behind political and economic relations, its history started in 1949, when Harry Truman announced his Four Point Program in which 'development' was made the new emblem of US policy and gained a currency it never had before.[4] Though a body of literature that has 'development' in its name proves the existence of a whole discipline, with its own theorists and practitioners, critics still question its tenets and motives. Rather than a matter of social fact or necessity, development, they forcefully remark, must be seen as a historically produced discourse or a *domain of thought and action*. This domain is defined by the 'forms of knowledge [...] through which it comes into being and is elaborated into objects, concepts, theories and the like; systems of power that regulate its practice; and the forms of subjectivity [...] through which people come to recognize themselves as developed or underdeveloped' (Escobar 1995: 10). In other words, development is a political construct comprising a set of representations that put different parts of the globe on various levels of the 'development ladder', defining them as needing a developmental intervention and sanctioning it. This ideology now penetrates human thinking to such a degree that it has become difficult, even for persons from regions diagnosed as 'underdeveloped', to think about their situation in terms other than those provided by this discourse. Seen in this way, 'development' fits Michel Foucault's concept of discourse: far from being 'just words' or a free expression of thought, it is a *practice* with certain conditions, rules, and its variously empowered actors. 'To speak is to *do* something' and thus a

3 On development, see Escobar 1995; Sachs 1992; Ferguson 1994; and Rahnema and Bawtree 1997.

4 On the history of 'development', see Rist 2003.

discourse is an apparatus not only of seeing, knowing, and expressing, but also of acting (Foucault 1972: 209).

The concept of development cannot exist without its antonym, underdevelopment. The latter is a product of the former and gains its meaning in relation to it (Esteva 1992: 11). Though it can be broadly said that underdevelopment is a state in which resources (of any kind) are not used to their full potential, with the result that the 'development' process is slower than it *should* be, it is clear that this is a definition that could pretty much be used to explain anything. This can be demonstrated by the example of 'poverty', another concept that is devoid of autonomous meaning and gains it only in relation with 'non-poverty'. This is defined according to the standards of wealth of more economically advantaged parts of the world. It is measured by statistical quantifiers such as GNP or per capita income and evaluated by institutions based in the 'developed' world, which use their own criteria of development and convert it into 'figures or combinations of elements disembedded from the cultural type of livelihood' (Rahnema 1992: 162). Poverty as an aspect of human existence has always existed, but in the 'development era' it became problematized as abnormal or pathological. Typically, this 'abnormality' had first to be diagnosed (or constructed), and only then could the development administrators approach the task of 'treating' or eliminating it. This is one of the moments in the biography of the idea of development when its artificiality is most striking: first, a consensus had to be reached that poverty (replace with: illiteracy, malnourishment, etc.) is a problem to be solved and then a cure for it could be sought (or, vice versa, first the cure was found, and then its use justified).

Poverty or underdevelopment has a relative character and is intrinsically tied to the other end of the scale – the ideal state of welfare or 'development'. This ideal is not defined either, and as the final destination is not fixed, the gap between those who call themselves 'developed' and those who try to catch up with them can only grow: the leaders *always* move faster, as Wolfgang Sachs has noted (2010: xvii). The scope and meaning of poverty changes, too, as it directly depends on what the institutions governing development identify as the world's needs: its measures inflate and are continuously redefined. Set in a 'counterpoint with wealth', the concept of poverty can be seen as 'a theoretico-practical support for the prospect of increasing abundance' (Procacci 1991: 154). However cynical it sounds, 'modern economized societies define their poor in accordance with the capacity to absorb the commodities and services they produce', as Majid Rahnema observes (Ibid.: 19).

The above discussion places development discourse in the global arena, where it is integral to the 'West' or the 'North', which sees itself as custodian and exporter of 'development', and is imposed on the rest of the world, which is in the process of 'developing'. But a similar debate can be moved to more localized settings where – sometimes within a single state – similar discourse is being employed. In a definition stating that development is a 'historical construct that provides a space in which poor countries are known, specified and intervened upon' (Escobar 1995: 44-45), 'countries' can be replaced with peoples, regions, social classes, and other units. The case of China is one of many that show how the state employs this developmental discourse to introduce hierarchical divisions and assign roles to its various population groups, making some of them guides on the path to development and others into passengers who need to be taken there. Various kinds of state projects described in this book, such as sedentarization programmes and economic reforms carried out in the name of improving living standards, security, and social satisfaction, or to protect the environment, are interventions that have their roots (and justification) in a diagnosis that some parts of the country and its population need help on this path to development from those who are more advanced.

With all this in mind, one can approach the topic of the transformations that are taking place in Golok. I deliberately avoid calling them development, as this would imply that they lead in the 'right' direction and arise between less and more complex systems. This linear teleology is a belief that I do not share. And I do not share it even though some of the changes observed in Golok can be taken as examples of the pastoralists' own developmental initiatives. Some of them match the state's vision of development and some counter it. In all of them, the pastoralists are active agents of change. The caterpillar fungus economy is an example of how actively the pastoralists engage with current market realities and how – with the gains from it – they change their life, society, and environment, 'upgrading' it to a model that they desire. But the pastoralists do not merely exploit the economic opportunity that the boom has created *for* them; they also work to enlarge this field in order to maximize their gains. This book shows that the authorities restrict some income opportunities created by the caterpillar fungus economy, and the pastoralists need to organize themselves in order to create the space in which they can most take advantage of this boom. This does not come without difficulties and the pastoralists need to have creativity and initiative to overcome them. Finally, they are actively trying to make sense of a phenomenon that they witness and participate in as well as to incorporate it in the wider setting of their experience and life in general.

Another concept that is important for this discussion is modernity, as the processes described in this book can be taken as an illustration of how a pastoral region such as Golok is becoming 'modern'. Unfortunately, modernity is also a notoriously vague concept that gains its meaning in contrast to what is believed to be its opposite: 'tradition' or 'backwardness' are the conceptual constructs that often serve as its antonyms. While these two latter terms have different connotations in a range of diverse contexts, 'modernity' on the level of politics and developmental discourse is usually valued positively. It is closely bound with development: it is through 'development' that societies reach this ideal state of 'modernity' and it is through 'modernization' that they reach the status of being 'developed'.

What it means to be modern is just as debatable as what it means to be 'developed'. In principle, 'modernity' refers to 'modes of social life or organization' that emerged in Europe from about the seventeenth century onwards and comprise a certain set of ideas about the world (e.g. that it is open to transformation by human intervention), and a complex of economic and political institutions (industrial production and market economy, nation state and mass democracy) (Giddens 1990: 1, 1998: 94). Though born in the West, 'modernity' became more or less worldwide in its influence, as Giddens observes. This leads to a question: how does it happen that societies (especially very diverse ones) become modern? According to one theory, this arises through a process of a culture-neutral 'operation that is not defined in terms of the specific cultures it carries us from and to, but is rather seen as of a type that any traditional culture can undergo' (Taylor 2001: 172-173). This theory is based upon the idea that modernity emerges through the dissipation of certain unsupported 'traditional' beliefs and practices. It implies that the paths of different societies converge: as they lose their 'traditional' illusions, they become modern and similar in their outlook. This vision of universal and linear 'modernization', happening as if by law of nature, has been criticized by many scholars who denounce its Eurocentric character and intellectual flaws. James Ferguson calls it a 'modernization myth', which on the one hand could be dismissed as false or factually inaccurate (in the popular usage of 'myth'), and on the other is 'a cosmological blueprint that lays down fundamental categories and meanings for the organization and interpretation of experience' (2000: 13-14).

Dilip Gaonkar reminds us that we should rather think in terms of alternative modernities and that modernity unfolds 'within a specific cultural or civilizational context and [...] different starting points for the transition to modernity lead to different outcomes' (2001: 17). Some of the changes may be similar, but different starting points ensure that new differences emerge in this process. A singular model of modernity enacted in endless settings should be replaced with alternative modernities or, as Schmuel Eisenstadt suggests, multiple modernities. He notes that 'the actual developments in modernizing societies have refuted the homogenizing and hegemonic assumptions of this Western program of modernity' and observes that institutional and ideological patterns seen in the world neither constitute a simple continuation of the traditions of their respective societies, nor follow the Western project. The key to understanding the world is to see it as a 'story of continual constitution and reconstitution of a multiplicity of cultural programs', which are 'carried forward by specific social actors in close connection with social, political, and intellectual activists, and also by social movements pursuing different programs of modernity, holding very different views on what makes societies modern' (Eisenstadt 2000: 2-3). Thus, Western patterns of modernity are not the only 'authentic' ones and modernity is not one but many.

The exact meaning or 'contents' of 'modernity' in different contexts are the result of a negotiation process between different visions held by different actors: groups and individuals. Visions of modernity, even those performed by the state, change with time. Since Golok became part of the People's Republic of China, the state undertook numerous modernization projects according to a model of what it deemed that this process should bring to the rural regions of the country. The sequence of reforms and other interventions aimed at making a modern economy or a modern person evince how the meanings vested in modernity had changed, for example in the field of the economy, where 'modern' meant something different in the people's communes, in the period of Deng Xiaoping's economic reforms, and with everything that came after. In Golok, at the time of my research, modernity was represented by spaceships and scientific laboratories pictured on billboards along the town streets and country roads. The doors to space or scientific conquests are not really open to Tibetan pastoralists, but the message of 'scientific rationality' that these billboards carry has a bearing on certain spheres of life that are closer to their experience. The policies regulating the use of the caterpillar fungus resources and some reforms of the pastoral economy discussed in this book demonstrate this well.

However, the state is not the only agent of modernity. The pastoralists in Golok, who are the target of the state reforms and modernizing projects, are not passive subjects who simply meet them with docile resignation. If they wanted to reject the state's modernizing or development projects (which bear some resemblance to what Stevan Harrell has termed 'civilizing projects') they do not have to retreat to 'premodern' forms of social and economic life.[5] The system is not bipolar and does not consist only of 'modernist' and 'anti-modernist' standpoints. The pastoralists can bring to life their own vision of modernity and make their own choices. As was said about development, here, again, their choices can agree or disagree with modernity as the state envisions it. But people are not uniform in thinking and neither are their decisions, and there are many actors in Golok who have different views and expectations of modernity. Some of them appear in this book – reminding us that every ethnography should be read as a polyphonic story.

Modernity is unthinkable without the market economy that is a crucial part of the economic ideology underpinning it (Coméliau 2002: 29). Market and money are important elements in this ethnography. Ultimately, the main role of caterpillar fungus in regions such as Golok is money related. This book asks what people do in order to earn money and what they subsequently do with these funds. Conversely, there is also a question about what this money is doing to them. This is an intellectual shortcut, since the technical properties of money alone are not powerful enough to change human beings, even if a popular cliché says the opposite. However, since people are not only creators of the socio-cultural realities they live in, but also the subjects upon which these realities work, one can assume that the presence of this income affects people in certain ways. What effects has the caterpillar fungus boom had on Golok society in fields other than those directly connected to the pastoralists' investments? How does having this money impact their other life calculations and choices?

The caterpillar fungus boom in regions such as Golok cannot be analysed separately from the main branch of the local economy, namely pastoral production. This economic boom did not take place in an economic vacuum, but in a society that, until recently, had relied mainly or entirely on livestock production. Before, pastoralists supported themselves through the sale of livestock and livestock products and occasionally added to their budgets with other income. Apart from selling livestock products, they also sold medicinal plants. In the 1990s, they dug and sold *pimo* (Lat. *Fritillaria* sp.), a medicinal tuber that elicited an attractive income at that time. Caterpillar fungus then overshadowed *pimo* and other medicinal species collected for

5 Harrell defines it as 'a kind of interaction between peoples, in which one group, the civilizing center, interacts with other groups (the peripheral peoples) in terms of particular kind of inequality' which 'has its ideological basis in the center's claim to a superior degree of civilization, along with a commitment to raise the peripheral people's civilization to the level of the center, or at least closer to that level' (1995: 4).

sale. The size of the income it facilitated not only led to relativization of other activities that no longer seemed profitable, but also created a situation where people did not need money from other sources. As its market price rose, caterpillar fungus pushed itself to the centre of the local economic stage. In terms of time, digging the fungus can be treated as a side-line economic activity, because the season for it is short. However, the income produced during this period gives this activity a different status. Where the border is between side-line and main sources of income and when exactly caterpillar fungus crossed it are interesting questions. But the central question relates to the links between the caterpillar fungus boom and the pastoral economy.

Does the caterpillar fungus economy synergize with the pastoral economy or are they in conflict? Can people combine them or does the presence of a novel source of income change their attitude to the pastoral economy? Do they need to reorganize their pastoral life? This book shows that when the caterpillar fungus economy takes a step forward, the pastoral economy takes a step back. This comparison oversimplifies the problem; the changes are more complex and are not simply about advancement of one economy and decline of the other. However, the question remains: what happens to the pastoral economy when it loses its leading role in the pastoralists' lives? This question will be answered in the coming chapters.

Apart from analysing the pastoralists' investments and the changes these bring to their lives and the pastoral economy, this study shows that the caterpillar fungus boom has had an impact on other areas of life that are not related to earning or spending money. I view this economic boom as a 'total social fact', whose importance and implications are felt in many spheres: economic, legal, political, religious, and more (Mauss 1966: 76-77). This economy is a historically new element in the Golok society, but one that is tightly interwoven with the other threads of the general matrix of social life, whether these are labelled 'economic', 'social', 'religious', or other. My first impression of Golok was that life revolves around *yartsa* or that *yartsa* is what turns the world – and not only the economic world *sensu stricto*. The fact that the caterpillar fungus boom has infiltrated even the field of cultural production, i.e., local singers record songs about it, shows that this phenomenon is too big to fit into a box labelled 'economy'. The criticisms voiced about the negative sides of this economy, which are also heard in this book, show that the market career of caterpillar fungus is analysed in Golok society on many levels, including those far-removed from this economy.

Despite the critiques and conflicting interests, my intention is to create a picture of accommodation. Karl Marx envisioned human societies as riven by the tension between 'a stabilizing principle, manifested mainly in the superstructural arrangements designed to contain conflicts arising from contradictions in society; and a destabilizing principle found in the theory of progressive intensification of contradictions' (Smelser 1973: xxiv). If, according to Marx, the dialogue between these forces is bound to end with the destruction of the old and the emergence of a new social order, my approach is dialectical in a different way: I see balance in imbalance and continuity where Marx sees rupture. Thus, the changes that this book describes can be seen as successful accommodations. The case of caterpillar fungus shows how 'throwing' a new element into the 'social laboratory' of a given society does not necessarily turn everything upside down and cause a revolution but reveals the hidden mechanisms of adaptation and the flexibility of existing structures. This new element not only finds its place in the system but is actively adapted to already existing forms. It does not disorganize the society, but rather reorganizes it. The fact that caterpillar fungus has penetrated so many areas of the pastoralists' lives can be seen as a reflection of its economic importance. It is possible that it is precisely because of its special economic status that such adaptational forces have been unleashed. However, in 'allowing' this new element into these diverse fields, the pastoralists had to integrate it into the existing structures of thinking and acting. They had to question, re-evaluate, adjust, and remodel these structures. The role of caterpillar fungus thus reveals much about human society or culture and the mechanisms existing therein.

It becomes clear from these considerations that my understanding of culture is not that of a stable and static reified entity, but rather of a constant flow of things, images, institutions, and meanings – an ever-changing social matrix existing *in* and *between* people. This resembles Clifford Geertz's definition of culture as a web of significance or a system of construable signs within which human behaviours, social events, institutions, or processes receive their meaning and can be 'intelligibly – that is, thickly – described' (1973: 14). Seen in this way, culture is created and continuously recreated by people through their involvement in social relationships and social interaction. I approach culture in terms of practices that are constantly transformed as people pick up new ideas and adapt them in the course of their life and interactions with their social and natural environment. That said, I generally avoid using the term 'culture', as – even if accepting polyphony within culture and pluralistic notions of it – this creates a risk of essentializing and building non-existing constructs, which, in the words of James Clifford (1988: 9), are a 'serious fiction'. Speaking about culture as a single concept or unit of analysis implies the existence of clear borders, cores, and peripheries, and a catalogue of fixed characteristics, embodiments, and representations. For me, culture is not an aggregate of such material manifestations, institutions, and practices but an ability of people to manage their lives, give them meaning, and shape their relations with their social and natural environment. This echoes Tim Ingold's argument about how human actors form their worldviews as a result of the interactions between themselves and their environment (1987: 9ff). Culture is about how people deal with things and what they think about them.

It is our destiny to observe events in the world from our own particular time perspective, and we often fall victim to a conviction that we live in a period of rapid change. Nothing actually justifies this assumption, especially if one accepts that human societies are in a state of constant flux and that this change is an unavoidable part of their existence. Seen in this light, the changes that societies undergo cannot be perceived as some cultural drama – as the proponents of cultural purity and preservation always wish. The changes described in this book are not deformations of some timeless cultural substrate, but episodes in a longer stream of changes that societies undergo and create. Although I use the word 'change' as it is handier to manipulate, I see the phenomena I describe as *transformations*. For me, this word carries a more complex meaning and designates processes stretching over longer periods, not necessarily having a fixed starting point or destination. They are *trans*-formations but, as I said, I believe that human beings live in a world of constantly changing forms. More importantly, 'transformation' carries less evolutionist luggage, which would imply a 'step' or 'transition' leading from one level to another. Evolutionism as a way of interpreting the history of human societies is gone from mainstream anthropological thinking but the evolutionary narrative still hides in the nooks and crannies of a large number of modern-day intellectual approaches to the world that are built around concepts of globalization, development, economic growth, modernity, and more. As Ferguson remarked, many 'influential recent critical analyses of the postmodern reinstate a teleological and Eurocentric evolutionary narrative, in which "postmodernity" becomes simply the next rung in a social evolutionary ladder' (1999: 17).

Finally, this book reveals something about the state, in this case represented by its low power registers governing administrative units at the province level: prefectures and counties. It tells of how effective or ineffective the state can be. But it also shows something more. Trade necessitates the existence of a network of contacts and the caterpillar fungus trade created such networks extending over thousands of kilometres and traversing regional borders. They bring together people whose collaboration is necessary to make the fungus travel from its place of origin to its destination. The section of the network that can be observed in Golok is only a small part of the caterpillar fungus trading world. However, the boom has given life to another type of network that connects pastoralists with people who arrive in Golok to dig the fungus there. This network functions in the shadow of the law and forms what can be called a parallel society, brought to life by different economic actors who coordinate their actions in spite of the state's attempts to control them. The case of the caterpillar fungus boom raises interesting questions about the state's (lack of) involvement in it. It shows how, at a certain point, the authorities decide to take action and try to subject this economic field to their legal regulations. It demonstrates what happens when the state takes action to regulate practices belonging to a sphere that has not previously been regulated by the state. However, the road from words to action is long and the state does not always interfere or does so only in a limited way. The question that arises in this context is: why? The answer will reveal itself in the coming chapters.

Some Important Notes from the Author

Every scholar writing about Tibet must clarify what this name means to her or him, as Tibet 'can encompass different geographical boundaries, depending on who is using it' (Childs 2008: 7). In this book, 'Tibet' refers to ethnic, cultural, and geographical areas in China where Tibetans constitute the majority of the population. These areas can be called 'ethnographic' or 'cultural', but in the context of the current map realities they form administrative units of so-called Tibetan autonomous status: regions (TAR), prefectures, and counties.

Although I write about Tibet and have no doubt that the protagonists of this book identify with this (currently geopolitically non-existent) country, I seldom use the ethnonym 'Tibetan' to refer to them. A single Tibetan identity developed in the course of political changes that the populations of the Tibetan plateau underwent in the twentieth century. As Tsering Shakya writes, it 'owes much to the nationality policies and ethnic categorization system', which 'categorized the people of the Tibetan plateau as a single *borig* and provided fixity to "Tibetanness", homogenising it typologically' (2012: 24).[6] This does not mean that there has been no sense of cultural, linguistic, or other affinity between different populations on the Tibetan plateau; nonetheless, it is the 'small homeland', or particular region of origin, which

6 *Borig* is the standard term adopted in China for Tibetans as an 'ethnic group' or 'nationality'.

was and often remains the primary marker of group identity. Moreover, the people in Golok recall in their written and oral histories that their region used to be inhabited by the Mongols. Some people still identify themselves as Mongols or are identified as such by their neighbours. In this situation, using the ethnonym 'Tibetan' would misrepresent people's complex identities. For this reason, to avoid the unnecessary 'ethnicization' or 'nationalization' of my informants, I use their local identifications rather than big-scale names and national constructs.

Tibet is depicted in most of the literature as consisting of three traditional provinces: Ü-Tsang, Kham, and Amdo.[7] Golok is conventionally said to be located in Amdo, which covers the north-eastern part of the Tibetan plateau. This prevailing view does not find support in the experiences of the inhabitants of Golok and their understanding of their history and geopolitical position. My informants did not think they live in Amdo. On the contrary, they stressed that 'Golok is neither part of Amdo, nor Kham. Golok is Golok.'[8] In this situation, using the province-umbrella of Amdo imposes geopolitical categories on them that they do not identify with. Instead of Amdo, I write about Qinghai, a province of the PRC that Golok is part of today. This is not a reflection of my political sympathies. It is evident that Golok has effectively become part of Qinghai only in the course of the twentieth century and – whether one likes it or not – it is to be found there today. On the other hand, identifying it with Amdo, a division that is traditionally Tibetan and that may seem more 'legitimate' in this context, can lead to misinterpretation of Golok's position and history.

The main actors in this book are pastoralists: this is the official, academic term for people colloquially called nomads.[9] The term 'nomad', as many scholars stress today, is imprecise and can be applied not only to *pastoral* nomads, but also peripatetics, traders, artisans, and other professionals who adapt their place of residence according to employment opportunities (Rao

7 The concept of the three provinces is a cornerstone of the politics of the Tibetan government-in-exile. But although 'it is deeply embedded in the political culture of the Tibetan diaspora [...] [and] enjoys universal support among the exile community, it has no recent historical base and it is difficult to assess the extent of support it might enjoy inside [...] Tibet' (Shakya 1999: 387).

8 After my fieldwork, I returned to the notes that I took at university during a lecture about Golok. I found there the same words quoted above. They were also repeated to me by John Reynolds, who heard them from Buddhist teacher Tarthang Tulku Kunga Gelek Yeshe Dorje (born in Golok in the 1930s; pers. comm., Berlin, 17 June 2013). This shows that they were part of an 'identity makeup' also observed by other scholars. Concerning the term Amdo, see Huber 2002 and Sulek and Ptackova 2017.

9 I use 'nomad' only in quotations from the interviews. My informants in Golok did not juggle with academic terminology and 'nomad' gives their words a colloquial feel.

1987, Berland and Rao 2004). Other reasons to abandon the term 'nomad' have been given by Caroline Humphrey and David Sneath, who observe that this term carries misperceptions that make us imagine 'nomads' either as embodiments of a romantic myth of freedom, or as 'low-tech, rapacious, disorganized wanderer[s]' (1999: 305). Both images hinder the proper conceptualization of the sophisticated technology and knowledge associated with pastoral life. Whatever term one chooses – pastoralist or nomad – it is important to ask what is the actual term that the people in Golok use about themselves. They call themselves, in Tibetan, *drokpa*. *Drokpa* denotes people of the *drok* or *drokki sacha*, that is, the ecological zone located above the uppermost limits of the agricultural areas and where the primary way of earning a living is through mobile livestock breeding. Yet, not all *drokpa*s have livestock. *Drokpa* is an identity category and many people born in *drok*, who are neither nomadic, nor have livestock, also use this name. Students, monks, officials, or intellectuals identify themselves as *drokpa*, although they live in town and do something else for a living. As Fernanda Pirie notes, Tibetans 'firmly refer to themselves as *drokwa* [...], as long as they still have a family tent in the pastureland', even if the routines of their daily life are far from those of the grassland (2005: 9). Finally, *drokpa* is a pan-local category. My informants stressed their affinity with *drokpa*s living thousands of kilometres away on the Tibetan plateau, saying that they share more with them than with their Tibetan non-pastoral neighbours in Qinghai.

The material for this book was collected during anthropological field research conducted between 2007 and 2010. Although this material has already acquired a historical value, for the sake of the simplicity of this narrative, in many general descriptions I use the 'ethnographic present' and, in more specific cases, sometimes write in the past tense. The reader should bear in mind that this 'present' refers to the time of my research. The ethnographic present, a literary device conventionally used in anthropological writing, has been criticized as reflecting an 'ahistorical or synchronic pretence of anthropology' (Crapanzano 1986: 51). However, similar to Kirsten Hastrup, I believe that – used consciously and cautiously – it does not have to be abandoned. Its use connects to the very particular character of anthropological field research, a participatory and very present-tense experience both for the researcher and other people taking part in it. Writing about this research later, I re-*present* the time of my encounters in the field and the state of the affairs as I observed them, or as they were narrated to me, but I do not claim to 'represent a truth about the timelessness of the others' (Hastrup 1990: 57). The situations and people's views change and the material presented here is showed in its historical as well as current political and economic framing to ensure that the reader is aware that they are reading about a *part* of social reality that was available to me as a researcher during the period of my research.

The State of the Research

Caterpillar fungus is famous among Tibetan and Chinese *materia medica*. Its name has appeared on magazine covers and articles about it have been published by leading titles of the international press. The biology of caterpillar fungus is bizarre enough to attract attention. And if you add a few other components, such as big money, a smattering of illegality and violence, and the sex factor (caterpillar fungus is often called 'Himalayan Viagra') you get a potentially bestselling topic.[10] Although the media 'like' caterpillar fungus, they were not the first to discover it. In academic literature, caterpillar fungus and the economic boom centred on it have been described by a number of scholars.[11] Daniel Winkler is the most important among them: his research covers many regions of the Tibetan plateau and most of what we know about this phenomenon comes from him (2005, 2008a, 2008b, 2009). Other authors include Francesca Cardi and Alessandro Boesi (Boesi 2003, Boesi and Cardi 2009), Andreas Gruschke (2011a, 2011b, 2011d, 2012), Michelle Olsgard Stewart (2009, 2014), Emily Woodhouse (2015), Kunga Lama and Emily Yeh (Lama 2007, Yeh and Lama 2013). Occasional remarks are found in works by other scholars: caterpillar fungus pushes itself into studies of contemporary Tibet, including those unrelated to the economy, and it frequently features in the footnotes or the margins of the author's main discussion.

10 Even a partial overview of the popular press articles on caterpillar fungus exceeds the capacity of this book. However, a brief look at the titles, which use such expressions as 'magical', 'mysterious', and 'black gold' and grab the reader's attention with messages such as 'Killer Fungus is Gold to Yak Herders' (*New York Times*), 'The Viagra of the Himalayas: Fungus Transforms Tibet with Huge Cash Influx' (*Huffington Post*), 'Caterpillar Fungus on Radar of International Smugglers' (*Deccan Herald*), 'Craze for Magic Caterpillar Fungus Damages Land' (*Shanghai Daily*), or 'Curse of Himalayan Annapurna Region' (BBC News) shows not only the journalistic ability to employ sensational language, but also the fact that the caterpillar fungus boom facilitates this sort of hysteria.

11 Research was also conducted in Nepal, Bhutan, and India; see Devkota (2006, 2010), Namgyel (2003), Negi et al. (2006), Sharma (2004), Garbyal et al. (2004), Singh et al. (2010), Bauer (2014), Childs and Choedup (2014).

It may seem that the caterpillar fungus boom has been on everyone's lips and that many people have written about it. Yet, for a phenomenon of such economic and social importance, it is surprising that it has not been the topic of a detailed book-length analysis. It has yet to be analysed in either China, or in the neighbouring countries, in its whole context, including its very diverse aspects and consequences. It has not been subject to a study discussing its economic side and social effects as well as its history and legal aspects – one that would place it within the matrix of the socio-cultural life of the people who engage in this harvest and trade. This book is an attempt to write such an analysis.

The focus of this book is the caterpillar fungus boom *in* the pastoral society of Golok. Different aspects of Tibetan pastoralism have been discussed by many authors. The economy and life of pastoralists have been analysed, first and foremost, by Melvyn Goldstein. His work on pastoralists in the Tibet Autonomous Region (TAR) was one of my first inspirations. His and Cynthia Beall's other works, based on long-term anthropological research, analyse specific aspects of the pastoralists' economic functioning and changes in their life (e.g. Goldstein 1994; Goldstein and Beall 1989, 1991). Valuable studies, this time from Yushu (a prefecture in Qinghai), also come from Andreas Gruschke (2008, 2011a, 2011b, 2012). Angela Manderscheid and Graham Clarke studied changes in the pastoral regions in Sichuan and the TAR (Manderscheid 1998, 1999, 2002; Clarke 1992, 1995a, 1995b). A significant contribution to our knowledge comes also from Nancy Levine, who worked in, among other places, Golok Serthar (a county in Kamdze TAP, Sichuan; 1990, 1995, 2015), as well as from Susan Costello (Golok; 2002, 2008) and Kenneth Bauer (2005, 2008, 2011, 2014). Bianca Horlemann, Fernanda Pirie, and Emily Yeh worked either in Golok, or neighbouring regions and discussed economic changes as well as other aspects of the functioning of pastoral communities (Horlemann 2002, 2012a, 2012b; Pirie 2005, 2008, 2012, 2013; Yeh 2003, 2005). Finally, scholars of my generation, such as Lilian Iselin (2012, 2014, 2017), Yusuke Bessho (2015), Jarmila Ptackova (2013, 2016), and Gillian Tan (2013, 2016, 2017, 2018), have been important for me because they conduct their research in similar economic and political circumstances and often in neighbouring regions.

The list of authors who have written about the pastoralists of the Tibetan plateau is longer. The works mentioned above provided me with comparative material and information about issues on which I had less expertise. However, a large part of the literature on Tibetan pastoralism has a very particular focus. In recent decades, Tibetan pastoralists have been targeted by numerous state interventions aimed at reforming the way they manage their economy and grassland resources. Fencing the land, controlling herd sizes by introducing selling and slaughtering quotas, and sedentarization programmes often involving some degree of coercion made many authors focus on the effects that these interventions had on the pastoral society. The over-presence of this topic in the literature as well as in media reports creates a unidimensional image of the Tibetan pastoralists as suffering under discriminatory policies of the Chinese state. There is no doubt that the state does introduce policies that can be seen as hostile to or lacking understanding of pastoralism and that are not locally consulted on, and perhaps not even well thought through. However, not all pastoralists are covered by the sedentarization programmes and not all are critical about the state policies. Furthermore, those who are affected by them do have some space to show agency in managing their lives in these often difficult circumstances. Seen in this context, my book stands out as telling a story about people who are doing well. It does not contradict reports by other authors but supplements them. Finally, it is a book about people who are not only actively managing their lives but are doing so in a way that is much less dictated by religion than we often assume when we think about Tibetans – people who suffer under the yoke of the stereotype that religion is the main determinant of their identity.

Structure of the Book

When conducting my research, I repeatedly found myself in a situation where my colleagues associated me with caterpillar fungus rather than with the pastoralists whom I worked with. This book is structured in such a way as to stress that it is the *pastoralists* who are its protagonists: we meet the pastoralists first and only later learn what caterpillar fungus actually is. This approach also mirrors the chronology of my encounter with the topic. The first time I visited Golok in 2007 was during the peak of the digging season. This is also how the reader encounters the pastoralists: when they are engulfed in the feverish work of digging the fungus. Only later does the book show what normal life in Golok looks like. It was also a conscious choice to start with scenes of people digging and only later discuss the legal side of this activity. This shows that my approach is 'from below', in other words from the *people*, and not from the position of the state authorities. Reversing this order could lead to writing a very different sort of ethnography.

Individual chapters of this book are thought of as autonomous units and the reader can start reading at any of them. After this Introduction, which has outlined the main argument of the book, explained the concepts used, and shown the place of this ethnography in the literature, comes Chapter 1, 'Golok: People and Places'. It introduces the region and its inhabitants and explains why this study was conducted there.

In Chapter 2, 'Digging', we travel to Golok during the caterpillar fungus season to meet the two families whom I lived and worked with. We see them and their neighbours during spring, when Golok faces an extreme mobilization of the labour force: one has a feeling that everyone who can goes out digging. Yet, at this point, it is not clear what causes this mobilization.

Readers wanting to learn about caterpillar fungus first should begin with Chapter 3, 'Fungus, Medicine, Commodity', which incorporates information about the biology as well as the medicinal and market career of this resource. Here, I answer whether the pastoralists use it in their domestic practices and show that, in Golok, caterpillar fungus functions more as a commodity than as a medicinal product. I also discuss the history of the trade, especially during the people's communes, and show how it developed into a proper economic boom.

Chapter 4, 'Market and Traders', is devoted to the caterpillar fungus market, understood as a physical space where transactions are made as well as the community of traders populating it. I discuss the legal position of the trade, show that it brings together different ethnic groups and that many traders resist state attempts to control them. I analyse the paths that caterpillar fungus travels between the moment it comes out of the ground and when it leaves Golok for bigger markets. I reveal its hidden market life and explain why this trade offers employment to such large numbers of people.

In Chapter 5, 'Market Operations', I return to the pastoralists and consider their strategies for selling caterpillar fungus. I describe different quality classes of caterpillar fungus and the criteria for evaluating it and I discuss the range of prices and their seasonal fluctuations. It thus slowly becomes clear just how expensive this fungus is. This chapter is illustrated with scenes of price negotiations and a description of a system of hand signs used in these negotiations.

Chapter 6, 'The Law in Action', focuses on the legal aspects of different activities that emerged with this economic boom and the state's attempts to regulate them. I discuss the pastoralists' opinions about them and show how the authorities' efforts to tighten control over caterpillar fungus digging have led to the emergence of underground illegal land leases. This chapter reveals the limits of the law and examines the social conditions responsible for its lack of efficacy.

Previous chapters show different ways in which the pastoralists earn their income from caterpillar fungus. Chapter 7, 'Money', answers how much they earn. Here, I not only explore the contents of the pastoralists' wallets and bank accounts but explain why this task is so difficult. I show that caterpillar fungus money has a very particular status, explain some moral dilemmas it creates, and the lines of conflict between the caterpillar fungus economy and people's religious beliefs.

Chapter 8, 'Pastoral Life and the Market', returns to the two families introduced at the beginning of the book. This time, I show them as livestock breeders. I introduce data on herd composition, migrations, and the pastoral work calendar. I also analyse the pastoralists' interactions with the market of pastoral products and show how the availability of income from caterpillar fungus has changed their approach to the pastoral economy.

What happens with the pastoralists' income is revealed in Chapter 9, 'Spending the Money', where I show the pastoralists as investors and consumers. I examine the rise of the house as a new status symbol and ask whether this is leading to increased sedentarization among this group. I discuss other technological changes in Golok and demonstrate that the caterpillar fungus boom has not only allowed pastoralists to improve their material standard of life, but also to undertake investments leading to infrastructural transformation in the highlands.

'Conclusions' gives us a chance to take a final look at the phenomenon under study and to ask questions about the future.

These reflections close the main part of the book but are followed by two more sections – 'bonus tracks', if you will. 'Afterword: A Note on Methodology' discusses the methodological toolkit for my fieldwork and reveals some problems that I encountered during my research. I leave this discussion to the end for two reasons: because this section has a partly personal character and because it can be taken as an independent essay. The methodological considerations are followed by an Appendix containing a translation of a Chinese state document outlining the rules governing access to the caterpillar fungus-producing land, which were in force in Golok during the time of my research.

1 Golok: People and Places

Golok is Heaven and Earth

Golok is not the most picturesque part of the Tibetan Plateau.[1] Its landscape cannot compete with the rocky canyons, alpine forests, and swift rivers in other regions, which find their way into wall calendars and photographic albums. However, the Golok highlands also possess some charm. The high, elevated tableland with the soft lines of the grassy hills of Gabde, dotted with distant lights from pastoralists' houses, is what instantaneously comes to mind when I think of Golok. Martod has a silent air, emerging from the still waters of its famous lakes, Hcharang and Ngorang. The dark peaks of Nyenpo Yurtse rise unexpectedly from the sea of grass in Jigdril. Green valleys around Amnye Machen, with marmots warming their fat bodies in the sun, give an impression of fertility and abundance. The night sky is sprinkled with myriad stars. The sky is big, the land is wide. As people in Golok say: *Sa Golok nam Golok*. Golok is heaven and earth.[2]

Golok lies at an average altitude of 4100 m. Its highest peaks include Machen Gangri in the Amnye Machen mountains (6282 m) in the north-west and Nyenpo Yurtse (5369 m) in the Bayan Har mountains in the south-east. Bayan Har form the watershed between the Drichu (Yangtze) and Machu (Huang He) or Yellow River basins: Golok forms the basin of the latter. The Machu flows through Golok until Gansu and Sichuan, then makes a turn to the north-west. Thus, Golok is easy to find on every map of the world: locate the sources of the Yellow River and follow it until its northward bend and this is where Golok lies. Entering Golok from the north or the south, one is struck by a sudden change in vegetation. In the foothills of Golok, in Rarja and Bamma, the climate is mild and small-scale land cultivation possible.[3]

1 Golok Tibetan Autonomous Prefecture (TAP) does not cover all the areas that formerly constituted the Golok region. The pastoralists stress that it included Serthar, Dzachuka (today in Kamdze TAP, Sichuan), parts of Ngawa (TQAP, Sichuan), and Machu (a county in Kanhlo TAP, Gansu). Scholars estimate that Golok used to cover over 100,000 km^2 (Don grub dBang rgyal and Nor sde 1992: 6).

2 *Sa Golok nam Golok* is a famous phrase in Golok. There are many stories connected to this, but few good ways of translating it. In the most popular version, a PLA soldier or a Communist leader asked an elderly pastoralist which is bigger, Golok or China. The latter said that, of course, Golok is bigger and used the above-mentioned phrase to mean that Golok is as big as the earth and as broad as the sky.

3 Farmland covers 0.02 per cent of all land usable for grazing and/or farming (Wu and Yang 2000: 187).

Although some forests are still to be found around Amnye Machen, the majority of Golok is treeless. It is a kingdom of alpine meadows, with sedges, shrubs, forbs, and grasses.[4]

Golok is large but sparsely populated. The prefecture covers an area of 76,442 km^2 and was inhabited, according to the census conducted in 2010, by 181,682 people (2.38 persons per km^2). Its population is relatively homogenous in its ethnic structure, with 93.43 per cent being Tibetan.[5] Other ethnic groups live mostly in prefectural and county towns, working in administration, public institutions, and as small entrepreneurs. Dawu, the centre of administration and trade, has the biggest percentage of non-Tibetan residents. The county statistics state this to be about 80 per cent, but they list only people with urban household registration (Chin. *hukou*). They are silent about Tibetans who live in the town but remain registered as rural residents.[6] In smaller towns, where there are fewer employment opportunities and living conditions are more rudimentary, Tibetans dominate. Regardless of their exact ethnic composition, Golok towns are important centres of trade and services for the pastoralists inhabiting the vast Golok highlands. The predominance of Tibetans as town-dwellers or visitors makes the Tibetan language the main medium of communication.[7]

If in towns non-Tibetans form a visible part of the population, the highlands are an exclusively Tibetan space. In Domkhok, a small township in Golok Prefecture and the place on which this book is centred, the last census recorded 99 per cent Tibetans, but in more remote townships the entire population is Tibetan. High-altitude rural townships cannot support any significant population employed in sectors other than the pastoral economy.

4 Joseph Rock, an Austrian-American explorer and botanist who visited Golok in 1926, was disappointed by its lack of botanical diversity (Walravens 2003: 60). About the types of alpine rangelands and plant species in Golok, see Hou 2011: 23.

5 There were 166,895 Tibetans, 11,934 Han, 1739 Hui, 107 Mongols, 247 Salars, 429 Monguors, and 331 members of other *minzu* (CNS 2010).

6 *Hukou* is a household registration system introduced in 1958 to control the mobility of Chinese citizens. It requires every person to register at birth as a rural or urban resident. This status determines the citizens' rights to state benefits, education, and employment. Although the system was gradually relaxed, changing a rural *hukou* to an urban one remains difficult (Solinger 1995; Chang and Zhang 1999; Wang 2005). Dawu is home to a large number of pastoralists, relocated during the sedentarization programmes, who remain formally registered as rural residents. Other pastoralists who moved to town independently of these programmes also kept rural *hukou*.

7 Many non-Tibetan residents leave Golok in winter and only the Chinese flags and Chinese street names remind the visitor that one is in China. During my research in winter 2008, there were hardly any non-Tibetan policemen in Dawu; they only reappeared around 10 March, the anniversary of the uprising in Lhasa in 1959, which is often an occasion for political protests.

Today, pastoralists in Golok do not move over long spatial distances. But their herds graze on the pastureland all year round, and people's lives show seasonal changes, adapted to the needs of their livestock and the state of the pastureland. The pastoral economy is the main, traditional source of subsistence and income; according to official statistics, over 70 per cent of Golok population is employed in it.[8]

Weather

Golok is the coldest prefecture of Qinghai, with an average annual temperature of 0.4°C.[9] It has a continental plateau climate, with high daily temperature amplitudes and high sun radiation. In summer, temperatures can exceed 30°C. In winter, they fall to minus 20°C and lower, but high solar radiation (almost seven sunlight hours per day) and the dryness of the air make this cold easier to bear. Winters are dry. Precipitation increases between May and September and reaches its peak in July.

The growth period for vegetation is short, lasting just three to four months.[10] Spring starts in April with its dust storms, but only in late May or June do grasslands slowly green up. Summer brings an explosion of life: in July and August the meadows are covered with a fragrant carpet of flowers. This is everybody's favourite season: pastoralists move to their tents, horse races are organized, and town people picnic in front of their houses. In August, a thin layer of snow appears on the higher mountain slopes, reminding of the coming autumn. In September, the grasslands turn brown and the herds start shivering in the cold and cutting winds. The cold and dry autumn is followed by winter, which starts three months later.

The vagaries of the high-altitude plateau climate and the stark contrasts between Golok and the lower altitude parts of Qinghai province are shown by the observations made during my research. Spring 2007, when I started my research, was unbearably hot in Xining. However, as soon as one crossed the mountain passes leading into Golok, the weather changed. It was warm during the day, but at night ice covered window frames, and before starting a motorbike one had to scratch a layer of frost off the vehicle. In mid-May, when I dug caterpillar fungus, the snow could blanket the land overnight,

8 According to the last census, the urbanization rate was 24.72 per cent with 136,762 rural and 44,920 urban residents (CNS 2010).

9 On this and the following data, see QPSY 2010.

10 Li and Li give its as around 99 days from 1961 to 1980 and 125 days since 1981 (2002: 33).

turning it into a truly winter-like landscape. During the day, it was burning hot in the sunshine and going out without a hat risked sun stroke, but it would be ice cold a moment later when the sun hid behind the mountains and a wave of hail came from an unknown direction. My hosts offered me six blankets at night, not only because they doubted my resistance to cold, but because, even inside the house, the cold was all-pervading.

Winter 2007/2008, when the second round of my field research took place, kept people waiting until January for the first snow. When the snow finally fell, it blocked mountain passes, cutting transport between counties. In Domkhok, people drove on frozen rivers. Every journey seemed endless, with the passengers leaving the car to throw sand and stones under the wheels, which were sliding on the ice. Fetching water was even more challenging: poured into containers loaded onto the back of yaks, the water froze immediately when spilled on the animals' hair. In Dawu's restaurants, the temperature was so low that the fat dripping from meat dumplings turned solid upon touching the table. In February, the temperature fell to minus 25°C and animals starved, unable to dig out grass from under the snow. The television showed animal carcasses and officials coming to help the pastoralists in times of crisis. In the year of the Olympic Games in Beijing the weather took on a political meaning. People commented that weather disasters often precede political changes. The mass deaths of livestock were believed to portend something; the coming months would show what.

After a long break caused by political protests in 2008 and restrictions imposed on travelling, I resumed my research in spring 2009 and continued until the end of summer under overcast skies.[11] In June and August, there were only ten days with more than two hours of sunshine. Clouds hung low over the ground, bringing rain, hail, and snowstorms. At the beginning of June, snow blocked the roads around Amnye Machen. A thick robe was a must, but even more necessary were rubber boots. Summers in Golok are humid but that summer was worse than ever. In Domkhok, it rained for twenty days without a break, complicating people's lives: cheese and yak dung did not dry and so fuel was in short supply. A bridge over the River

11 Beginning in March 2008, Tibet witnessed a wave of dramatic protests: street demonstrations, peaceful marches as well as attacks on symbols of the Chinese rule organized by Buddhist nuns and monks, students, farmers, pastoralists and other Tibetans. The authorities responded with a military crackdown, restrictions on freedom of movement, patriotic education campaigns, detentions, and prison terms. The 2008 protests and ensuing crackdown led to a tragic series of self-immolations in many regions and across social classes, age, and gender groups. See ICT 2008 and 2009.

Domchu was destroyed by the raging waters, making people prisoners in their houses. The pastoralists talked about 'man-made' rains, a part of the state environmental conservation programme.[12] They complained about the government's lack of understanding of the local ecology: 'The results will come to them', someone prophesized, 'They don't know the law of *karma*'.

The last round of research ran from spring to autumn 2010. Caterpillar fungus season was late and new diggers were still arriving in Golok at the end of June. On the 1st of June, Dawu was flooded after torrential rains in the mountains: the streets turned into rivers of black water. The water level fell, but memories of the earthquake in Yushu in April the same year were still fresh. People recalled that the earthquake was preceded by a flood.[13] Owners of cell phones on the China Mobile network received messages urging them to refrain from panic: 'In the coming days torrential rains are expected in many parts of the province. The water reservoirs are functioning properly. Do not listen to the rumours and continue your usual life.' Landslides in other parts of China seemed for many people to be another result of weather engineering. The flood contributed to a general disorder: the pavements in Dawu were torn up and the ground dug with ditches for electric cables. These construction works coincided with changes in the local authorities. 'Why is it that every time the leaders change, they start repairing the road?', I heard someone asking. 'I'll tell you why. That's how they make the money disappear. It's because of the one-party system. If we had more [parties], there would be more control', another person answered.

Administration and Travel

Golok has the status of a Tibetan Autonomous Prefecture (Map 1).[14] The prefecture consists of six counties (Tib. *dzong*, Chin. *xiang*): Gabde (Chin. *Gande*); Darlag (Chin. *Dari*); Jigdril (Chin. *Jiuzhi*); Bamma (Chin. *Banma*);

12 This programme was not a figment of my informants' imagination. A leaflet of the Meteorological Academy of Golok informed that the 'man-made rains' (Chin. *rengong zengyu*) help against retreating glaciers, receding water levels in lakes and rivers, as well as against avalanches, desertification, and air pollution.

13 The mood was so grim that we – me and a friend whom I stayed with – packed our most important belongings and slept with the lights on, ready to evacuate. This is also what many people in Golok did directly after the earthquake in Yushu.

14 This status dates back to 1955. A year earlier, in 1954, Golok Tibetan Autonomous Region was founded.

Map 1 Golok TAP in Qinghai Province, China

Map 2 Administrative divisions within Golok

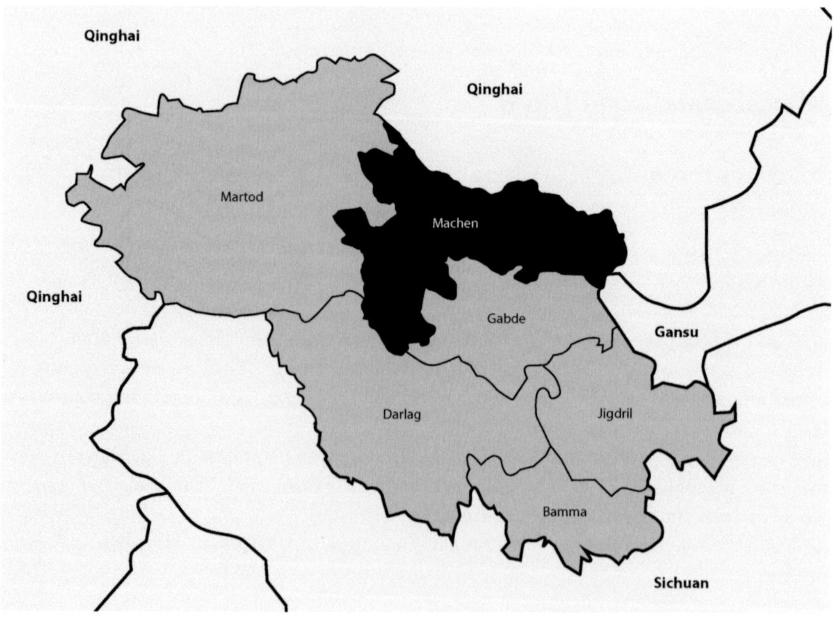

Martod (Chin. *Maduo*); and Machen (Chin. *Maqin*) (Map 2).[15] The counties are split into rural townships (Tib. *shang*, Chin. *xiang*) and towns (Tib. *chongrdel*, Chin. *zhen*).[16] In Machen County, where I conducted most of my research, there are eight townships: Domkhok; Xueshan; Chamahe; Yigzhung; Danzhung; Danlag; Dawu Zhuma; Dawu Gongma; and two units of town status: Rarja and Dawu (Map 3).[17]

Travelling to and through Golok is not easy, partly because of the large distances and the insufficient transport network. It is further complicated by the restrictions placed on access to many parts of the prefecture. During the period of my research, only three out of six Golok counties were open to foreigners: Machen, Jigdril, and Martod.[18] Visiting the remaining three was possible only upon obtaining a so-called Alien Travel Permit. Foreigners could apply for one of these at the prefectural police station. It allowed them a stay of up to one week in whichever county the permit covered. In 2008, the year of the Olympic Games, as well as 2009, travelling to counties that were normally 'open' was also restricted: in Machen, stays were limited to one week. The situation relaxed in 2010. Yet, in Rarja it remained tense: because of the political protests that broke there in March 2009, foreigners had to apply to three different institutions to obtain the necessary documents to visit.[19]

15 The first four counties were founded in 1955, and the remaining ones three years later (Xing 1994: 66). The county borders changed and parts of Gabde and Darlag went into Machen and Martod in 1958. There were also shifts between Golok and other provinces (see Levine 1998, Yeh 2003). Dawu has been the prefectural capital since 1962.

16 'Town' is a standard translation of Chin. *zhen*. As with a township, it is a fourth-level administrative unit (under province, prefecture, and county). It normally designates a small town and its surrounding rural area (often of considerable size). Because of this predominantly rural character, to translate *zhen* as 'town' is misleading. A better rendering would perhaps be 'urban township' (in contrast to *xiang* or 'rural township'). In Golok, the main difference between *zhen* and *xiang* is that *zhen* are located along bigger roads and their administration hubs are more populous.

17 The name Dawu simultaneously denotes a *dewa*, an administrative unit of zhen level and the biggest town in Golok. In order to avoid confusion, when Dawu as administrative unit is mentioned it is followed by the word *zhen*.

18 Golok officials in charge of tourism stressed that travelling is restricted in many parts of China. As they said, this does not necessarily have political but infrastructural reasons: some areas with major tourist attractions have better infrastructure and others are not 'ready' to receive tourists. This logic seemed to apply to Golok. All three open counties had tourist attractions: Amnye Machen in Machen; Nyenpo Yurtse in Jigdril; and Hcharang and Ngorang lakes in Martod. All three underwent major processes of investment and offered better services than the others, where no hotel had a licence to accommodate foreigners.

19 This was a result of the suicide of a monk from Rarja monastery and ensuing political protests, see RFA 2012a and CD 2012.

Map 3 Administrative divisions within Machen County

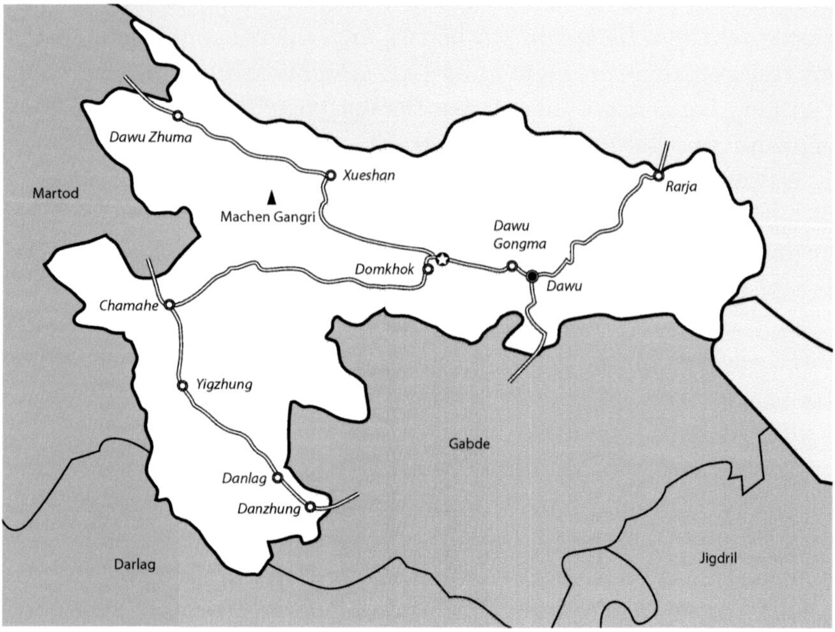

road
sacred mountain
Sumdo checkpoint
prefectural capital
administrative centers of townships and towns (zhen)

Although open to foreigners, Machen is not a popular travel destination. Golok in general is not. It has a reputation for being a region that is remote, isolated, and difficult to reach. This picture was partly true, especially at the beginning of my research, when reaching Dawu from Xining, the capital of Qinghai, meant spending a whole day on a bus.[20] But this 'remoteness' was not only a function of physical distance: in the field of tourism, Golok even lost out to Yushu, another, more remote pastoral prefecture (its capital, Jyekundo, lies 820 km from Xining, while Dawu is only 440 km away). Golok's 'remoteness' is a mental construct supported by stereotypes that depict the region as wild and its people as bandit-like. This reputation is supported by tourist guidebooks and other reports that contrast the beauty of Golok's landscape with the

20 In 2014, there were already a number of *marshrutkas* that covered the distance between Xining and Golok within six hours. During my fieldwork, getting a ticket for a sleeper bus to Golok could take days and so travellers often rented seats in private cars leaving from Runda Hotel.

reputed hostility of its inhabitants.[21] This idea has a long history: already in the nineteenth century, Western authors characterized Goloks as bandits terrorizing neighbours and travellers (Horlemann 2007). Today, the image that Goloks have among their neighbours is not much different.[22] As a result, tourists – both from China and abroad – are a rare sight in Golok. I was often welcomed with startled looks: 'Oh, a foreigner!' I realized how automatic this reaction was when, after several months of fieldwork, I myself saw a foreigner and exclaimed, to my embarrassment, the same words. But this was already 2010, when Golok 'adopted' a number of tourists who would normally have gone to Yushu, which was then struggling with the effects of an earthquake.[23]

The People

Golok Prefecture takes its name from the Golok pastoralists, who form the dominant part of its population. The prefecture, however, is not inhabited only by Goloks. The pastoralists from Domkhok, about whom this book is about, belong to the Metsang *dewa* and claim different origins than the Goloks, speak a slightly different language, and stress their distinct identity. The Metsang are not the only non-Goloks in the prefecture. They are neighbours to several other non-Golok groups: Dawu *dewa* (after which the main town in Golok is named); Guru Tsodruk; Yunzhi; and others. In fact, large parts of Machen are inhabited by non-Goloks: Xueshan; Dawu Gongma; Dawu Zhuma; Dawu; and Rarja are all non-Golok townships and towns.[24]

The term *dewa*, which the Metsang pastoralists use to describe themselves, denotes a unit of social organization consisting of people inhabiting

21 The reasons for Golok's unpopularity are worthy of a separate study. Several foreigners in Xining to whom I suggested visiting Golok believed it was dangerous. Chinese trekkers expressed their admiration for the Golok landscape and, in the same breath, an aversion to its all-pervasive dirtiness. They were shocked that someone might stay there for more than a day. When I asked my Tibetan colleagues in Xining why they did not visit Golok, they said it was too far away. Opinions such as Dawu being 'a town of such grimness that even prefectural capitals elsewhere in transformed Tibetan areas cannot rival its bleak artificiality' (Marshall and Cooke 1997: 2217ff) do not help in promoting it as a tourist destination.

22 To name some examples: my friends who studied in Xining said their classmates called them 'Golok enemies' (*dra Golok*). A man from Yushu remarked that when he was a child his parents threatened him that if he was naughty 'Goloks will take him away'. Another man from Yushu recalled that pastoralists there scold unruly yaks with the words 'Fall into Golok hands!' (*Golok lag jog*).

23 Within five months after the earthquake, the number of tourists in Golok had doubled, as a prefecture official in charge of tourism informed me in 2010.

24 The four Machen townships with predominantly Golok populations are Chamahe, Yigzhung, Danzhung, and Danlag.

a certain area and sharing a common past or ancestor. The *dewa* is loosely organized, but it either has a leader (called *huonpo*) or is led by a group of elders (Hermanns 1959, Pirie 2005). In Metsang, the *huonpo*'s position is hereditary, but in exceptional cases a person can be chosen for this function based on his (the *huonpos* are usually men, though history knows a few exceptions) abilities, respect, and oratorical skills.[25] The role of the *huonpo* is crucial in local politics, allocating pastures, but also in mediating conflicts.[26] The *dewa*, being of relatively small size, is usually part of a bigger unit called a *tsowa*, which comprises a number of *dewa*s inhabiting a larger territory and tracing their origin back to a more distant or mythical ancestor (Levine s.a.). Common descent is an important but not indispensable condition for a *tsowa* affiliation. Families and larger groups can migrate for demographic, environmental, or other reasons and be admitted into another *tsowa*, often for strategic purposes.[27]

The Metsang live at the southern edge of territories that stretch from Golok up to Koko Nor Lake, from which Qinghai province (Chin. for Koko Nor) takes its name. These territories are inhabited by a large group of pastoralists whom the Metsang call Wranakh. This is also how the Metsang describe themselves: they say that they belong to Metsang *dewa a*nd Wranakh *tsowa*. The Wranakhs were mentioned by a number of authors who travelled through the Tibetan plateau in the nineteenth and early twentieth century. They pictured the Wranakhs as a distinct group or alliance of pastoralists (Przhevalsky 1876; Rockhill 1894; Grumm-Grzhimailo 1899, 1907; Grenard 1974 [1904]; Ekvall 1968 and Roerich 1933, 1958, 1961).[28] Nikolai Przhevalsky provided the first notes on their history (1876: 151-153), William Rockhill estimated their population number (1894: 115), and George Roerich listed their tongue as one of the three Tibetan dialects spoken in Amdo (1958: 6). Finally, Rockhill provided the first photographs of the people and Grigorii Grum-Grzimailo the first glossary of the Wranakh language (1899: 417-420). The common identity that the

25 About the female leaders of the Archung Huonmo or Ralo *dewa* immortalized in Western literature as 'Golok Queen', see Horlemann 2007 and 2015, Tugs chen skyabs 1991, Yang 2006.

26 On the art of mediating conflicts, see Pirie 2005, 2012.

27 Similar approaches were described inter alia by Ekvall (1977), Norbu (1997), and Thargyal (2007).

28 Some scholars have argued that the name Wranakh ('Black Tent') does not refer to any distinct group of pastoralists but includes them all (Hermanns 1949: 45). Other authors contradict this: 'within the aBrogpa [i.e. pastoralists] are a number of subcategories: the sBa nag (black tents), an agglomeration of tribes in Amdo; the mGo Log [...]; and the Byang Pa [...], nomads of the high northern Plain' (Ekvall 1968: 20). This confusion stems from a belief that Tibetan pastoralists always lived in black tents; but several informants told me that the Goloks lived in white tents before. For more on the Wranakhs, see Sulek 2010b.

Metsang share with other Wranakhs finds its expression not only in verbal self-statements, but in a number of cultural practices. My informants from Domkhok often chose their spouses from among the Wranakhs living north of Golok. The Metsangs' language is closer to that of their northern neighbours rather than that spoken by the Goloks. In their religious sympathies, they are predominantly followers of the 'new' school of Buddhism – the Gelug – rather than the older schools that the Goloks prefer.[29]

The Metsang pastoralists trace their history back to Central Tibet, which – as they say – their ancestors left during the time of Sakya Pandita Kunga Gyeltsen and his nephew Drogön Chögyel Phagpa.[30] The connection to these famous Buddhist hierarchs, who played a major role in the history of Tibet, is a source of pride for many Metsang pastoralists, who stress that they have historical origins rather than mythical ones. This alludes to the origin narrative of the Goloks, which starts with a relationship between a wanderer from another part of the Tibetan plateau and a daughter of the *zhibdag* or territorial deity of Nyenpo Yurtse.[31] The migration from Central Tibet is not the only one in Metsang history. Before they arrived in their current settlements, their ancestors lived in Rongwo Lonchu in Rebkong (a county in Malho TAP, Qinghai). The reasons they left Rongwo Lonchu are explained in different ways. According to a local historian, the explanation hides in the name Metsang, which alludes to an internal conflict (Tib. *me* or *nangme*) as a result of which one of the embattled sides had to leave the area. This is said to connect to military conflicts between Tibetans and Mongols, which tore up this part of the Tibetan plateau in the

29 Other differences express themselves in marital customs, kinship terms, clothing, dietary habits, etc. Although some of them were perhaps slightly exaggerated by my informants, who, living in Golok, felt a need to stress that they are *not* Goloks, they are interesting material through which to study the processes of mutual stereotyping and self-representation.

30 Sakya Pandita Kunga Gyeltsen (1182-1251) was a Buddhist scholar of the Sakya school. When Tibet stopped paying tribute to the Mongols after Genghis Khan's death, Godan Khan sent his troops to Central Tibet, creating an urgent need for negotiations. This job fell to Sakya Pandita Kunga Gyeltsen, who took up the role of agent of Mongol policy in Tibet and had a position as a feudatory chief under Mongol suzerainty (Shakabpa 2010: 210ff). He is credited in Tibetan sources with big successes in preaching Buddhism to the Mongols and remained among them for the last years of his life. He seems to have been followed by a whole group of people who served as auxiliary troops in the Mongol army (Petech 1990: 8, 13). He also took along his nephew Drogön Chögyel Phagpa (1235-1280). This Phagpa Lama is known as the one who established a preceptor-donor or *chöyon* relationship between Tibet and the court of Kublai Khan (following an example set by Sakya Pandita and Godan) and authored the so-called square script to be used across the whole multilingual Yuan dynasty empire.

31 On this narrative, see Rock (quoting Stein) 1956: 125, Kornman 1997, Don grub dBang rgyal 1991. My Golok informants said that the man arrived from Ngari, in western Tibet. Other sources mention different places.

Figure 1 Domkhok Township Seat

eighteenth century (Pad ma Tshe ring 2004: 25-26). The pastoralists whom I interviewed were divided in their opinions as to why they had to migrate. Some admitted the existence of the past conflict; others mentioned reasons of economic or demographic character. In either case, they agreed that the migration took place six generations ago, counting in lifetimes of the leaders of the *dewa*. In 2010, Ada Sodruk, the sixth *huonpo* counting from the time of this migration, was in his eighties and lived in Domkhok Township.

The Metsang make their first appearance in European literature at the beginning of the twentieth century. Vsevolod Roborovskii, describing his expedition to Golok in 1893-1895, mentions people called Rtau-metsen who, as he says, are 'impossible robbers' and fall under the authority of the Khangsar Goloks (1900: 384). Similar information is given by Joseph Rock in his monograph about Amnye Machen. Rock writes that the Metsang are 'a Tibetan tribe who lived in constant fear of the Go-log robbers, and as one Go-log does not rob another, they joined the tribe of the powerful Khang-sar Go-log chief in order to gain his protection' (1956: 125).[32] Both authors allude to a political alliance that existed between the Metsang and

32 The Metsang appear also in Thomas (1948: 21). I thank Wim van Spengen for this information.

Goloks in which the Metsang were the 'outer *dewa*' or *pshilde* of the Goloks.[33] This arrangement benefited both sides. The Metsang (together with the Dawu *dewa*) supported Khangsar Goloks in their military campaigns. For this support, they received from the Goloks the land on the sunny side of the Machu, including the site of today's Domkhok Township (Pad ma Tshe ring 2004: 30).

Research Locations

The main part of the material for this book was collected in Domkhok Township in Machen County. This township was chosen for a number of reasons. Domkhok is small, but its fame is big. It is well known, both in Golok and outside, for producing caterpillar fungus of superb quality. Consequently, it has a reputation as a particularly prosperous area. For research focused on the effects of the caterpillar fungus boom on the pastoral society it was a perfect destination. Moreover, Domkhok lies not far from Dawu. The pastoralists can reach the town, access its markets, and use the facilities it offers relatively easily. For a study showing the pastoralists' contacts with the town, this proximity mattered. Traffic on the Domkhok-Dawu road also goes in the opposite direction: diggers, traders, and officials controlling the use of caterpillar fungus resources can easily reach the township. For a study showing cooperation networks between pastoralists and diggers, as well as the problems of access to the caterpillar fungus-producing land, this was not insignificant.

There were also a number of practical reasons to select Domkhok. The township is inhabited by a relatively homogenous community belonging, with a few exceptions, to one *dewa*. These common origins and historical experiences reduced the risk of too many diversifying factors influencing the research. Domkhok is also the smallest (780 km^2) of the three townships in Machen that produce the best caterpillar fungus; the other two are Xueshan (1350 km^2) and Dawu Zhuma (1641 km^2).[34] This spatial compactness carried a hope that I could build a picture representative of the township as a whole. Its short distance from town meant that I could commute between the

33 This organizational structure was mentioned by Gelek in his study about Golok Serthar: the 'inner *dewa*' (*nangde*) shared a common ancestor, while the 'outer *dewa*' had different origins and was less 'organically' tied with the rest (1998: 51), see also dPal bzang 1991: 240.

34 Xueshan functions in the local language under its Chinese rather than Tibetan name Gangri. Another example is Chamahe, which the pastoralists seldom call by its Tibetan name of Chongmar.

Figure 2 Soglung Valley in winter

Figure 3 Summer camp

township and the town, and conduct interviews in both of them, which was crucial for a study about pastoralists *and* the market. This also allowed me to interview pastoralists who had moved to town and who are only guests in the highlands. Lastly, Domkhok is officially open to foreigners. This reduced the risks that could arise for the local inhabitants from participating in my research if I had conducted it elsewhere. For me, it was a guarantee that I would be able to carry out my research without stumbling on too many obvious legal obstacles.

Domkhok is located in a valley formed by the Dom River (Domchu), a small tributary of the Yellow River. Its name, so the pastoralists say, comes from the time when this area teemed with wildlife: deer, wild yaks, and especially Tibetan brown bear. It is from the latter (Tib. *dom*, Lat. *Ursus arctos pruinosus*) that the river is believed to take its name. *Khok*, in this context, means a valley where a river spreads its waters wide in the mountains. The Domkhok Township seat lies precisely in such a place.

The township seat is a small place with a village-like feel. Rows of low buildings run along one street. There is a small clinic, a primary school, police station, small restaurant, a few shops, a bank, and a motorbike repair shop. Township administration offices were distributed around a small courtyard: all of them rather rudimentary, nothing like the multi-storey glass architecture of county towns. Private houses, enclosed with mud walls, spread in a disorderly fashion around this village hub (Figure 1). There are few inhabitants: mostly elderly pastoralists who enjoy the comforts of permanent housing and look after their grandchildren attending school. Township officials have their flats and offices here, too, but they usually spend more time in the town. Under a large parasol, in front of the restaurant, the same group of neighbours gathers every day: a local doctor, a veterinarian, and a restaurant owner spend sunny hours on a worn-out sofa. Sometimes, a group of cyclists stops here on a tour around Amnye Machen, causing amusement with their outfits and expensive bicycles. But on the average day little happens. One of the reasons is that Domkhok is only 32 kilometres from Dawu and people prefer to go there to find better shops and entertainment.

Around the township seat in all directions spread rolling hills of grassland framed by rocky silhouettes of mountain peaks. In the valleys created by the small tributaries of the Domchu, on both sides of its course, dwell the pastoralists who are the main protagonists of this book. These valleys are divided into administrative units called *richen* and these, in turn, into *richung*, as the smallest administrative units in the township are known.[35] I conducted research in different *richungs*, which meant studying different valleys and working with their communities. These valleys differ in size, population, and quality of their resources. In some, the breath of the town is felt in people's lives and a string of cars pass in front of their windows. Others lie in more secluded corners: reaching them during bad weather can be difficult, if not impossible. The smallest valley was inhabited by only two and the biggest by eighteen households.[36] In some valleys, all families continued a rural way of life. In other valleys, some people had moved to Dawu and employed herders to manage their livestock back home. In some valleys, people boast of their caterpillar fungus resources. In others, they complain that their land is less fertile. However, all this was relative and did not change the fact that Domkhok was known for the good quality of its land, which attracted diggers and traders from distant regions.

Although I conducted research in several valleys, only one of them appears in this book. This valley is called Soglung.[37] It is a fictious name. Soglung exists, but I have concealed its name in order not to make identifying it too easy.

Soglung is a long valley with a small stream murmuring on its stony river bed. The water swarms with tiny fish and there are swampy meadows with frogs. In lower Soglung, the pastoralists have their houses, all connected by a road that runs in a sinusoid-like wave (Figure 2). When wet and muddy, these ups and downs are bone-chilling: once in a while, a car slides down the slope and dirt, ashes, and gravel are thrown under the wheels when one needs to go to town. Travelling in winter is easier: all traffic in Domkhok is done on frozen rivers. The summer pastures in Soglung are located higher up in the valley, where people camp in tents.

35 *Richen* (Chin. *xing zheng cun*) is translated as 'administrative village' and *richung* (Chin. *zhi ran cun*) as 'herders' group', 'herders' association', or 'natural village' (Rong 2011). Before 1984, when the pastoralists were organized in the people's communes, these units had different names: *richen* was a production brigade (Chin. *dadui*) and *richung* a production team (Chin. *xiaodui*). These names are often still in use. Speaking about a meeting held by the *richen* administration, the pastoralists would often say that there was a '*dadui* meeting', and in explaining where they live they often give a *xiaodui* number. Thus, the names of the people's commune units outlived the system they were part of.

36 During my research, the number of households was 369 and the number of residents did not exceed two thousand (MCSY 2009: 11ff, MCGW 2010).

37 Soglung means 'Mongolian Valley'. Many place names in Golok contain 'Sog' ('Mongol' or 'Mongolian') and allude to a time when the area was inhabited by the Mongols. Some toponyms have a meaning understandable only in Mongolian, such as Amnye Wayin whose name comes from Mongolian *bayan* 'rich'. On Mongolian names in Golok, see Sulek 2014.

Yaks and sheep graze on the meadows covered with yellow *metok serchen* (Lat. *Viola biflora*). Smoke rises from the tents, cheese dries in the sun, fat cats rub themselves against the tent ropes (Figure 3). It is a good place to live, say most pastoralists.

But Soglung is not paradise and people have their problems. One of them is growing populations of plateau pika and the larvae of the moth *Gynaephora alpherakii*, which destroy the grassland.[38] Another problem is mining. In a neighbouring valley, a large mining company made its headquarters; its impact on the environment is a source of concern for the pastoralists. People observe that the quality of the land is deteriorating and the animals produce less milk. Some attribute this to climate change and others to mining. Some of them joke, wistfully, that not only the government, but also nature does not want to support the pastoralists. Finally, the level of security has decreased too. Although there is much more order now and feuding and raiding are not as common as in the past, as my informants observed, new material wealth that people have accumulated attracts thieves. At night, when the howling of wolves resounds in the air, the pastoralists let their mastiffs loose: the ghostly sound of their chains gives this otherwise very pleasant valley a hellish atmosphere.

38 The *Abra* or plateau pika (Lat. *Ochotona curzoniae*) is a small burrowing mammal commonly blamed for contributing to grassland degradation. Some experts argue that pikas help sustain biodiversity (Foggin and Smith 1999). Concerning *Gynaephora alpherakii* (Tib. *bu thuwa*), see Zhou et al. 2012.

2 Digging

My first days in Domkhok were like diving into deep water. Everyone in the family was up and about well before sunrise. As soon as the alarm clock rang, Tsering Drölma would run to milk the female yaks (*dris*). When the herd left to graze, she collected dung for fuel. Tendor made a fire in the stove. After having a bowl of tea and a few bites of old bread, we set off for the mountains. The path led along a small stream, which in May was still covered with ice. It took me some time to get used to climbing at over 4000 m altitude and running across fields of sleet covering the slopes. I made even slower progress in digging caterpillar fungus. Caterpillar fungus is buried in the ground, only its 'head' (*go*) or 'horn' (*ra*), which Tibetans say to refer to its stroma, sticks out, but it has the same brown colour as everything else covering the pastures. It was like looking for a needle in a haystack. All around me, I heard cries of joy. People waved their hands to show what they had found, but I had hardly got anything. I must have overlooked many fungi, because just a few metres to the left and right people found dozens. My companions scanned the ground with enviable efficacy. No fungus could escape their attention. Then they made two moves of a hoe to dig it out of the ground with a lump of soil covering it: a quick move of the hand to break the soil and extract the fungus; another move to put the soil back into the hole. The fungus goes in their pocket and the digger moves on.

It was May, but at this altitude in the highlands the first signs of spring are shy. Patches of snow lie in shady parts of the valley and snowstorms are common. High on the mountain range, wind blows the hat off your head. Sun burns your skin during the day, but in the evening the temperatures fall below zero. The slopes are steep and you must be fit to reach all the places where caterpillar fungus grows. Spending whole days walking on all fours or bent over with your face close to the ground is exhausting, too. The breaks, when we met for lunch or some snacks, were cherished by everyone. In the evening, we would run down the slopes with newfound energy to reach the house as quickly as possible. But there would be many tasks to be accomplished before dinner. Someone has to bring the yaks home and tether them for the night. Someone has to fetch water. Waiting until a dried yak leg boils for supper, we would watch television and clean the fungi. One person notes how many fungi were found; another chooses the best specimens to take to market. The lights in the house go out late at night. Taking a good rest is necessary before a new day arrives and everyone sets off into the mountains again.

Figure 4 Caterpillar fungus digging

Diggers

Our digging team consisted of the extended family of Tsering Drölma and her husband Tendor.[1] Tsering Drölma was 21 when I met her for the first time, Tendor was 27. They had been married for six years and had a four-year-old daughter. Also part of the team was Jigmed Dorji, Tsering Drölma's maternal uncle, who lived in a nearby house and shared Tsering Drölma and Tendor's pastureland. Jigmed Dorji was 46. He was married and had three teenage daughters. Apart from Jigmed Dorji, his wife, and children, we were joined by his brother and sister-in-law, who lived in another less fertile valley and preferred to dig caterpillar fungus in Soglung. Another sister-in-law of Jigmed Dorji was there too. Her husband had recently taken religious vows to become a tantric practitioner and so stopped digging. The woman did not want to work alone and so joined her in-laws. Tsering Drölma's siblings

1 He was a *magpa*, a man who marries into his wife's family. This happens when the woman's family does not have a male heir (or when all sons married and moved out or chose monastic life) or when parents have a special emotional bond with one of their daughters (often the youngest) and want to keep her at home. Economic reasons play a role too: when a man does not have land or livestock, he can settle with his in-laws.

also came from the town: her sister from Dawu, where she studied at the Tibetan high school, and her brother from Xining, where he was trying to make it as a pop singer. Tsering Drölma's mother was with us as well. She had remarried after Tsering Drölma's father died and left the land and house to her daughter. She now lives in Xueshan and her new husband was a township official who did not have any land. Finally, we were joined by two men who worked for Tsering Drölma and Jigmed Dorji as herders and lived in houses owned by the family.

We were a very diverse group of people. Tsesojid, the student, and Norbu, the singer, brought with them a city-like air. Tsesojid had pink fingernails, wore a Hello Kitty jumper and a baseball cap. She was too shy to speak English but learned it at school and hummed some American hits when no one was listening. When she was not helping her sister with work, she spent time reading a Chinese girl's magazine. Norbu sported a Korean-style haircut, a North Face jacket, and leather trousers. After work, he watched a television show about amateur singers: he dreamed of taking part in it. My presence embarrassed him: he excused himself for the fact that his sister could not cook and that Tibetans are 'backward'.[2] Compared to this young generation, Jigmed Dorji's sister-in-law looked like something from a photographic album about Tibetan folk costumes. She arrived in Soglung wearing an elaborate robe trimmed with a fake leopard skin and a high felt hat. The next day, she was barely recognizable in her working clothes: dirty jeans, worn-out trainers, and thick glasses on her nose. This working dress code levelled differences: all members of our team adjusted to it and donned trousers, outdoor jackets, scarves, and gloves. I also left my Tibetan robe at home: it was too heavy for this work. The only thing I lacked was a face mask, which the women wore to protect their skin against the sun and wind.

Time and Tools

The caterpillar fungus digging season in Golok starts in May. This is later than in southern and warmer areas of the Tibetan plateau, in Sichuan, but

2 He also spoke about the British military expedition to Tibet of 1903-1904. In this act of imperialistic aggression, he said, tens of thousands of Tibetans were killed – this is contradicted by Western sources, but described as such in China. In a moment of irritation, I told Norbu that he had 'no brain' and that I was not going to talk to him anymore. I immediately realized how unfortunate it was to offend a person from my host's family. However, this episode actually became a widely cherished anecdote that subsequently opened many doors for me.

earlier than in higher-elevated and drier parts of the TAR.[3] The season lasts several weeks, around forty to fifty days.[4] The first weeks are the time when the greatest number and the best quality fungus are found. The later the season, the less fungus is found, but people keep digging even in June and July.[5] The timing of the season and its quality depend on the weather during winter preceding it. After long and cold winters, the season starts late, but cold winters, when the earth is frozen for a long time, serve the fungus well. Mild winters are considered bad. Spring weather also has an impact. Heavy rainfall inhibits the fungus's growth, whereas light rains carry a promise of abundant harvests ('It's raining with money', as one trader said to me when it began to rain). Heavy snowfalls in spring – something not unusual at this altitude – are also bad for the fungus, but light and fast-melting snow plays a positive role ('Today there's snow and tomorrow you go digging!', as Tendor said). Other factors play a role, too, as Tendor explained:

> *Yartsa* grows best on the winter land (*gunsa*). It grows better in shady places (*shybsa*) than in the sunny (*nyinsa*) ones. And in the shady ones it grows better in the high places (*mthosa*). In spring, if it rains a little and then the sun comes out, it's really good. But if it snows, it can be bad. When the snow covers the ground [too long], *yartsa* can't grow and goes rotten in the soil.[6]

The 'winter' and 'summer land' refer to two areas of Tsering Drölma and Tendor's pastureland. We worked on the winter land and commuted between the house and the winter pastures every day.[7] In Soglung, the short distances allowed this, but in other regions of the plateau, in Sichuan or Yunnan, people leave their homes and move their herds to higher pastures and camp in tents. Tsering Drölma or Tendor could not imagine doing this. They weren't about to give up their home comforts. Taking yaks to the summer pastures in spring is impossible and spring is also calving time. Although both families employ herders, Tsering Drölma wanted to be near her herds.

3 In Sichuan it starts in April (Boesi 2003: 32) and in the TAR as late as June (Winkler 2008a: 294).

4 This length was given by my informants. It is also a period for which digging licences were issued in the past.

5 Winkler also observes that 'a sharp elevation gradient can extend a locality's season by several weeks since the higher slopes tend to fruit later than the lower ones' (2008a: 294).

6 Boesi quotes his informants as saying that 'when the spring season is cold and snowfalls are abundant, it is very difficult for the grass to grow upwards because of the thick layer of snow that covers the ground' (2003: 32).

7 The relation between the character of the terrain and the practice of setting up camps in the mountains is clear in Nepal or Bhutan, where people trek for several days to reach the elevation at which the fungus grows and carry with them tents and provisions for the weeks ahead.

Caterpillar fungus digging is considered as having relatively low barriers to entry (Winkler 2008b: 18). This is one of the reasons why it is often compared to a 'gold rush': it is open to a large number of people and requires little in terms of preparation or tools. Good eyes and a fit body are a prerequisite. Some knowledge about the fungus and the terrain is also needed. But every person fulfilling these basic conditions should be able to engage in this work.

The verb used to describe gathering the fungus, or *kuwa*, reveals much of the process: one does not *collect* the fungus, one *digs* it. The main tool is a hoe with a 25-30 cm long blade (Figure 5). In Domkhok houses, this hoe or *kakle* was normally used for chopping dung and digging edible or medicinal roots. One or two hoes suffice for these standard uses, but now several extra hoes are kept for relatives and other diggers.[8] Apart from a hoe, little else is needed. Caterpillar fungus, when extracted from the ground, is covered with a thin layer of soil. It is then packed, together with this soil, in a plastic bag and put into the digger's pocket. The people I worked with used instant noodle soup bags: they are a handy size and the soups were popular during the digging season when people have little time to cook. At home, people clean the fungus with old toothbrushes and spread it out on cardboard to dry. Dry fungus is then stored in cloth bags or shoe boxes. Specimens that are to be sold fresh are wrapped in moist toilet paper and toothpicks and matches are used to fix broken parts. This is the basic toolkit: with a dose of ingenuity, these tools, borrowed from other activities or even 'recycled', gain a new life.

The digging season is a period of mass mobilization of labour. Every morning, groups of people armed with metal hoes walk up the hills and, every evening, they return the same way. If we saw, from our window, Jigmed Dorji's family going upstream, we knew we had to hurry. In the mountains, when we saw them going downhill, we knew it was time to go home. Our working day lasted eight or nine hours. The exact length of time the pastoralists spend digging depends on the weather and the amount of fungus growing in a given area. As Tendor explained: 'If there isn't much to find, we get home at 2-3 pm. But if there's a lot, we return even as late as 6-7 pm.' This meant ten or eleven hours of work! During peak season, people spend all their time in the mountains, and spend the least time needed for managing household tasks. If one could look for fungus by torchlight, there would certainly be people ready to extend their working hours into the night.

8 The popularity of this hoe can be measured by its *absence* from the shops. When I searched all the Dawu hardware shops in June, I could find only one hoe in the whole town. The sellers looked at me puzzled from behind piles of horse saddles and rows of metal bells: 'You came too late'. The same thing happened in Xining: 'Come next year *before* spring', the sellers said.

Figure 5 Freshly dug caterpillar fungus

Caterpillar fungus digging takes up a large part of the pastoralists' time, but it is only one thing on their list of daily chores. Every morning and evening, people need to tend their herds, milk the *dris*, send the herds out to graze, and bring them back home. Some *dris* need assistance during calving and weaker calves need extra care. If the household has sheep, someone has to take care of the flock. But there is also other work: processing milk, fetching water, collecting dung, and cooking all take time. Spring is also when oats and barley for winter fodder are sown. All these jobs are pushed to the margins of the day when the caterpillar fungus harvest is in full swing. The *dris* are milked before dawn and women fetch water after dusk. The days are stretched to the maximum. Although it may seem that digging caterpillar fungus is squeezed in between the other tasks that structure the working days in the highlands, the opposite happens. It is the other tasks that are subordinated to the digging routines: the money-producing potential shapes the new hierarchy of work. Only occasionally, when snowfalls make trips to the mountains impossible, do people do the things they usually don't have time for: Tsering Drölma would do the laundry and Tendor repair a motorbike. As soon as the snow melted, they would take their daughter to the neighbours and set off to the mountains to look for fungus again.

Does Everyone go Digging?

For Tsering Drölma and Tendor, digging caterpillar fungus is pleasant work. They enjoy being in the open and spending time with their relatives, many of whom now live elsewhere. The pastoralists' enthusiasm for digging caterpillar fungus is well observed. Empty towns and villages is a common trope in the literature about spring in pastoral Tibet where 'everyone who can walk goes digging'. However, even if it is good and brings in decent profits, this work is physically strenuous and someone unfit would have difficulties performing such a task day after day.

Do all pastoralists dig caterpillar fungus? My survey showed that two thirds of people in Domkhok do. Those who stay home are usually persons whose health or age does not permit them to perform such strenuous activity. People over sixty do not normally dig, and the same goes for children younger than ten: their parents usually said they would hinder rather than help the work. Needless to say, there are exceptions. Jigmed Dorji's aunt was 74 when I met her. She was well known for her talent in digging and she proudly told me that she had found over a thousand fungi in less than a month.

Scholars have observed that Buddhist monks, nuns, and other religious specialists refrain from digging caterpillar fungus for faith-related reasons (Hyytiainen 2011; Yeh and Lama 2013). Domkhok does not provide many examples confirming this, partly because there are few monks and religious specialists in the township. There is no monastery there, nor has there ever been one, and parents are reluctant to send their children to more distant places where they have fewer contacts. In the whole township, several boys were enrolled in Aku Chöyon's monastery in Dawu.[9] They divided their time between the monastery and their parents' home, where they helped with work, sometimes also with digging. There were several elderly monks from Rarja monastery, who had returned to their families due to age and weak health. They seldom left home and certainly did not go digging.[10] There were also two retired commune cadres, who had taken religious vows and committed their life to religious practice, and one tantric practitioner (*ngagpa*), Jigmed Dorji's brother.[11] Only he did not dig caterpillar fungus for reasons explicitly related to religion.

9 The monastery was founded in 2004. Its full name is Lhari Tashi Thondröl Dokha, but it is usually referred to after its founder Jadrel Chöyon Rangdröl.

10 Rarja (Rarja Gon Ganden Tashi Jungne Shesub Darje Ling) is a Gelug monastery on the Yellow River founded in 1769.

11 Tibetan religious culture is populated by many kinds of religious specialists. In contrast to fully ordained monks, *ngagpas* are non-celibate tantric practitioners who often enter the path of religious life in their adult years.

The degree of a household's engagement in digging depends on its size and age profile. The best adapted are households consisting of a couple in their thirties or forties and teenage children: each of them can be a digger. Couples with small children are less fortunate, but larger households are not always in a better position. A bigger household generally means a more complex structure. A household of seven usually comprises three or four generations living together, of which the youngest and eldest members cannot dig. Having a small child or elderly parent means that someone has to stay home with them. Jigmed Dorji's household illustrates this. In 2007, he and his wife were in their forties and their daughters were between twelve and sixteen years old. Before my research ended, one of the girls got married and the young couple moved in with the girl's parents. They were in no hurry to start a family and they continued digging. However, as soon as they had a child, they would have to decide how to combine childcare with work. One person would have to stay home or they would need to employ a babysitter. This is what Tsering Drölma did: together with a neighbour, they hired a man who looked after their children when they went digging. The man was unable to dig himself because of his poor eyesight. Working as a babysitter offered him a perfect chance to earn additional money.

A household reaches equilibrium of sorts when grandparents help with the children, but the situation becomes complicated when the eldest family members require care. So, it was with the family of Tamkho, our neighbour in Soglung. Tamkho's mother was in her nineties. She suffered from dementia, did not recognize her relatives, and spent days consumed in prayer. Tamkho stayed home with her, while her husband, daughter, and son-in-law went digging. Tamkho's two grandsons stayed at home, too. The family had decided that they were too small to work and should learn to read and write instead, which they both did very well. But Tamkho did not want to be idle. After thinking over the problem for a time, she took in a daughter of a distant relative: when she went to dig, the girl stayed with Tamkho's mother, and when Tamkho was home, the girl went digging. The fungus she found acted as payment for her work.

These examples show that the pool of human resources in the household shrinks the smaller the children and the older the seniors are. The maximal mobilization of labour force is possible only when the family is not burdened with taking care of its youngest, eldest, or disabled members. But optimal situations are short-lived. Households must seek creative solutions for how to employ their workforce in digging caterpillar fungus without losing much of the harvest or neglecting other duties. In each case, a decision about who stays home and who goes digging is based on the calculation of gains and losses. If there are enough able-bodied diggers, they take turns at the work. Otherwise, the person who is least prepared to endure the hardships of such work, or has other skills required at home, stays behind. Finally, persons who have other income earning prospects can also be released from digging.

Women and Men

Different people in the household engage in digging caterpillar fungus to different degrees. While some work every day, others are only occasional diggers. The latter are mostly men, who – as in the case of Jigmed Dorji – sometimes worked with us for the whole day, sometimes joined us in the afternoon, or did not come at all. Jigmed Dorji owned a small shop in the township. He hired a shopkeeper but visited the shop to check up on the situation and to supply new products. Other men also had extra jobs. One neighbour worked in the township administration. He had to attend official meetings, which prevented him from working with his family every day: he helped them as much as he could, but on some days he fulfilled his other duties. Another man drove tourists to Amnye Machen. Arguing that it was too far to walk to the mountains to dig caterpillar fungus for half a day once he had finished his job, he sat in the cold and empty kitchen, smoking cigarettes and waiting for his wife to get home.

Men's participation in digging is governed by their other income generating opportunities. Having another job is sufficient reason to cease work in the mountains. But men also go to town on other errands: to do shopping, pay electricity bills, and sell caterpillar fungus. Compared to them, women are less mobile. They are tied to the pastureland and domestic work with fewer flexible bonds and their role as caterpillar fungus diggers is more fixed. Women do not have any economic alternatives that bring in money, or other duties outside the household. My female informants could barely remember when they had last taken a break from work. It usually happened when their child was ill, more seldomly when they themselves were. Sometimes, but rarely compared to men, they might take a day off and accompany their husband to town. On most days, though, women follow a daily routine, tending to their usual household tasks in addition to digging caterpillar fungus.

It was a common opinion in Domkhok that the best workers are women and children and that the loss of a female digger in a family is a bigger setback than losing a male one. While children owe their success to their 'good eyes and being close to the ground', women – so my informants said – are particularly industrious and get the best results. These better results should not be a surprise given that they spent more time doing this work than men. Praise of women's skills and industriousness can be used by men to justify why women should continue to dig, whereas men can stop digging with little effect on the family budget. However, women's industriousness may have another source. The usual work that women perform in the household does not translate into direct access to cash. Digging caterpillar fungus provides a rare occasion to earn their own money. Women's readiness to climb the mountains day after day can be connected to the sense of economic empowerment that this work brings. By performing work that contributes so much to the family budget, women can negotiate a stronger position in the household, but they can also hide some fungus to sell it later and thus have their own money.

In classical anthropological studies, gathering has often been labelled part of the female economy. This applies to Tibet, as well. Robert Ekvall, who lived among Tibetan pastoralists during the 1930s and 1940s, notes that hunting and raiding, performed away from home, were male occupations, while gathering, done close to home, was an 'exclusively female role'. He admits that it was a work 'of some significance, bringing into the economy both foodstuffs and cash crops'. Women dug for wild garlic, leeks, and small tubers called *droma*, but also 'unidentified mushrooms, which pastoralists call "soil gold", because of their colour, and never eat them, but sell them to Chinese' (Ekvall 1983: 56).[12] This simplified model of labour division is questioned by many scholars, who argue that it rarely conforms to reality. They show that in many societies men go gathering and women hunt game, also alone and on a daily basis (e.g. Lee and DeVore 1968; Estioko-Griffin and Griffin 1981; Willerslev 2007). This flexibility of economic roles is, however, obscured both by the scholars' own bias to see some roles as 'male' and others as 'female' and by their informants' opinions about what jobs *ideally* belong to whom (Ingold 1987: 87; Jarvenpa and Brumbach 2006: 6).

The case of caterpillar fungus shows that gathering in present-day Golok is not a 'female' activity and men perform it as well. Men's tendency to evade their digging duties can be interpreted as indicating that some negative value is ascribed to it as 'female', but no one ever expressed such

12 *Droma* are the tubers of Silverweed (*Argentina anserina*) and are eaten as a delicacy, with rice or yoghurt, and in other dishes.

an opinion to me.[13] Caterpillar fungus is also not the first and only species that men gather. They collect other commercial species: matsutake and morels, culinary mushrooms exported to Japan and Europe, as well as marsh orchids, snow lotuses, and *pimo* (Winkler 2008b: 5).[14] In the past, they also gathered rhubarb, the trade in which was popular in nineteenth- and early twentieth-century Tibet (Norbu 1997: 68; Winkler 2008b).[15] If some prejudice against men engaging in the gathering economy really existed in the past, the income from selling these commercially important species could have played a role in relaxing this prejudice and widening the gap between the model division of the economic roles that scholars observed and the actual practice.[16] What distinguishes caterpillar fungus from other species collected for sale is, of course, the sort of income it brings. Thus, although the emergence of the caterpillar fungus economy is not the first moment when Tibetan men have engaged in gathering, the financial incentive perhaps explains why they engage in it on such a scale.

13 Seen from the perspective of Tibetans from farming areas, the 'female' character of this activity looks even more doubtful. There are both female and male diggers in Golok, but a farming family would rather send a man on such a trip, especially if they did not have an agreement with the pastoralists on whose land the person was going to dig. This makes it similar to the phenomenon of 'going for income', when men leave their villages in search of non-farm work (Goldstein et al. 2008).

14 These species are: matsutake (*beha* or *behing hamo*), Lat. *Tricholoma matsutake*; morels (*guguhamo*), Lat. *Morchella esculenta*, *Morchella conica/elata*; marsh orchids (*wanlag*), Lat. *Gymnadenia* ssp.; snow lotuses (*ganlha metok*), Lat. *Saussurea medusa*. Digging *pimo* was popular in the 1990s when one *jin* cost 200-300 yuan. During my research, it was dug only by people who found themselves in a difficult financial situation. One family who I lived with were a good example. Their son had killed a man during a pasture fight and so they had to pay 'blood money' or compensation to the victim's family. This left them bankrupt and they were forced to look for any jobs bringing even the smallest income, even digging up *pimo*. My other informants laughed when I asked whether they dug *pimo*: it was too cheap to be worth the effort. On blood money, see Ekvall 1964a; Pirie 2005 and 2008.

15 Trade in rhubarb (*hchum*, Lat. *Rheum officinale*), well-known in many parts of Tibet in the past, did not seem popular in Golok. One of the *dewa* leaders (b. 1926) said that it was a reason for violent conflicts between pastoralists and in-migrants prior to the 1950s.

16 During my research, one *jin* of matsutake cost 60-120 yuan. Fresh morels cost 70 yuan per *jin* and dry ones 700-800 yuan per *jin* (Winkler 2010: 22).

3 Fungus, Medicine, Commodity

In the Tibetan language, caterpillar fungus is called *yartsa gumbu* or 'summer grass winter worm'. This name captures the complex nature of this organism and the seasonal metamorphosis it undergoes. As the pastoralists say, in winter it is *mbu*, or an insect, and in summer is *rtsa*, or grass. Tsering Drölma explained:

> In winter, it's a worm, a very small one. When the soil thaws in spring, the worm grows bigger and a horn grows on its head. If you don't dig it, in summer you'll see a flower growing on this horn. The flower withers and the wind disperses the seeds. They fall on the ground and the whole process starts again.

This organism is an entomophagous fungus called *Ophiocordyceps sinensis*, which parasitizes the larvae of ghost moths, largely from the genus *Thitarodes*. These caterpillars (the 'worm' from the quotation above) are attacked by the fungal spores, the fungus develops inside the host organism and then causes its death. Upon reaching maturity, it produces a stroma, or fruiting body (the 'horn'), which protrudes from the larva's head. On the top of it a sporacia, or spore-producing tissue, develops (the 'flower'), containing spores that attack more larvae. A large part of this process takes place in the soil, where the larvae spend the winter. In spring, caterpillar fungus already has the hybrid look of a larva with a blade- or horn-like stroma. The larva has an almost lifelike appearance but is completely filled with mycelium. It rests vertically in the ground. The only part visible above the ground is the stroma; it is by locating this that the caterpillar fungus can be found.

This sort of caterpillar fungus is endemic to the Tibetan plateau and the Himalayas. It grows in China, Nepal, Bhutan, and India. In China, its distribution reaches up to the Qilian Mountains in northern Qinghai and to Nagchu Prefecture in the TAR (Map 4). It covers five provinces or autonomous regions: the eastern part of the TAR, Qinghai, south-western Gansu, western Sichuan, and north-western Yunnan – which in total accounts for one tenth of China's territory (Yao 2010). Three factors shape the extent of this distribution zone: altitude, humidity, and temperature. The fungus grows at altitudes between 3000 and 5000 m above sea level (the highest confirmed

Map 4 Caterpillar fungus distribution zone (after Daniel Winkler)

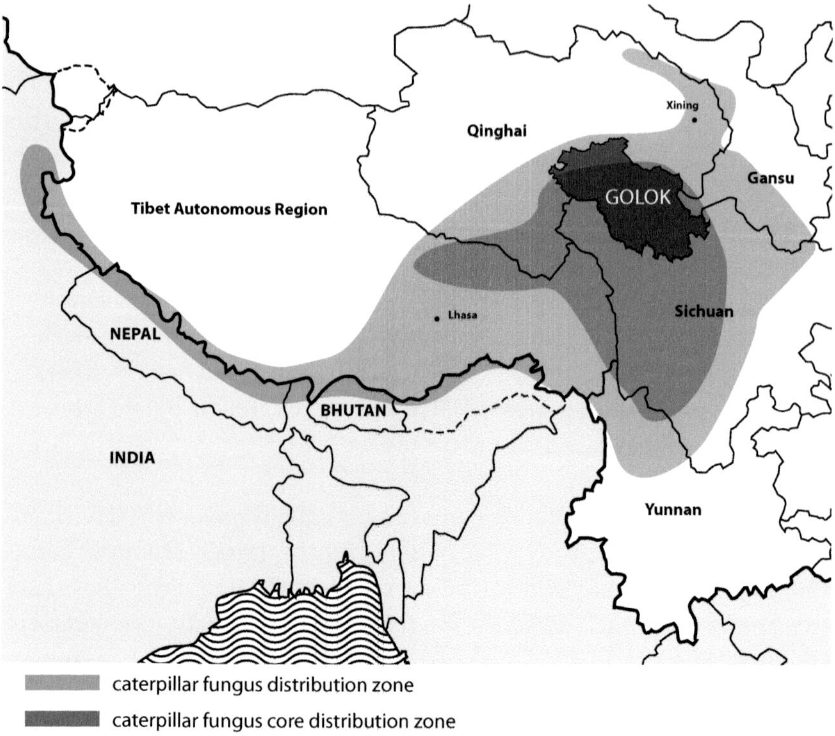

altitude is 5200 m), but most of it is found at 4000-4700 m.[1] It requires a minimum precipitation of 350 mm per annum and relatively moist soil (36-48 per cent). The optimum temperature for its growth is 15-18°C, but it can already start growing at 1.2°C and survives winters in the ground when the soil temperature falls below zero (Yang 2010).

Medicine

The beginnings of caterpillar fungus's medicinal career are not easy to ascertain. This task is additionally complicated by the contemporary explosion of interest in this resource as well as advertising strategies that envision its usage as reaching deep into the past. A popular narrative reproduced in advertising materials says that caterpillar fungus had been 'discovered'

1 Due to climate change, this distribution zone is moving higher (Yang 2010). Moreover, pastoralists observed that the fungus can be found at higher altitudes than before.

1500 or 2000 years ago: Tibetan herders observed that their yaks became particularly vigorous after eating this fungus. Thus, having recognized its unique properties, they started using it as a medicine.[2]

What we know for sure is really much less than that. The earliest identified Tibetan text containing a description of caterpillar fungus (without referring to it by name) comes from the fifteenth century. In a work entitled *Ten Million Relics* (*Bye ba ring bsrel*) by Nyamnyi Dorje (1439-1475), the fungus is described as follows: 'In the summer it is a blade of grass [growing] on a worm similar to the leaf of mountain garlic. The flower resembles a silken green sedge. The root resembles cumin seed at the end of autumn' (Winkler 2008b: 32-33).[3] Tibetan doctors in Golok say that caterpillar fungus is also mentioned in the most important (and earlier) Tibetan medical treatise: the *Four Medical Tantras* (*rGyud bzhi*).[4] The name *yartsa gumbu* is not mentioned there either though. These doctors say that the fungus hides behind the name *rtsabshid* or *rtsa tabshid*. *Rtsabshid* is said to be similar in its powers to *tabshid* or *kang tabshid*. The latter two names denote so-called snow frog, whose meat is used as a medicine against 'cold disorders of the waist and kidneys' and the urine of which is a potent aphrodisiac (Clark 1995: 179).[5] *Rtsabshid* is said to be a cheaper equivalent of snow frog. Some authors state that *rtsabshid* and caterpillar fungus are synonymous (dGa' ba'i rDo rje 1998: 109, Byams pa 'Phrin las 2006: 707). Tibetan doctors in Golok stress that *rtsabshid* is a broader category that cannot be reduced to one species. It includes many substances of plant, mineral, or animal origin; caterpillar fungus is merely one of them (Boesi and

2 This narrative appears in a lot of advertising materials and some academic works (e.g. Devkota 2006: 48; Holliday et al. 2005: 2).

3 For a longer excerpt, see Winkler 2008b: 32-36.

4 *rGyud bzhi* or *Bdud rtsi snying po yan lag brgyad pa gsang ba mang ngag gi rgyud* is the most important work in the canon of Tibetan medicine. Its importance explains why debate about the use of caterpillar fungus in Tibetan medicine focuses on whether this fungus was mentioned in it or not. *rGyud bzhi* is conventionally portrayed as a translation of a Sanskrit original, which was been brought to Tibet, translated, and adapted to local conditions. Another school of thought holds that it was originally written in Tibet by Yuthog Yontan Gonpo the Younger (1112-1203). On the origins of *rGyud bzhi*, see Karmay 1998b; Gerke 2011: 89ff.

5 The medical dictionary of *dGa ba'i rDo rje* identifies 'snow frog' (Tib. *ganwel*) as the stream salamander (Lat. *Batrachuperus pinchonii*) which lives at lower altitudes in Sichuan and Yunnan (1998: 376). It could also be alpine stream salamander or Tibetan mountain salamander (Lat. *Batrachuperus tibetanus*), which lives at up to 4250 m on the plateau (thanks to Li Li for a discussion on this topic). The doctors in Golok explained that 'snow frog' refers only to male frogs that live on snow-clad mountains. During the mating season they migrate to lower altitudes and need to be caught before mating: only then does their flesh have medicinal value. Many young doctors claimed that this is all pure fantasy, but other doctors, including such respected figures as Abu Karlo, said they used snow frog in their practice.

Cardi 2009: 5). This confusing web of interrelations illustrates the difficulties of investigating the place of caterpillar fungus in Tibetan medical sources.

Although caterpillar fungus is not explicitly mentioned in the *Four Medical Tantras*, knowledge of it is believed to be much older.[6] Some doctors in Golok believe that its properties were known during the reign of Tri Songtsen (d. 649), the founder of the Tibetan Empire. His marriage to the Chinese princess Wencheng explains how China learned about this medicine. Other doctors argue that the fungus has been known since the reign of Tri Songdetsen (742-*c.* 800), who is credited with calling a 'medical congress' in Samye that was attended by medical practitioners from different parts of Asia.[7] Caterpillar fungus, according to my source, was one of the topics discussed during the congress. It is unclear whether it was the local oral tradition that suggested such beginnings to the medicinal career of caterpillar fungus or if it was the question itself that provoked these speculations. It is also possible that the current popularity of caterpillar fungus has enticed people to reconstruct or reinvent its history. Such reconstructions must be placed in the context of the ongoing competition over the discovery and symbolical ownership of this resource. Chinese literature calls it 'Chinese caterpillar fungus', which is partly justified by the scientific name *Ophiocordyceps sinensis*, but the adjective 'Chinese' gains additional dimension in the context of the Sino-Tibetan political conflict. In any case, caterpillar fungus is famed as one of the three biggest treasures of Chinese pharmacopeia (alongside ginseng and deer antlers) and it has been proposed that it receives the title of National Fungus of China (Zhang et al. 2012, Ma et al. 2010: 9). Some sources claim that it was described by Chinese authors well before the Common Era, which would give China the lead in the race as to who discovered the medicinal properties of this resource (Holliday et al. 2005: 2). Needless to say, any remarks about the use of caterpillar fungus appear in the Chinese medical literature much later.[8]

The lack of attention paid to caterpillar fungus in Tibetan medical literature is no great surprise given how marginal a role this fungus plays

6 For more information about caterpillar fungus in Tibetan medical sources, see Boesi and Cardi 2009.

7 This congress is a famous (mentioned in Tibetan histories as well as by Tibetan doctors today) event that very likely never took place. However, there is little doubt that exchanges between medical practitioners from Tibet and other countries really did happen at this time and in earlier periods (see Martin 2011).

8 About caterpillar fungus in Chinese medical sources, see Li et al. 2011: 913; Zhu et al. 1998: 289. The first source they identify is *Ben Cao Bei Yao* (*Essentials of Materia Medica*), composed in 1694. Other sources are *Ben Cao Cong Xin* (*A New Compilation of Materia Medica*) (1757) and *Beng Cao Gang Mu Shi Yi* (*Supplements to Compendium of Materia Medica*) (1765).

in Tibetan doctors' practice. Those I interviewed did not differ from their colleagues from other regions who 'do not frequently employ and do not highly value *Cordyceps* claiming to know other herbs that, once mixed, have the same properties but give more effective results' (2009: 6). I heard a similar opinion in Golok, where the doctors said that caterpillar fungus 'is good, but certainly not as good as people say' and 'its price is better than its quality'. In the most popular pharmacy in Dawu, caterpillar fungus was an ingredient in only four out of hundreds of medicines: three were Chinese biomedical products and the fourth was an alcohol-based tincture with *pimo*, deer antler, and other ingredients. A similar situation was observed in Bhutan, where general recognition of the medicinal properties of caterpillar fungus did not translate into its use: local doctors have only started using it recently, clearly due to the impact of the fungus's new popularity (Boesi and Cardi 2009: 6). Similarly, patients in Golok have started requesting prescriptions for medicines containing caterpillar fungus. The doctors were reluctant to comply with such requests. They were mildly sceptical about the qualities of caterpillar fungus and surprised or even perplexed by its rise in popularity, which they view as unexpected, if not undeserved.

In the Chinese pharmacopoeia, caterpillar fungus has been officially classified as a drug since 1964 (CPCMH 1964: 77). It is recommended as a tonic for strengthening the immune system as well as a remedy for a long list of ailments. It is prescribed for lung, liver, and kidney disorders. It supposedly improves respiratory function, and combats hepatitis and cirrhosis. It is advised for use against cancer, to neutralize the side-effects of chemotherapy, and in cases of organ transplants. It is said to help against high blood pressure and to have anti-aging properties; it stimulates hormone production and inhibits the oxidation process. It is also believed to enhance women's fertility and combat male sexual dysfunction. Whether used against a particular disease or in prophylactics, caterpillar fungus can be taken as a highly processed pharmaceutical product or in raw and unprocessed form. It is also popular in so-called herbal cuisine. Customers of such pricy 'herbal restaurants' are assured that food is inspected by a Chinese Medicine doctor and adjusted to their body constitution. Cooking with caterpillar fungus is also popularized by cookbooks, which recommend dishes against insomnia, asthma, atherosclerosis, and even memory loss.[9]

9 Such cookbooks were available in the more elegant caterpillar fungus shops of Xining (e.g. *The Sanjiangyuan Cookbook* proposed a chicken soup with caterpillar fungus, which 'increases body energy, helps against premature ejaculation and sore back', DCXC [s.a.]: 12). Some were also available in Golok, and in Tibetan language, and proposed recipes (e.g. yak tongue with caterpillar fungus) adapted to the palate of Tibetan customers.

The body of literature describing the medicinal qualities of caterpillar fungus in China is growing proportionally to its popularity and price, with many studies confirming its unique values. The major active ingredient of caterpillar fungus is called cordycepin, but the exact chemical composition of caterpillar fungus, especially compared to other *Cordyceps* species, appears to be inadequately studied (Zhu et al. 1998: 293). Some authors argue that Chinese scholars overlook the fact that many of the chemical ingredients of this fungus are actually typical to a wider group of fungi. They contend that much of the medical research on this topic shows contradictory results and is not peer reviewed to international standards (Dong and Yao 2010: 1032; Cannon et al. 2009: 2269). Some authors call the enthusiastic reports about the fungus unsubstantiated and criticize the advertising of it as unethical, giving patients empty hopes of finding a cure for cancer and other serious diseases (Paterson 2008: 1489). Although pharmaceuticals derived from other *Cordyceps* species are used in biomedicine, caterpillar fungus is not, at least not in mainstream healthcare in the West (Cannon et al. 2009: 2269). It is, however, available in both official and unofficial circulation. Advertising materials attract customers by calling it 'Himalayan Viagra'. The mooted anti-cancer and anti-aging properties are underlined: the fungus is even claimed to offer the secret of eternal youth and is advertised as a cure for the consequences of the stress of living in technologically developed societies (Paterson 2008: 1470).

Caterpillar fungus is a wild product of Tibetan grasslands, but a number of companies, in China and abroad, cultivate it in artificial conditions.[10] This means cultivating fungal strands (hyphae) on artificial substrate (such caterpillar fungus is often called 'artificial', in comparison to the 'wild' one, growing in nature). Cultivating it on infected larvae, as happens in the natural environment, remains difficult.[11] One of the reasons why cultivating caterpillar fungus is advertised as important is a growing insecurity

10 Xinhui Chinese Caterpillar Fungus Cluster in Guangdong claims to cultivate 50 tons of artificially grown fungus per year (Fan and Lin 2007: 1), Qinghai Everest Aweto Pharmaceuticals aims at producing 110 tons (EP [s.a.]: 16), and Aloha Medicals from the USA ('The World's Largest Producer of Organic Medicinal Mushrooms' as its website informs) declares that it produces 2 tons of the fungus which is 'genetically and analytically bio-identical' with the wild one and grown in conditions identical to nature (see AM 2012). Shen Nanyin, who conducted research on caterpillar fungus and its cultivation since the 1950s, did not believe that it would ever be economically viable. Interview, Xining, 14.06.2010; see also Stewart 2009: 83.

11 Winkler notes that growing it on larvae is important in Asia where 'people have an appreciation for the wild fungus so they want to see the caterpillar. In the West, people don't want to see the caterpillar. [...] So, the artificially produced *Cordyceps* is just perfect for the Western market' (2008c: 4).

about the quality of the wild fungus. Because the fungus is sold by weight, the traders purportedly increase its weight using methods that are not entirely health neutral. Stories about such practices were told to me by the pastoralists and by traders in Golok, who mentioned whole specimens made of plaster and wheat flour. These stories were skilfully used (perhaps even produced) by the pharmaceutical industry, which benefits from the moral panic that such reports cause. In 2010, CCTV aired (and repeated many times) a film made as if with a hidden camera showing a gang of criminals who sprayed fungi with heavy metal powders, soaked it in alum solution, and inserted a lead wire into the larvae – as the voice-over explained.[12] The manager of a large pharmaceutical company in Xining told me that no fungus available at the market is 'safe': because 'wild' caterpillar fungus is less natural than it seems, artificial cultivation is more than needed.[13]

Domestic Uses

The pastoralists often refer to caterpillar fungus as *menrtsa* or 'medicinal grass'. This, however, does not mean that they use it in their own domestic medical practice. Asked about whether there were any domestic uses of caterpillar fungus, my informants usually shrugged their shoulders and said, 'We actually don't use it'. Some people recalled how, in their childhood, they would bet with their playmates to see who could eat the most fungi, *pimo*, or *droma*, but – as one woman stressed – this was not considered healthy. Observations from Golok confirm those from other Tibetan regions: that caterpillar fungus was seldom used in people's domestic practices (Boesi and Cardi 2009: 4). In comparison to other regions, in Golok, caterpillar fungus was used even less, a fact that is not surprising if one considers that caterpillar fungus was also less established as a commodity or trade item there. In Lithang, for example, Tibetans made an alcoholic drink with caterpillar fungus (Boesi and Cardi 2009: 6).[14] In

12 According to Shen Nanying, a bigger risk comes from replacing caterpillar fungus with similar-looking species, such as *Ophiocordyceps militaris*, which not only lack medicinal properties but can be damaging to health. Interview, Xining, 14.06.2010.

13 Interview with Zhang Xuefeng, chief product designer at 5xVery Grass, Xining, 20.06.2010. See also Holliday and Cleaver for photos of caterpillar fungus specimens with a metal wire inside (2008: 223).

14 The drink is made by dipping a few fungi in *arak* (liquor produced from barley or rice) in a proportion of 3-5 fungi for each half-litre. After storage in a cool place for 2-3 months the drink is ready: the longer it 'ages' the more potent it becomes (Boesi and Cardi 2009: 6).

Golok, this use was mentioned to me by a few persons, but they could not give me any details. Other, more sophisticated uses known from the Himalayan regions include a milk-*yartsa* drink (the fungus is boiled with milk and butter and drunk daily at least for a week, Devkota 2010) and a tea-*yartsa* drink (caterpillar fungus with tea, milk, and honey; Lama et al. 2001: 56; Panda and Swain 2011: 10). None of them had ever been heard of by my informants.

In fact, the material from Golok adds only one point to this rather brief list of domestic uses of caterpillar fungus, but in this case, it is not used for people but for livestock. A pastoralist and former PLA soldier reported that caterpillar fungus is a good fodder supplement, especially for race horses. He recalled feeding horses with a mixture of dried fungi and *rtsampa* or fungi boiled with *rtsampa*, tea leaves, and salt.[15] A similar use was mentioned to me by a Han trader in Xining, a well-established figure with long experience of trade with Tibetans:

> I met a nomad, perhaps eighty years old. He said he had seen piles of *chongcao* in his life, but he never ate it.[16] He was convinced about its powers and fed it to his horses. It was a long time ago, in the 1960s or 1970s, when people still travelled on horseback. At that altitude, at 5000 m, not only people suffer, but animals too. This nomad had two horses. When one of them was exhausted, he rode the other one. When he gave his horse water with *chongcao*, the horse recovered and half an hour later he could ride it again.

This use of caterpillar fungus would fit in well in the Golok context, where livestock – especially in winter when animals are weaker – are given fodder supplements of tea leaves, wheat, *rtsampa*, and salt.[17] However, my informant, quoted above, said that the 'horse recipes' he knew were neither used in his family in the past, nor in his *dewa*. He learned them in the army. Another man explained: 'In the communes, *they* told us to feed horses with *yartsa*'. Today, though, nobody would feed caterpillar fungus to horses, as he underlined. What one might use is actually the soil covering the larvae: water boiled with this soil can be given to livestock.

15 *Rtsampa*, roasted barley flour, is a Tibetan staple foodstuff with a rich symbolic meaning. '*Rtsampa*-eaters' is sometimes used as an evocative designation of all Tibetans; see Shakya 1993, Ramble 1993.

16 An abbreviated version of the Chinese name of caterpillar fungus, *donchong xiacao*, which is normally used in colloquial language.

17 A similar use of caterpillar fungus is confirmed from Nepal (Pohle 1990: 37).

As one woman remarked: 'People can drink it too. But no one would eat the whole fungi, maybe only some old or broken ones. Otherwise it would be too expensive.'[18]

Trying to access a large group of informants from different parts of Golok and gather comparative material, I conducted a questionnaire at Golok Tibetan High School in Dawu and asked students about the uses of caterpillar fungus in their families.[19] A number of students admitted that the fungus is used at home. Four (out of 69) declared that it is given to yaks and horses, eight that their parents take it with alcohol, four that it is taken with food, and another four that it is used as a medicine. A problem arising here concerns where the border between a medicine, food, and alcoholic drink is: caterpillar fungus taken with alcohol can also be classified as a medicine. These four partly overlapping uses were mentioned by the students; none of them was particularly well represented. The majority of students (36) replied that in their families caterpillar fungus is not used at all, eleven gave no answer, and two replied that it is used but did not know how.

These answers show that caterpillar fungus has not generally been popular in Golok pastoralists' domestic practices. Moreover, the positive answers given – those stating that it *is* used – do not offer information about the history or regularity of a given practice. Caterpillar fungus has become so acclaimed for its medicinal value and is so closely associated with a particular standard of life, that it is difficult to tell whether the ways people use it now are a continuation of a formerly existing practice or part of a new lifestyle. Whereas caterpillar fungus could have been used as a fodder supplement in the past (although some informants suggested that this use also is rather new), alcoholic drinks with caterpillar fungus may well be a recent invention. Finally, consuming caterpillar fungus with food is most likely inspired by dietary practices from China's urban centres, which are popularized by cookbooks and illustrated magazines available also in Golok.

18 This soil is sold at the market in Xining. During one of my interviews, a Han couple waited until the women employed in the shop had cleaned newly delivered fungi. The couple used this soil for their houseplants. There are two types of soil obtained from cleaning the fungus: the outer and the inner one, closer to the larva's body. The latter is used on small injuries to stop bleeding. This explains why the fungus is often cleaned in two phases: first the outer layer of the soil is removed, then the inner one.

19 It covered first- and second-grade students of between 15 and 20 years old, from all counties and mostly from pastoral families (63 out of 69). The age differences resulted from the different times at which they had started their education.

Trade

In some parts of the Tibetan plateau, the trade in caterpillar fungus was already practised in the seventeenth century.[20] Western explorers and scholars mentioned a 'worm plant known as the Shar-tsa gong-bu' in Lithang (Rockhill 1984: 361), a 'grass worm' in Bathang (Shelton 1921: 318), and 'insect grass' in Dartsemdo (Coales 1919: 244).[21] None of them mentioned Golok, a region located far from important trade routes and seldom visited by foreigners. When I asked the local pastoralists how old this trade was in Golok, they said it had a short history. Before Golok became part of the PRC, the trade practically did not exist. As one elderly man explained: 'In the old society, there were no Chinese and grass belonged to yaks, sheep, and horses. There was no other use for it.'[22] This opinion was often repeated: many pastoralists said that the trade started only 'with the Chinese'.

The trade in caterpillar fungus in Golok seems to have a short history, but this history is nevertheless longer than one may assume: the recent economic boom did not evolve from nothing. This trade was also carried out during the period of the people's communes, where it was part of the command economy and was controlled by the state. The earliest date of confirmed, organized state purchase of caterpillar fungus in Golok is 1962, as both the former commune cadres and pastoralists recalled. This locates the trade in the period following the crushing of the pastoralists' armed resistance against the Communist invasion, forced collectivization, and Great Leap Forward.[23] This dramatic period was followed by a short political thaw when people could again practise religion, at least to a degree, monks returned to the monasteries, and some properties confiscated in the initial phase of collectivization were returned to the owners. This was also when the trade in caterpillar fungus was introduced in the communes. The outbreak of the Cultural Revolution in 1966 brought it to a halt. The former commune cadres in Dawu recalled that the trade was resumed soon after Mao Zedong's death and continued until the end of the communes.

20 The main suppliers were Tibetan areas of Kham, today in Sichuan. During the late Qing dynasty they exported 10 t per year, and in the Republic of China even more (see Liu 1994 and Wang 2000, after Winkler 2005: 75).

21 Lithang and Bathang are today counties in Kamdze TAP, Sichuan. Dartsemdo (Chin. Kangding) is the name of a county-level city and seat of Kamdze TAP.

22 The 'old society' or the 'old world' (Tib. *jigten nyangwa*) is how people in Golok often refer to the times before Golok became part of the People's Republic of China.

23 Perhaps the most dramatic description of this period in recent literature is Li 2016. It also agrees very much with what people in Golok who remember those days recall.

Digging caterpillar fungus was one of the tasks performed by the pastoralists working in the people's communes. Alongside typical pastoral work, such as herding livestock, milking, producing dairy and other pastoral products, they logged trees, hunted wild game, produced sod walls and bricks, built roads, and collected *materia medica*. The list of medicinal plants, minerals, and fungi that the pastoralists collected differed between different parts of Golok, but caterpillar fungus was common throughout most of the prefecture, only the amounts that people dug varied. It also had a special place among other medicinal species: digging it was obligatory, whereas most other species people just collected on a voluntary basis. A former commune cadre recalled:

> In those days, we had a planned economy and received orders from the province. Then, we divided them between counties. If there was an order to produce 1000 *jin* of wool, it would go to the prefecture and be divided between the counties.[24] With *yartsa* it was similar. The prefecture divided the plan among the counties, giving bigger obligations to Machen, Gabde, and Darlag, because they produced more. So, people dug *yartsa* according to where they lived. For example, if we had to deliver 100 *jin*, it was divided among the households and each of them had to get one *jin* or maybe three or four *shang*.

The pastoralists in the people's communes supplied caterpillar fungus to the so-called trade agencies (Tib. *chitsog nyotson*, Chin. *waimao gongsy*), which bought pastoral and other products, such as yak hides and hair, wool and sheepskins, but also animal pelts and medicinal species. For supplying the required amounts of caterpillar fungus, they would receive payment. This, again, gave caterpillar fungus a special position. For other work, the pastoralists were remunerated in work-points, called *karma* (Chin. *gongfen*), which were converted into cash in proportion to the output of the production team the person belonged to. It is an interesting and little-known fact that the situation with caterpillar fungus was different: its price was fixed and it was paid for regardless of the production team's overall work results. This means that a direct link between caterpillar fungus and cash income was already recognized in those days, which were otherwise characterized by a shortage of cash.

24 One *jin* (Tib. *jama*), is equivalent to 0.5 kg. It is the main weight unit used for caterpillar fungus. I use it in the book instead of converting all values into the metric system, to avoid 'half kilograms'.

The 1980s brought the economic reforms initiated by Deng Xiaoping and with them the end of the people's communes and the state monopoly over caterpillar fungus trade. The communes were dissolved and the Household Responsibility System was introduced, which made pastoral households into autonomous production units. Households now carried 'responsibilities' towards the state: they had to meet quotas and fulfil tax obligations. But they could also make their own decisions about how to manage their land and livestock and how to use their income. The economic liberalization of the 1980s opened the gates to private trade and allowed the pastoralists to sell the surplus of their production (everything above the quota) on the private market. The trade agencies did not disappear: they were now responsible for purchasing the quota products. They also fought for surpluses of pastoral production, trying to encourage people to remain within the state trade channels. Whereas prices for the quota sales were kept at a low, under-the-market level, for delivering anything above the quota a supplier got a higher, 'negotiated price' (Chin. *yijia*). Another incentive that the agencies used was a system of prizes (*gartah*). It had previously been used in the people's communes and entitled particularly successful production teams or brigades to buy extra rations of fertilizer, clothes, or fodder (Oi 1991: 51; Potter and Potter 1991: 147). In the shortage economy, this privilege had a special value. In the 1980s, when the bonus system was brought in in Golok, the principle was similar: particularly successful caterpillar fungus suppliers were entitled to buy certain attractive and sought-after products, such as felt hats and otter pelts, at a lower price. Many pastoralists fondly remember such enticements to this day.[25]

Pastoral regions such as Golok had limited access to private markets during the 1980s, but even there private trade slowly developed, opening up new possibilities for those who resisted the incentive programmes of the trade agencies. The dominant picture of the beginnings of private trade in Tibet assigns the key role to Hui and Han migrants from other parts of China, who successfully colonized the emerging entrepreneurial niche on the Tibetan plateau (Gladney 1998: 1; Hansen 2005: 4). Although the Huis have played a key role in caterpillar fungus trade since the 1980s, Tibetans also engaged in it. Particularly active were Tibetans from Ngawa in Sichuan, who were one of the ice-breakers of private trade in Golok. Local pastoralists

25 Otter pelts were traditionally used to trim Tibetan robes. They fell out of use after the Dalai Lama, in 2006, criticized Tibetans for wearing endangered animal pelts. Soon after, Tibetans in China started publicly burning animal pelts torn off of their clothing. See Bassini 2012; Yeh 2013b.

were active as well. Some of these pioneers are still possible to identify. They were often individuals who had previously worked as people's commune cadres and continued their work in the new administration system when the communes were replaced with the townships, with their own administrative apparatus. This work allowed them to establish contacts, gather know-how, and accumulate or access capital. A son of one of these pioneers in Domkhok recalled:

> My father was the first to start business here. People got profit from it and he profited too. He traded with everything: butter, cheese, yak hides, sheepskins, and *yartsa*. When *yartsa* was 1 yuan my father paid 1.50, and people earned from it. It was similar with butter. When the market price was bad, people were doomed to lose. But we paid more. We also paid in advance, if people needed cash.

It is tempting to think that developments in Golok resembled those in other parts of China and that some unofficial private trade in caterpillar fungus existed on the margins of the command economy. Studies of rural economies in China show that private trade survived the command economy and intensified during periods of political relaxation in the 1960s and 1970s (Eyferth 2009; Ruf 1998). If some unofficial trade existed in Golok too, it must have been done either by the commune cadres or whole commune units – production teams or brigades – who sought better income or barter exchange outside of state-controlled channels.[26] Such a 'hidden economy' was tolerated in many parts of China: the more remote or struggling for subsistence a commune was, the more the authorities turned a blind eye to its hidden economic life, which was necessary for the commune to survive (Chan and Unger 1982). If such unofficial trade in caterpillar fungus existed in Golok, its structures could have helped the private trade to emerge during the 1980s or formed a blueprint for future private trade structures.

Freeing the trade from the constraints of the command economy triggered changes in the caterpillar fungus market. The decades following the

26 Eyferth shows how the leaders of the production teams in Sichuan made their private deals instead of selling to the state: 'those "with guts" sold on the black market, whereas "cowards" sold to the state' (2009: 181). Occasional reports suggest that such unofficial trade existed in Tibet, as well. For instance, the villages around Yamdroktso Lake in the TAR bartered fish for agricultural products with the neighbouring commune units (Diana Lange, pers. comm., Berlin, 8 February 2012). There are signs that in Golok it could be similar. A man, who in the 2000s traded used motorbikes and cars, said of the 1980s: 'we could finally start a normal life, instead of being afraid that doing business secretly would bring on us some terrible punishment'.

economic liberalization of the Deng Xiaoping era brought growth to the trade in caterpillar fungus, measured by the number of people engaged in it as well as prices, market demand, and volumes sold. From the 1980s, prices rose year on year. Between the 1960s and the late 1970s, one *jin* of fungus in Golok cost 4-12 yuan, depending on its quality.[27] At the end of the 1980s, prices reached four-digit values for the first time. The trade agency employee remembered:

> Before 1978, *yartsa* was twelve yuan [per *jin*]. Later, the price changed from one year to another. In 1981, they paid twenty-five yuan, next year forty, and a year later sixty. There were years with 150 yuan, 125 yuan, and 250 yuan per *jin*. In 1986, it was 400 yuan. In 1987, it was already 600 yuan. And in 1988, it was 1000 yuan.

This was still the beginning of the process, and today the 1980s and 1990s are remembered by the pastoralists and traders as a time when caterpillar fungus was 'very cheap'. The traders who started their careers in that period recall almost in disbelief that in 1990 one piece of fungus cost 7-8 *jiao*. Some pastoralists remember even lower prices of 3-4 *jiao*. Taking into account the inflation rate during this period, the increase in caterpillar fungus's value was smaller than the prices suggest.[28] However, the relation between the value of the fungus and prices of other goods and services shows how this change was felt by the population. As my informants recalled, in the mid-1980s, one fungus was worth 1 *jiao*, equivalent to *one fifth* of the simplest dish in a local eatery. Twenty years later, it cost 30-40 yuan, which could pay for several dishes in a proper restaurant.

Caterpillar fungus prices in different parts of the Tibetan plateau increased at different rates. In Golok, this was a slower process compared to other regions. Within the same period, the prices of caterpillar fungus in Lhasa reached double the values seen in Golok (Winkler 2009: 299). This situation continued for a long time, and in the year 2000, the best caterpillar fungus in Dawu fetched only 5000-7000 yuan per *jin*, while in Lhasa it cost double or more (Horlemann 2002: 262; Winkler 2009: 299). Only after 2000 did developments in Golok pick up speed. This was due to improved transport links, which facilitated the region's closer integration with the rest of the country. This was when the caterpillar fungus boom in Golok commenced.

27 A similar price (21 yuan/kg) was reported from Xining in the 1970s; see Wen 2004, after Winkler 2008a: 298.

28 For a discussion of inflation and price development, see Winkler 2008: 298.

The Boom

The market demand for caterpillar fungus started growing rapidly during the 1990s, and in the 2000s it was already said to exceed the price of gold (Dong and Yao 2010: 1027; Zhang et al. 2012). According to some statistics, it became the most expensive medicinal product exported from China (Mu and Zhu 2005: 3).[29] A number of reasons explain the growth in popularity of caterpillar fungus: rising consumerism of the new Chinese middle class and growth of the market for luxury goods (Stewart 2009: 75; Mu and Zhu 2005: 3), growing concern with health (Ma 2010: 7), and the progress in scientific research that made people recognize the fungus's medicinal value (Shen 2010). A 'fashion for Tibet' has perhaps contributed as well. In China, as in Europe, Tibet is often envisioned as a place of unspoiled nature and spirituality (Kapstein 2004; Sulek 2006), which makes it a popular tourist destination and attracts donors who support Buddhist monasteries as well as customers who buy products associated with this image (Lama 2007: 51; Hillman and Henfry 2006: 257).

Apart from these general phenomena, several other factors helped to market caterpillar fungus. One of them was a series of successes by Chinese athletes during Chinese and international sport events. This started in 1993 during the 4th World Championships in Athletics in Stuttgart where Chinese track runners broke several world records. This unexpected success led to speculation about the use of illicit drugs by the Chinese team, but the coach announced that his only secret is high-altitude training (on the Tibetan plateau) and a 'stress-relieving tonic' with caterpillar fungus (also from Tibet) (Steinkraus and Whitfield 1994: 235). This story circulated in the international media and was skilfully used by the marketing industry in China. This scenario repeated itself during international sport events in subsequent years, further strengthening the association between caterpillar fungus and health, stamina, and success.

The outbreak of the Severe Acute Respiratory Syndrome (SARS) epidemic in China in 2002-2003 helped increase interest in caterpillar fungus as well. In the TAR, the epidemic saw customers turning to Traditional Tibetan (and Chinese) Medicine. Pharmaceutical companies in Lhasa ran out of

29 Although it is sold to Japan, Korea, Southeast Asia, and other regions with large Chinese communities, it is mostly consumed in China. This also includes caterpillar fungus imported or smuggled from other countries. Hywel-Jones et al. (2002) estimate that 2 t of fungus are trafficked to China from Bhutan every year (after Namgyel 2003: 8). Negi et al. (2010) describe how the fungus is trafficked from India.

stock within days (Craig and Adams 2008: 3). A parallel rise in demand for caterpillar fungus was observed in Golok and Yushu (Lama 2007: 49). The epidemic created a sense of fear, which the pharmaceutical industry used to increase the sales of their products: caterpillar fungus, being a medicine against almost everything, was perfectly suited to being marketed as a protection against SARS. A Tibetan doctor from Dawu explained to me that people's interest in this product came from a combination of fear caused by the epidemic, emergence of new, health-oriented behaviours in society more generally, and people's growing affluence: 'People have money and want to have a good life. Men want to retain their sexual potency and *yartsa* can help them. People want to stay young, as if they were always thirty. The SARS epidemic also reminded them to take care of their health.'

The market career of caterpillar fungus is also connected to the transformation in gift culture in China. Apart from being bought for the customer's own medical use, the fungus is purchased as a present. Many shops sell elegant carved boxes with an assortment of particularly well-shaped fungi displayed on red velvety cloth. During my research, such gift sets cost around 29,000 yuan (100 g) and even up to 76,000 yuan (250 g), depending on their size and the quality of the fungus. An elegant boutique in Xining offered them tailor-made and recommended its best fungus for 150,000 yuan per *jin*. This fungus was kept in a crystal vase in a locked cabinet: 'We call it "soft gold" [Chin. *ruan huangjin*], but it's worth more than that', the shop manager said.[30] He added:

> Before, people gave each other alcohol or cigarettes, but after the outbreak of SARS they understood the importance of health and started buying *chongcao*. Who do you give it to? Parents, friends, colleagues ... Well, let's be frank, you give it to your superiors. Corruption is a tremendous problem. People despise it, but they have no power to change it.

The relation between caterpillar fungus and *guanxi* or 'connections culture' is widely made and traders admit that caterpillar fungus is given to upkeep social relations, show appreciation or gratitude, and to 'push through' certain

30 The shop belongs to Qinghai Sanjiangyuan Pharmaceutical Co. Ltd. A similar price (21 yuan/kg) was reported from Xining in the 1970s; see Wen 2004, after Winkler 2008a: 298. The company uses the Tibetan ethnicity of the founder as a guarantee of the quality of the product: the trust between the pastoralists and the owner 'symbolizes the quality of caterpillar fungus processed in the company' (SQSY 2009: 16).

decisions.[31] One trader in Xining remarked: 'You can't give anyone 40,000 yuan in cash, but a box of *chongcao* of that value? Absolutely!' Asked about who gives such presents to whom, he said: 'Lower officials give them to higher officials. Anyone does when he wants to see his matter positively settled.' The more expensive caterpillar fungus gets, the more popular it is as a luxury gift, as the traders commented, because – as one of them said – 'this is exactly what the officials want: luxuries!'

The last area where caterpillar fungus makes its presence known is the tourism industry, where it is sold as a 'medicinal souvenir'. This has been observed in such areas as Shangrila (Chin. Xianggelila) County in Dechen Prefecture in Yunnan, one of the most popular tourist destinations on the Tibetan plateau (Crouch 2010: 10).[32] Also in Golok, which is not a popular travel destination, tourist brochures use caterpillar fungus to attract visitors. There has even been a proposal to organize a 'caterpillar fungus festival' in Golok, during which tourists would watch dance performances, taste local specialities, and buy the fungus. They could also visit the pastoralists' settlements to see how people dig caterpillar fungus and perhaps do it themselves (Li et al. 2010: 35).

Caterpillar Fungus Production in Qinghai and Golok

Any figures on caterpillar fungus production are broad estimates and are based on fractional volumes of the fungus as sold and processed, which are declared by traders as well as transport companies and the pharmaceutical industry. However, the caterpillar fungus trade remains largely beyond state control; traders often pay no taxes and are not registered. Efforts are being made to regulate the trade, to give it uniform shape and centralized administration. Before this happens, though, the trade and social circles involved in it remain opaque to data collectors and official figures representing the quantity of fungus passing through the market must be taken with caution. Additionally, not all caterpillar fungus from a given

31 *Guanxi* is a Chinese term which my informants used in this context. It means 'relation' or 'relationship' and refers to social connections or networks which exist for instrumental and other purposes. According to its critics, *guanxi* is fuelling corruption, but other voices say that it is not that simple. In any case, *guanxi* is closely linked with a gift economy, and its attendant rites, rituals, and rules. See Gold et al. 2002.

32 The county was previously called (in Tib.) Gyalthang. It was officially renamed after the fictional land Shangri-La from James Hilton's novel *Lost Horizon* in 2001 in efforts to promote tourism.

year enters the market immediately but is sold later. All this limits the reliability of the statistical data that appear in the literature and official sources. Still, it is interesting to study them in order to realize that although caterpillar fungus is expensive and in high demand, it is by no means a scarce commodity.

The official sources estimate the total annual production of caterpillar fungus in China at 150-200 t (Sun 2008: 5). The main producers are Qinghai and Tibet Autonomous Region. The usual estimates for Qinghai speak of about 90-120 t per year.[33] Regardless of its exact contribution, Qinghai is an important caterpillar fungus producer, and it is new to this position. In the 1980s and 1990s, the TAR was the leader, although the values for both provinces were quite similar (TAR fifty per cent, Qinghai forty per cent; Sun 2008: 5). Within Qinghai, the main production zone is located at the headwaters of three rivers: the Machu (Yellow River), Dzachu (Mekong), and Drichu (Yangtze). This is where the Sanjiangyuan National Nature Reserve (SNNR) or the Three Rivers Nature Reserve is located. Some authors claim that this nature reserve produces 80 per cent of all caterpillar fungus in Qinghai (Cai 2010).[34] Such numbers, regardless of their accuracy, carry a message that the most fertile caterpillar fungus parts of Qinghai are Yushu and Golok, the two prefectures that contribute the largest tracts of land to the reserve.

In Golok, most if not all counties produce caterpillar fungus. A contentious place for scholars is Martod: some authors say that the fungus has never grown there, and others state that it did and perhaps still does.[35] In sum, the fungus is found in 36 townships and the area where it grows covers more than 20,000 km^2 (Zhang 2003: 33). The data concerning caterpillar fungus production in Golok are subject to similar problems as those for Qinghai – and maybe even more so, since only a tiny minority of traders are officially registered and the digging of caterpillar fungus goes on

33 Cai et al. (2010: 6) and Ma (2010: 9) mention 100 t, and Sun (2008: 5) 140 t; 90-120 t was given by Zhao Jinwen, General Manager of Xinqian International Fungus Co. Ltd (pers. comm., Xining, 14 June 2010).

34 The problem with the SNNR relates to the ambiguity of the terms Sanjiangyuan Reserve and Sanjiangyuan Area. The Sanjiangyuan Reserve initially covered 152,300 km^2, but the Sanjiangyuan Area is bigger and covers 318,000 km^2 (Foggin 2005, Li and Li 2002). Many sources use the term 'Sanjiangyuan Region' without stating what it refers to, and so it seems that the SNNR grows and shrinks depending on the occasion and purpose. In 2012, the SNNR was officially enlarged to 395,000 km^2 (Meng 2012).

35 Yang (2010) and Li et al. (2010) claim that Martod produces caterpillar fungus (they even provide figures). Other authors say that caterpillar fungus has never grown there (Li et al. 2011: 917). According to the pastoralists, Martod still produced large quantities of caterpillar fungus during the 1980s.

largely outside any control or regulation. The way the data are gathered also invites doubts about their accuracy. The data collectors extrapolate general production quantities from the turnout reported by a small number of traders. The problem is that caterpillar fungus changes hands many times before it leaves Golok; it is sold and bought between different traders on the same market. In this situation, collecting reliable data would require a profound understanding of the market and a much more elaborate data collection scheme.

Official publications provide two sets of figures: one is the production capacity of Golok and the other the amount of caterpillar fungus that is actually dug. The production capacity is usually estimated as 23-30 t per annum (Li et al. 2010: 32, Zhang 2003: 33, Zhou 2004: 37). Some sources speak of 40 t (Qiu 2010). The greatest production capacity is in the county of Gabde (7.4 t per year). Other county figures are: Machen 6.4 t; Darlag 4.7 t; Bamma 3.1 t; and Jigdril 3.4 t. The amounts of caterpillar fungus actually dug are smaller: in Machen 2.3 t; in Gabde 2.1; in Darlag 1.7; in Jigdril 1.2; in Bamma 1.1; and in Martod 0.6 t (Li et al. 2010: 32). This gives 9 t of caterpillar fungus dug in the whole prefecture, approximately one third of its supposed production capacity (Zhou 2004: 37).[36] Considering that Qinghai is said to produce 120 t per year and that Golok is one of its main suppliers, the above quantities are surprisingly low.[37]

These numbers, even if highly imprecise and questionable, show that Golok is not homogenous and not each of its counties is equally fertile. In Machen County, three townships produce the best quality caterpillar fungus and the greatest amount of it. Every trader in Dawu can list them in one breath and always in the same order: Xueshan, Domkhok, and Dawu Zhuma. All of them are located a short distance from Amnye Machen and it is this mountain that the pastoralists credit for giving their land its fertility. The rains and snow and the streams whose sources are within this mountain system, so many people say, nourish the land and provide it with wealth. The fact that Amnye Machen is one of the most sacred mountains in the north-eastern Tibetan plateau plays a role as well. As one woman commented: 'People in Domkhok have the blessing of Amnye Machen. That's why they have so much *yartsa*.' Other townships in Machen are less lucky. Some of

36 This is much less than is officially estimated for TAR, which is said to produce 65-70 t per year, of which 40-50 t is collected (Mu et al. 2011: 169).

37 These quantities could be true only if, as some sources say, Qinghai produced only 20-50 t (Gruschke 2011e: 69). However, most sources give much higher values. According to some, only Yushu (one of the prefectures in Qinghai) produces 12-22 t (Mu and Zhu 2005: 1).

them are even said to produce no caterpillar fungus at all.[38] No information about caterpillar fungus production in particular townships is to be found in the county statistics, but an official from Domkhok estimated for me that his township sells 0.5-1 t of the fungus per year. If Domkhok really yields 1 t of caterpillar fungus per year, it would be one fifth of what the whole of Machen County produces. The official estimate for Machen is 2.3 t, but in private interviews Tibetan county officials said that the county produces 5 t. Upper estimates for both Domkhok and Machen (1 t and 5 t) are more realistic than lower ones (0.5 t and 2.3 t), especially compared to the results of my survey.

Commodity

The above discussion shows that in today's Golok caterpillar fungus is more a commodity than a medicine. It is a wild natural product that became commoditized to such a degree that its market value overshadowed its other uses (if such were known before). The paradox consists in the fact that caterpillar fungus also grew in Golok before the boom and was known to the pastoralists. However, subsequent developments turned it into a market celebrity incomparable to anything else that the grassland produces.

In the simplest definition, a commodity is a socially desirable thing with a use value and an exchange value, but, seen in a socio-cultural perspective, none of these attributes is its intrinsic property. A commodity is not merely a material object, but rather its marked social form and an embodiment of an order of values and meanings projected on it by various actors. The constructed character of these values, 'forms of social consciousness' that build 'invisible chains linking relations between things to relations between people', is clearly visible in the case of caterpillar fungus (Gregory 1997: 13). It is a natural product of the Tibetan grassland, but its commoditization was stimulated by processes taking place outside the Tibetan plateau. Its market value comes from the outside in a double sense: it does not intrinsically 'belong' to caterpillar fungus and it comes from beyond the area where the fungus grows. The pastoralists in Golok have observed within their own lifetime how the price and market demand for this product have grown under

38 Four townships that are declared ecologically degraded (Yigzhung, Danzhung, Danlag, Chamahe) do not produce caterpillar fungus, as Golok officials claimed. The traders, however, bought fungus from there and the pastoralists who had relocated from these townships to Dawu returned to their land in spring to dig it.

the influence of processes in mainland China, where the main trade centres and consumers are. The scale of this sudden growth and the fact that the prices are dictated from beyond the local society, cause the fungus's market value to be perceived as something 'added' and not deriving from the fungus itself.[39] Although some of the medicinal powers of caterpillar fungus are recognized in Golok, at least by medical practitioners, its economic value is something that the pastoralists do not fully identify with. They could not explain why this fungus is so expensive. Its high prices and their seasonal changeability made people ask: What would remain if caterpillar fungus lost its economic value? One of the few women who used to trade in caterpillar fungus, said: 'This trade is like gambling. When the price declines, you're left with nothing. If *yartsa* one day becomes cheap, what will we do with it? It's *just* grass. We won't eat it.'

Caterpillar fungus has a high use value for its consumers, but for the pastoralists in Golok on whose land it grows and whose hands literally excavate it from the ground to introduce it into the world of commodities and trade, this use value is nearly identical with its exchange value. It is the saleability of the fungus that makes it useful.[40] Arjun Appadurai defines the 'commodity situation' in the social career of a given thing as one in which 'its exchangeability for some other thing is its socially relevant feature' (1986: 13). Caterpillar fungus barely exists outside this exchangeability context. It has been socially produced or rather *re*produced as a commodity so successfully that this commodity status nearly exhausts its identity (Gregory 1980: 641). Caterpillar fungus is dug mostly, if not exclusively, for sale. It is a commodity by *destination* or an 'object intended by their producers to be principally for sale' (Appadurai 1986: 16). If it is consumed in Golok at all, it is when it is damaged and cannot be sold. Even this my informants did more out of curiosity than conviction.

In Golok, caterpillar fungus is so closely associated with money that its price stands out as its immediate attribute. Being sold and bringing in income appears as its main function. What is caterpillar fungus good for? 'It's good for earning money', as one pastoralist stated. There was more than a grain of truth in his words. The close relation between caterpillar fungus and cash (and this relation is older than the boom itself, as shown above)

39 This reminds me of the words of a Tibetan mastiff trader: 'It's not the dog but the process that costs. The market is doing that. For me, dogs worth 10,000 and 100,000 yuan are the same.'

40 Byg et al. (2010) conducted an interesting study in which they showed rural Tibetans samples of plant and fungus species and asked about their names and uses. Four of them were identified by almost everyone: all four were commercially important.

is visible in my informants' statements that eating this fungus would be 'a waste of money'. One woman, when I asked if she ever ate it, exclaimed: 'Never! I'd feel like eating money!' This comparison is less metaphorical than it may seem. The caterpillar fungus to yuan 'conversion rate' is widely known, especially for large specimens that are sold fresh and per piece. During the caterpillar fungus season, prices of other goods and services can be calculated using 'one *yartsa*' as a cash-replacement unit. I observed cashless transactions where itinerant sellers of wristwatches, binoculars, or electric shavers accepted payment in caterpillar fungus. I was repeatedly told that prostitutes in Dawu also accept payment in caterpillar fungus. In all such transactions, caterpillar fungus is used as 'occasional currency': it is its convertibility into cash or its capacity to 'produce' cash that allows it to take this role.

Searching for other than commercial uses of caterpillar fungus showed that, at least in Golok, there were few. But if a commodity is an 'impersonal bundle of use value and exchange value' (Carrier 1995: 18), this bundle can be remodelled and new elements can be added to it. The fungus can be 'inscribed' with new meanings and functions: new ways of using it can penetrate into the fungus-growing regions under the impact from China's urban centres. These new uses can be perceived as imported and lacking original character. Asked whether people drink alcohol with caterpillar fungus, one man remarked: 'Few people do. *Yartsa* is so expensive that it's better to sell it.' Then he added: 'Well, maybe some people do. They copy the Chinese.'

The process of commoditization happens in a particular context, 'at the intersection of temporal, cultural and social factors' (Appadurai 1986: 16). These factors create conditions for making caterpillar fungus a complete commodity with its own commodity culture encompassing 'knowledge, attitudes, tastes, and meaningful actions involved in their [the commodities'] making, exchanging and use' (Cook 2004: 9). The socio-economic history of caterpillar fungus can be divided into two parts: the first when it was relatively unimportant, one of many species of the grassland, and the second, when it became the main product of the Tibetan grassland, dug and sold on a massive scale. Seen from this perspective, caterpillar fungus has (so far) had two lives: before and after it became commoditized. As Igor Kopytoff observes, objects move in and out of the commodity state within their 'lifetime' (1986: 67). Caterpillar fungus does this as well. When it comes to individual specimens, no significant shift in their biographies is visible. As long as it stays in the exchange space of Golok, the fungus remains a commodity – the fact that some people store their savings in such a 'fungal bank', as the following chapters show, shows only that the moment of the realization of this exchange value can be delayed. However, because it is a product of the natural environment and not of human hands, one may ask at which moment in an individual fungus's biography does it turn into a commodity. In fact, all fungi become a commodity as soon as they leave the ground because people dig them out with the sole purpose of selling them. More accurately, all caterpillar fungus growing in Golok soil is – from the diggers' perspective – a commodity, it is just that not all of these commodities can be found by the diggers.

4 Market and Traders

Caterpillar fungus trade in Golok has many faces. In the high-altitude pastoral townships people trade under the open sky. The traders and sellers sit on the road, bringing foldable stools and tables or squat on the ground weighing caterpillar fungus, sorting it, cleaning, and haggling. Drivers honk their horns to push their way through the crowd, while traders slowly move their baskets and pastoralists their motorbikes. The traders gather in the afternoon when the first pastoralists return from the mountains and they disperse after dusk to return the next day. Outside the digging season, it is hard to tell where these markets take place: the suddenly appearing trading crowd resembles a flash mob and leaves no trace when it is gone.

In smaller county towns the trade conquers the pavements and public squares. In each town there is an enclosed area allocated for a marketplace, but caterpillar fungus is seldom traded there. Instead, the traders occupy the stairs leading to shops and restaurants, sit on the street or on the lawns. Customers and onlookers gather around them. Someone is cleaning the fungus. Another person offers scales for rent. Pastoralists walk among the shops, spending what they have just earned. Someone ties a frozen sheep carcass to a motorbike. Another person carries a sack of rice. More significant traders stand out in the crowd: their posture, clothes, and a circle of people around them reveal who the biggest trader is. Even though some of them own shops where they could meet their suppliers, they prefer to trade on the street.

Finally, there is Dawu, the prefectural capital (Figure 6). Of all Golok towns, it has the most urban texture. The prefectural administration is located here. The Communist Party buildings shine with their glass walls. There is a television station and a football stadium. Dawu is the main commerce centre of Golok and many threads of the caterpillar fungus trade network meet here. The town attracts both pastoralists from the highlands and traders from faraway cities. When a weather-bitten pastoral family finishes their lunch in a dusty eatery, a group of neatly dressed men walks into a hotpot restaurant whose opening was announced with firecrackers. Tibetans whisper: 'These're bosses from Shenzhen. They've come to buy *yartsa*.' Just as people from different levels of the trade network cross paths in Dawu, so meet different styles of doing business. One can strike a business deal sitting on a curb or visit a caterpillar fungus shop with a counter and a small selection of products displayed in a glass cabinet.

Figure 6 Dawu, view of the town from the direction of Domkhok

Dawu during the caterpillar fungus season and the rest of the year are two different towns. Someone who has been here in winter would not believe that its half-empty streets can be blocked by traffic and its pavements swarming with people. It is caterpillar fungus that is behind this transformation. At any other time, Dawu is visited mainly by pastoralists coming down from the highlands for shopping or entertainment. There are several periods when life in town intensifies. In autumn, pastoralists transport livestock for sale and come with trucks to buy provisions for winter. When Tibetan New Year approaches, the monastery attracts pilgrims, tailor shops get crowded with people who want to welcome the new year in a new robe, and supermarkets are stormed by customers buying delicacies for family feasts. But spring is different. In spring, Dawu attracts not only pastoralists, but also people who otherwise have no reason to be there: traders from Qinghai, Gansu, Sichuan, and other provinces. Police regulate the traffic, all rooms are taken in both the most expensive and most basic hotels, and you have to wait for a table in even the filthiest restaurant.

Two parts of Dawu are especially closely associated with the caterpillar fungus trade: Koja and Darlag Street. Koja is a quarter on the eastern side of town, in the direction of Gabde. Crowds of people from the highlands rarely dwindle there and all sorts of business are settled (Figure 7). Horses doze

Figure 7 Koja, general view

tied to electricity poles, motorbikes parked in the mud wait for their owners. Pastoralists spread fresh yak hides on the pavement waiting for a buyer and unemployed-looking women standing in front of their little rooms wait for men with some money to spend. If you need to get a motorbike fixed, this is the place to go. If you want to rent a car to go to another county, this is the place to check. But first and foremost, this is where caterpillar fungus is sold. A line of shops calls attention to this fact with signboards saying, 'Caterpillar Fungus Purchasing Shop'. Traders stand in groups waiting for their suppliers or scout the street asking passers-by in a low voice: 'Do you have any *yartsa*?' They skilfully spot potential sellers and hurriedly lead them to a shop or to one of many flats concealed off the yards. One does not need to worry about finding a trader. It is enough to appear on the street with a plastic bag in hand: the traders will be there in an instant.

The second caterpillar fungus trade hub is Darlag Street in the centre of Dawu. Compared to Koja this area has a more urban feel and many of the town's flagship buildings are located here: Agricultural Bank of China, China Telecom, and Xueyu Hotel – one of more elegant buildings in the town. There are restaurants, boutiques with fashionable imports from Xining, a discotheque with a DJ playing a mix of international hits, and even a manicure studio. Between these nests of modern consumption, a

Figure 8 Darlag Street, caterpillar fungus shop

Figure 9 Anywhere is good for business: street scene in Dawu

Figure 10 Gesar Square

number of small, one-room shops stand with their doors wide open. To an uninformed onlooker they look mysterious: they seem empty of any goods for purchase but are full of people (Figure 8). As in Koja, these shops are also the traders' living quarters: even if the traders have a flat in the town, they sleep in their shop to guard it and be there any time suppliers arrive.

These two areas cannot contain the whole trade done during the peak of the caterpillar fungus season: traders flow out from the shops and cover the streets (Figure 9). A popular place among them used to be Gesar Square, the main square in Dawu, with a statue of King Gesar of Ling from the famous Tibetan epic (Figure 10). The square's spacious lawns created ideal conditions for the trade until it was forbidden. This was the first case where the town authorities stepped into action to regulate the expanding street trade. Together with initiating several construction projects around the square, including the first public library for the town, the authorities decided to evict the traders from the area. Posters warned that people violating the ban would be charged a fine of 5000-10,000 yuan. Thus, Gesar Square was 'returned' to the town residents, who could again picnic on its lawns. The officials were proud of their success. 'In spring, this town becomes impossible to live in! We have to prevent traffic jams and gatherings. We have to look after Dawu's appearance!', one of them told me.

Regulating the street trade is not easy, as another initiative shows. The town authorities attempted to build a market hall, where all caterpillar fungus trade could be done in comfortable and – as the officials stressed – hygienic conditions. This initiative did not win the traders' support and even the investors withdrew. The traders preferred their old working places, regardless of how imperfect these were. In fact, caterpillar fungus traders, both in Dawu and in smaller county towns, avoid official marketplaces. Between Koja and Darlag Street, there are two large market squares: one with textiles, hardware, and everything else that can be useful at home and in animal husbandry, and another selling food products. But these are retail markets and caterpillar fungus is traded in wholesale quantities. Although small-scale purchase for one's individual needs is possible, the trade in Golok supplies primarily wholesalers and not individual consumers. It is not the sellers who wait for the buyers, as is the case at a usual marketplace where groceries or textiles are sold. In case of caterpillar fungus, it is the wholesalers who are the buyers and who wait for the suppliers to arrive. However, there is another reason why the traders prefer to remain outside the officially designated market squares: the trade in caterpillar fungus is, more often than not, done rather unofficially.

Traders' Official status

The caterpillar fungus trade creates an employment market of enormous size. However, during my research, very few shops in Dawu were officially registered. In the whole town only 28 shops had this status, and they employed 67 people. This is a minute fraction of the number of traders who work in the town. This number comprises not only local residents, but also people who visit the prefecture seasonally. According to the officials from the Industry and Trade Bureau whom I interviewed, at least 5000 traders come to Dawu between April and June each year. The traders themselves believed that this number reaches 7000. 'It's difficult to count them, it's such a crowd!', as an official from the Bureau remarked.

Why so few shops were registered as specialized in the caterpillar fungus trade is partly explained by the fact that the registration procedures were new: they were only introduced in 2009. The officials of the Industry and Trade Bureau assured me that there ought to be no other reason: the whole process is simple and cheap. Only a few documents are needed: an identity card, a rent contract for the shop, a trade permit from the county government as well as one from the Agriculture and Animal Husbandry Bureau, and a certificate in Chinese and Tibetan Medicine. A trade licence with a four-year validity period cost 80 yuan, a sum rather insignificant compared to the traders' income.

If a licence is cheap and easy to obtain, why do the traders not apply? First, if they want to 'jump' into business, earn quickly, and swiftly withdraw from it, applying for a licence is too big a commitment. Second, opening a shop requires capital. If someone cannot afford to buy large amounts of caterpillar fungus, they do not open a shop but work on the street. Furthermore, many traders fear the state control associated with the official status. This status, they believe, is more of a hindrance than a help, as no one knows when the policies will change or new regulations come in. Finally, keeping an unregistered status carries the hope of evading taxes. Thus, staying away from the sight of the authorities is for many traders a convenient way to minimize the expenses of running a business and avoiding the complications they associate with state control. But there are also factors that pull other traders into the registration process. Officially recognized status spares them any stress caused by unexpected controls and fines, with which they could be charged if they were caught without a licence. 'If you don't have *yartsa* in your papers, you shouldn't trade it', as one official explained – stating that unregistered traders, if caught trading caterpillar fungus, have to pay a fine proportional to the value of the fungus they have on them. He mentioned 2000-20,000 yuan for individuals and 80,000-100,000 yuan for companies, that is, other general stores whose personnel can be caught trading the fungus on the side. This sort of money is not insignificant and the traders fear controls and denunciation by competitors. Thus, whereas small deals are often made on the street, bigger ones are usually made behind closed doors.

Taxes, in the context of caterpillar fungus business, are a complex topic. Traders, both in Golok and Xining, often claimed that their work is tax free. They explained this with the 'Open up the West' campaign, which had been launched to boost economic development in China's western regions, as well as with the state's support for ethnic minorities. As a young Hui trader said: 'We're an ethnic minority and the government supports ethnic minorities. That's why we don't have to pay taxes.'[1] The officials from the Tax Bureau in Golok contradicted this. Officially, the traders have to pay Individual Income

1 Chin. *shaoshu minzu* denotes 55 officially recognized 'ethnic minorities' or 'minority nationalities' that, together with the Han, constitute the population of the PRC. There are at least 15 other groups aspiring to this status (see Gladney 2004).

Tax, Value Added Tax, as well as Urban Construction and Maintenance Tax and Educational Surtax.[2] But they also admitted that extracting these taxes is difficult. Traders are supposed to pay taxes in the place where their business is registered: those running shops in Golok should do it in Golok and those who are only visiting the prefecture should do it where their business is based. However, finding out a trader's residential address is difficult and many remain fiscally invisible. As one Tibetan official complained: 'We don't know these people's names and have no energy to chase them. We have no people to do it.' The very fact of carrying out caterpillar fungus trade is also difficult to prove. The shops have many apprentices, scouts, and middlemen and are usually full of people. This makes it difficult to say who is legally responsible for the trade done therein. Unless one is caught red-handed in a situation where caterpillar fungus is exchanged for money, it is not easy to prove who bought and who sold the fungus. This crowded character of the trade field, as well as the fact that the shops are usually registered in several people's names (traders in Xining openly told me that they do this to 'reduce risk') help the traders evade the law and make them elusive for the state organs. A Tibetan official lamented:

> There is no way to control these people. What if one day you catch a man with a plastic bag full of *yartsa*? Next day he's gone. It's the same with the shops. You see a person sitting in the shop. 'Is this your shop?' 'Yes.' But you never see him again. The shops are meeting points of lots of people. Who do you punish if you can't get hold of anyone?

The lack of cooperation between different state institutions does not facilitate implementing the fiscal regulations either. The officials talked about a joint project of Tax Bureaus, Industry and Trade Bureaus, and other state institutions in the counties that would coordinate their actions to improve control over the trade. However, all interviewed officials were sceptical about it. In fact, some of them saw implementing the tax policies as potentially hurting the interests of the local population or as detrimental to the state security. A close relation between taxes, people's income, and the state security was mentioned by several of them as a reason why the policies are not enforced. Another official stated:

2 In 2010, officials from the Bureau informed me that in Machen County the Individual Income Tax was 1 per cent of total income; VAT was 3 per cent, including 5 per cent for Urban Construction and Maintenance Tax and 3 per cent for Educational Surtax.

> If we start extracting taxes, it will directly hit the nomads' income and they will protest. Generally speaking, the problem is that there're so many ethnic groups in Qinghai. Since we have the policy of the harmonious society, we need to make sure that the society really does stay harmonious.[3] So we don't tax Tibetans. I think that Tibetans should be allowed to earn this money. It's our best local business and people deserve to profit from it.

The Informal Economy

Caterpillar fungus traders, staying outside the official registration system and evading tax duties, can be classified as belonging to the realm of the informal economy. In its usual definition, this term denotes economic activities with minimal linkages to the 'formal' industrial or bureaucratic sector (Hart 1973: 63).[4] It encompasses enterprises characterized by low capital, moderate physical facilities, possibility of easy entry and exit, reliance on kinship or other non-contractual working relations, and lack of written records. These features are considered to be a result of a lack of access to resources, capital, or training as well as a method enabling the 'informal' entrepreneurs to retain autonomy and evade state control and taxation (Clark 1988: 3). The caterpillar fungus trade has some of these features: the possibility of entering and leaving the trade quickly, reliance on kinship and other social relations ('it's impossible to do this business without them', as one trader noted), and reliance on word of mouth ('having written contracts among friends would be an insult', another trader said) all make this trade a good candidate for inclusion within the ranks of the informal economy.

Although the term 'informal economy' has gained wide currency, there are several problems associated with it. First, it covers a broad field of heterogeneous phenomena, ranging from cases of self-employment through activities that take place outside of state regulation but do not openly contravene them, to activities characterized by various degrees of illegality. 'Informal economy' lacks clear conceptual borders that would facilitate

3 'Harmonious society' (Chin. *hexie shehui*) is a concept introduced in 2002 by PRC president Hu Jintao in response to increasing social injustice and inequality in Chinese society as a result of unchecked economic growth. It was a key feature (together with 'scientific development') of Hu Jintao's signature ideology, see Chan 2009; Brox and Bellér-Hann 2014.

4 On the history of the concept, see Hart 2009.

viewing all activities designated as 'informal' as cohering in one economic sector. The plurality of understandings as to what the informal economy means is shown by a range of terms used to describe it: 'non-observed', 'irregular', 'second', 'hidden', 'shadow', 'parallel', 'unmeasured', 'unrecorded', and 'non-structured'. Another problem is that the definition of 'informal', 'hidden', or 'shadow' economy has as its starting point its 'better' counterpart, which is conceptualized as 'regulated', 'formal', 'first', and which consists of mainstream economic practices authorized by the state. The informal economy is thus defined by the negation or *absence* of certain features that its counterpart is believed to have. This negative definition, which stresses deficits or operational weaknesses, coexists with another one based on a number substantive attributes that are said to be typical of this economy. These are no less questionable and some of them (e.g. reliance on kinship) are found not only in 'informal' economies, but also in 'formal' ones and even in the state institutions; this makes the distinction between the two even more problematic.

A more fitting designation of a large part (but not of the whole) of the caterpillar fungus trade would be an 'unofficial' economy. The term 'unofficial' refers to the *current* legal status of a given enterprise and not to its alleged modus operandi. A trader can change his or her status and hence this 'official' or 'unofficial' designation does not capture any inherent qualities of the trade as such. The methods of doing trade on the two sides of this conceptual division do not have to change. In Golok, no significant qualitative differences were observed between the work of traders with a licence and those without: they worked shoulder to shoulder and frequently in the same shop. The fungus (a product that as such is legal) was passed from 'official' to 'unofficial' hands or vice versa. The traders were all part of one network. It was the distance from the registration procedures that decided whether the status of a trader or an enterprise could be called official or unofficial. In real life, these worlds mix and one person can follow both paths. Depending on the location and time, an 'official' trader can make transactions that are not taxed or are done in a space that is not authorized for it.[5] A person can move to the other side of this conceptual division when his or her trade licence expires; by the same token, someone who has never had a licence can apply for one.

5 It is noteworthy that the caterpillar fungus street market in Xining is always busy, while the shops are empty. The traders explained that they prefer working on the street, but open shops for prestige reasons.

Describing the caterpillar fungus trade through its official or unofficial status helps to understand the place that this trade takes and its relations with the state organs. But defining it from the perspective of the state is problematic because this trade existed before the government introduced the registration procedures. Thus, these two designations can be used only as long as one remembers that the notions of 'officiality' or its lack are a recent innovation deriving from state attempts to regulate the trade. This also applies to calling some part of this trade 'illegal': such a designation would relegate it to the same category as trade in goods forbidden in private trade or trade in general, such as foreign currency or endangered animal species. But none of the officials in Golok, not even those who complained about this trade the most, called it 'illegal'. They spoke about it with a tone of resignation, sometimes even understanding; they often showed an attitude that could perhaps be described as benign neglect. A similar attitude is observed in other contexts where 'the state is weak, where it is undergoing transformation and lacks an adequate infrastructure, trained personnel, techniques, facilities or funds for monitoring, or where state personnel themselves derive benefit from the opportunities it [i.e. the trade] supplies' (Smith 1988: 190).

The caterpillar fungus trade can sometimes also be called 'grey', but again this is conditional: 'white' channels for this trade did not exist during my research, as the licenced and non-licenced traders worked within the same market structures and cooperated with each other. They did not form parallel worlds, they were *one* world. The non-registered traders did not fill any spaces left by their registered colleagues who, inadequately covering the needs of the population, invited 'grey' entrepreneurs to fill the gaps. This has not been the case in Golok, where no alternative to the usually practised forms of the caterpillar fungus trade existed. During my research, this trade was only gradually undergoing a state-initiated process of regularization. This could lead to diversification of the trade field and creation of parallel markets. The campaigns against 'counterfeit' caterpillar fungus, in which the traders were accused of artificially increasing the weight of the fungi in ways detrimental to health, can be seen as an attempt to 'blacken' the traders' reputation and direct consumers to channels that the state or pharmaceutical companies recommend as 'safe'. As a result, clear 'black', 'white', and 'grey' markets could emerge, but in the course of my research such divisions were not observed.

Tibetans and Others

The trade in caterpillar fungus acts as a social glue that brings together members of several ethnic groups.[6] The most active of these in Golok are Huis, Salars, Tibetans, and Han. The first two groups are most widely associated with this trade, which is often described as being controlled by Muslims.[7] Muslims, and especially Huis, were active in the trade in the north-eastern Tibetan plateau in the past and thus brought into contact communities living many kilometres apart (Ekvall 1977). Many commodities passed through their hands: sheep, wool, pelts, salt, and other products of the plateau as well as goods from farming communities and urban centres: grain, tea, cloth, firearms, tobacco, hardware, pots, and many others. While the role of the Huis has been widely acknowledged, less attention is paid to other ethnic groups.

The exact participation of different ethnic groups in the caterpillar fungus trade in Golok is difficult to estimate. The leading role of Huis and Salars is no surprise, since a similar situation exists in Xining. According to the Industry and Trade Bureau in Dawu, Huis and Salars constitute 70 per cent of all caterpillar fungus traders active there. The remaining 30 per cent are Tibetans and Hans. The traders whom I interviewed saw it differently. The Huis believed that as much as 95 per cent of the market is in their hands and the Hans argued that 30 per cent is 'theirs'. Each group inflated its own influence, but two facts remain undisputable: that the Huis play the major role and that Tibetan traders' participation in the market is relatively small.

For the Industry and Trade Bureau, the number of Tibetan traders is particularly difficult to judge. While Huis and Salars with their straw hats and white skullcaps can be easily recognized as involved in the trade, more inconspicuous Tibetans are difficult to differentiate from other inhabitants of the town. The officials in Dawu could only estimate the situation in Dawu or in Machen County, but this is not representative of the whole prefecture.

6 Chin. *minzu* has several meanings. In *shaozhu minzu* it means 'ethnic minority' or 'minority nationality', in 'Han *minzu*' it is 'nationality', and in *zhonghua minzu* it is 'Chinese nation' not reducible to any single nationality, but encompassing all of them. Concerning the term *minzu* and its meanings, see Bovingdon 2010: 16.

7 The Salars are a Turkic-speaking, mostly Muslim people, numbering around 130,000, who inhabit the Qinghai-Gansu border region (CNS 2010). The Huis are more than 10 million and live in many parts of China, with bigger groups in Qinghai, Gansu, Ningxia, and Xinjiang. In contrast to other Muslim groups, who speak Turkic, Iranian, or Mongolian languages, the Huis do not have a distinct 'non-Chinese' language and their commitment to Islam is considered their main identity marker (Gladney 1996: 26-36).

The situation at different markets depends on their location. While Dawu is in the trade orbit of Xining, where the Huis are the main market players, in southern Golok the situation is different. In Jigdril County, which lies just a two-hour drive from Ngawa, an important centre of Tibetan commerce, only 40 per cent of the caterpillar fungus traders, as my informants there estimated, are Huis and Hans. The rest are Tibetans, both from Golok and Ngawa (ten and fifty per cent respectively). In the highland pastoral townships, where the population is almost entirely Tibetan, Tibetans dominate. Their purchasing powers are smaller than their colleagues from the town, but they are nevertheless the first-contact buyers for pastoralists selling their fungus and are an important link in the trade network.

The number of Tibetans involved in the caterpillar fungus trade is small, but the traders observed that it has grown since the early 2000s. This reflects a wider tendency visible in rural Tibet to seek income outside the rural sector (Goldstein et al. 2008; Childs et al. 2010). But it is also a result of several decades of this trade going on in the region, during which Tibetans have accumulated capital, acquired market know-how, and established contacts. Caterpillar fungus trade offers new job opportunities, particularly attractive where there is limited employment outside of the pastoral economy. The trade stands open to people with capital and courage to bear risk. Interpersonal contacts and experience are helpful, but no formal education is needed and knowledge of caterpillar fungus and the market can be gained by observing and cooperating with others. The trade is especially attractive for people who, for various reasons, cannot continue their previous life, who left pastoralism, are displaced from their earlier environment, or want to start a new life in the town or in lay society. A visible group of men active in the trade are former monks who have renounced their monastic vows and returned to the lay life, but who neither have livestock, nor a wish to return to pastoralism and lack relevant education that would allow them to find employment in the public sector. Their interpersonal, often transregional, contacts and sometimes experience from the monastic trade may additionally encourage them to engage in this business. The caterpillar fungus trade is one of a handful of good options that such men have.

Coming from the caterpillar fungus-producing areas, with direct access to relevant people and the resource itself, Tibetans should have a 'natural advantage' in this business (Fischer 2005a: 166). However, their position at the market – as the traders said – is far from satisfactory and although many Tibetans try their hand in the trade, few ever gain prominence. There are several factors that the traders blamed for making their careers difficult. Almost all traders whom I spoke to named insufficient knowledge of Chinese and lack of experience with bureaucratic procedures as their biggest problem. Lack of capital is another difficulty. It forces Tibetans to trade in small quantities and to constantly move between buying and selling without having a chance to accumulate larger amounts of the fungus. Tibetans accused Huis and Salars of blocking competitors from the market. As some traders said, the Huis support and credit each other, something that Tibetans do not do. Several Tibetan traders admitted that they had taken out loans from the Huis to start their own business. Moreover, many traders believed that Tibetans are less commercially minded and lack the experience, self-assuredness, and courage of the Huis and Salars. Finally, there is one more factor in play. Tibetan traders frequently fall prey to a moral quandary in which they are trapped between the 'moral economy' of their community, who expect fair prices and preferential treatment, and a need for profit. In choosing the former they risk bankruptcy, with the latter they expose themselves to social ostracization. Tibetan traders signalled that their suppliers expect preferential prices, sometimes below the level of cost effectiveness, and often do so in the name of ethnic solidarity. However, as the traders said, the pastoralists are not necessarily driven by a similar sentiment and make very different choices in choosing the people who they trade with.

The above reasons explain why Tibetans often work as part-time traders. They trade in caterpillar fungus during spring and winter, and otherwise run shops or restaurants, work as school teachers or state officials, or return to their pastoral life in the highlands. A Tibetan owner of an 'ethnic products' shop explained: 'It's safer this way. In a good year I can earn enough from *yartsa*. But in a bad year, it's good to have the shop. People always buy robes.'[8] This part-time engagement can lead to a narrower specialization. None of the Tibetan traders who had a well-established position in Dawu during my research had started out with caterpillar fungus. All practised rotational systems before: they traded in caterpillar fungus in spring and in other pastoral products during the rest of the year. How many such part-time traders from the time of my research will have become full-time traders in the future is difficult to tell. But it is important to note that many Tibetans engage in this trade seasonally. Thus, even if the number of Tibetan traders is growing, it is growing with people who do this trade as a lucrative but only a side-line job.

8 Tib. *mirig thondzi*, a common term for 'ethnic assortment' including clothing, jewellery, music VCDs, and religious articles: prayer beads, prayer flags, etc.

The Multilayered World of Caterpillar Fungus Traders

The world of caterpillar fungus traders splits into many categories. The traders occupy different places in the trade network and differ in the ways they operate. Some open their own shops. Others work from home. Still others, who come to Golok seasonally, invite their suppliers to a hotel room or work on the street. The following section offers descriptions of several traders and their shops and shows how different the traders' worlds and paths to business are.[9]

Trader no. 1

A shop in Koja, under a big green signboard, the door wide open. A bench covered with a Tibetan carpet, a few stools, and a counter. On a bench lies *Military Affairs & Weapons*, a magazine, which the shop owner, Mr. Liu, likes reading in his free time. On the shelves wooden gift boxes for *yartsa*. Mr. Liu says he keeps them for decoration; they don't sell here, anyway. A large mirror covers the wall. Under a ceiling a scroll of red paper features the names of guests who visited the shop opening to give their congratulations and good wishes. On the counter a dollar plant and Christmas Star. Behind a glass wall the trader's living quarters: two beds, a table, and a television. Two Tibetan youngsters, in blue jeans and outdoors jackets, try to sell *yartsa* from a North Face backpack. Without success: they leave the shop disappointed. The next visitor, a man in a thick robe and dirty boots, is more cordially welcomed. Mr. Liu takes him directly to the private part of the shop and treats him to a cigarette. Mr. Liu, a Han, has been doing this business for fifteen years, but he has lived in Golok longer. He came here to dig *yartsa* and knows this business from its lowermost level. He used to dig *yartsa* in Domkhok and the contacts he struck up there are still paying him back today. The majority of his suppliers are from Domkhok (like the pastoralist puffing his cigarette behind the counter). Another man sits on the bench next to the door holding a plastic bag of *yartsa*. It has been over one hour since he came, but he has still not agreed on a price with Mr. Liu. So, there are two customers sitting in the shop at the same time: the one with the plastic bag in the front, and the one with a cigarette in the back. The counter is the border between the more official and the more

9 All traders described are men. I interviewed one Tibetan female trader from Ngawa, met female traders in Xining, and read about them in Yushu (Lama 2007: 70), but I have never met any female caterpillar fungus traders in Golok.

private space. I suggest taking a photo of this Han-Domkhok friendship. Playing out a scene of bargaining, they both laugh into the camera. Old-time relationships pay back well.

Trader no. 2

I meet Mr. Ma on the street. He takes me to his flat on the first floor with a window overlooking the road. The flat is also his shop. 'I don't have a visiting card, but people know me', he says. Mr. Ma (a Hui) is 37. His wife is 28. Both are from Gansu, and that is where their sons go to school. They miss their boys, but the telephone is a remedy for longing. Mr. Ma seats us behind a table and fetches painted tea cups. He produces the ingredients of Eight Treasures tea from his drawer and crushes two pieces of sugar with a metal weight.[10] Mrs. Ma sits on a little stool embroidering a pair of insteps (which, on another day, I will receive as a gift). Later, she serves us – me and Sonamjid, my friend and Chinese language translator – potatoes with chilli, but she does not eat any: 'My husband likes potatoes. I don't.' A black veil covers her hair and a white blouse with an open neck shows a golden necklace. She is no doubt a good homemaker. There is not a trace of dust in the flat (and this is in a town where dust is an integral element of the air!). What is called a flat is actually a one-room space. There is a table, two beds with carefully folded blankets, clothes in a plastic cover, a stove, and a cupboard. There is also a metal safe, which Mr. Ma opens to show us several kilogrammes of caterpillar fungus packed in plastic bags. When his customers arrive, the table turns into a counter. Two pastoralists from Gabde unpack their *yartsa*. Long discussions about the price ensue. Mr. Ma is proud to show how fluent his Tibetan is. He has lived in Golok for ten years and started with digging caterpillar fungus in Rarja. 'Is that ok?', the pastoralists ask. 'I don't have such money', Mr. Ma answers. The disappointed customers will go to another shop, but Mr. Ma says that they will return soon: 'They're my long-time suppliers, the connection matters.' When the pastoralists leave, Mrs. Ma wipes the dust from the table and refills the cups.

Trader no. 3

Mr. Hu is playing cards with a group of men in one of Koja's backyards. He wears 'donkey-ears' – long stripes torn from a plastic bag which hang from

10 Chin. *ba bao cha*, a composition of green tea, ginseng, chrysanthemum flowers, jujube, and other ingredients.

his ears. 'We don't play for money, so if you lose, you wear your cap inside out or put these donkey ears on', his playmates explain. Our interview is also a form of entertainment, so the game stops when we are talking. Mr. Hu wears a tight white shirt exposing his round belly and a straw hat on his neatly shaved head. It starts drizzling and water washes away my notes, but Mr. Hu does not invite us into his home. He occupies a room in this yard right in front of us. Through a half-open door, a very basic space can be glimpsed. There is a metal bed, a chair, a plastic bowl, and a safe. The room cannot be more than six square metres. There are at least five rooms like that on this side of the yard and maybe a further ten in the two-floored building on the other, all taken up by traders like him. Mr. Hu comes to Golok in spring and stays as long as the business goes well, usually until August. He does not know how long he is going to stay this time. There are good days, when he gets up to a hundred customers. But there are also bad ones, when he does not get any. He is 40 years old and started coming to Golok some 20 years ago. At that time, he bought everything: sheepskins, cheese, and caterpillar fungus. When the market for *yartsa* improved, he dropped livestock products. He learned some Tibetan, but only as much as is required for business. Why does he not open a shop? Mr. Hu shrugs his shoulders. What he buys, he sells on to bigger traders. Sometimes he takes the fungus to Xining, but it is more convenient to sell it here. Can he earn enough? Yes, if he manages to sell two or three hundred *jin* in a year. But that is only if the year is good, like in 2007. And not like in 2008, when the economic crisis struck the market. 'Anyway, this trade is a losing game', he says and he does not mean cards.

Two More Traders from Dawu

The biggest Tibetan traders in Dawu are Pakpajab and Lorden. Their shop is right next to the Agricultural Bank of China. It looks like any other *yartsa* shop; there is a bed, a couch with the headrest covered with a towel, a safe, a coffee table with a jar of chilli paste, and a calendar on the wall. It is the bare bones needed to run a business. Both traders used to hold official positions in the township administration. They moved to Dawu in the 1990s: sold their livestock, bought houses, and brought their family to the town. They come from the same *dewa* and if one examined their family histories, they would also appear related by blood. The two men are not alone in business: they have a partner, a Salar. 'We took him on to do the paperwork for us', they explain. Like many other Tibetans, they complain that their knowledge of Chinese is not good enough to let them be independent. While

I conduct my interview with Pakpajab and Lorden, a man brings in three *jin* of dry fungus. Their Salar assistant, in round, saucer-like glasses, squats on the floor to sort the fungi. A while later a lady with a mask on her face brings along a few specimens of *yartsa* in her handbag. She leaves the shop disappointed. She hoped for a higher price. It is now lunchtime and we go to a Sichuan restaurant, leaving the Salar at work. The talk continues over beef soup. It starts raining. This means that more *yartsa* will grow and one will not need to buy prayers in the monastery. 'Do people ask monks to pray for rain so that more *yartsa* will grow?', I ask. 'No. One can ask for grass for the livestock, but for *yartsa*, which is a private gain, one shouldn't', answers one of them. 'In practice it can be different though', the other one adds. Both traders regularly go to Xining (I met Lorden there: in a felt hat, grey suit, and tinted glasses, surrounded by his Han colleagues he looked like a real boss). They even went on holiday to Malaysia a few years ago. What will they do if the market declines? 'We're prepared for that', they say.

A Trader from Ngawa

A sunny afternoon at Gesar Square. Cows graze among people eating watermelon and ice-cream. Someone takes a nap under an umbrella. A Han lady walks with a loudspeaker mounted on her bicycle: '*Lamian, lamian*' – the voice advertises chilli noodles that she sells from a plastic bucket. Some ten paces away from me and my two Tibetan friends, a group of young men sit around a basket of *yartsa*. At some point they come to talk to us. One of them turns to me and says in good English: 'How are you doing?' All three men are from Ngawa and came to Golok to earn money. 'Did the business go well?' 'Not really…', Gendun, the English-speaking one, says. He was a monk in Labrang before.[11] Now, since he has returned to lay life, he not only has to improve his Chinese, but also to help his parents to support his five sisters, all of them at school. He tried his luck with *yartsa*, thinking it would bring an easy income, but in two years he did not achieve much. His story shows that the beginnings, for amateurs like him, are not easy:

> ES: Is it the first time that you've done business in your life?
> Gendun: No. It's actually the second time. I sold second-hand books on the street in Lhasa before.
> ES: How much money did you have to start out with *yartsa*?

11 Labrang Tashi Khyil is a Gelug monastery founded in 1710 in Sangchu County, Kanlho TAP, Gansu Province.

G: Not much. My aunt gave me 5000 and a cousin 3000. So, I had 8000 yuan.
ES: Is that enough?
G: For a good business it's not enough. It's altogether not enough. Other people have much more, 20,000-30,000 yuan. With that you can do business. But what can you do with 8000? Nothing. It will never suffice to buy, let's say, 1000 *yartsa*. So, I have to buy small quantities and sell small quantities, but this way one can't make big money.
ES: Why did you choose Golok? Have you been here before?
G: My cousin dug *yartsa* in Domkhok. I was there once too, but only for a short time.
ES: How would you describe the situation of the market in Dawu?
G: Without connections it's very hard. And one needs big money as well. These are the main things.
ES: So, you didn't know anyone here?
G: My father has a friend here. Apart from that I didn't know anyone.
ES: Did this person help you?
G: Well, he didn't give me money, but he at least told me what *yartsa* to buy and how to deal with the traders. He showed me how to divide *yartsa* into groups. He said that if I buy more of the good ones, I should mix them with those of not such good quality. If you have more of those of high quality you get a good price. Among those of low quality you also see [pieces of] different shape and colour. They have a different price. He also said I should buy hard and heavy ones. This helped a lot. I really didn't know much about it before.

Traders and their Suppliers

The different positions that the traders have and the different degrees of their integration into local society were highlighted in a questionnaire I conducted with traders from Koja and Darlag Street.[12] This questionnaire, in which I asked about the traders' careers, turnover, suppliers, and methods of work, showed that the caterpillar fungus trade is less spontaneous than it seems when one judges it by the informal character of the settings where it is done. Many traders rely on recurrent suppliers from one area: in the shops covered by the questionnaire as many as half of the suppliers return

12 This questionnaire covered 12 traders whose ethnicity, business size, and degree of integration into local society differed: some lived in Golok and others visited seasonally. I conducted the questionnaire in 2010, so the amount of caterpillar fungus these traders bought refers to 2009.

there every year. This means that the pastoral townships had *their* shops and the shops *their* townships. The traders prefer long-term modes of exchange over short-term ones: thus, one-off exchanges tend to be repeated and so develop into enduring relationships. The bonds formed between the traders and their suppliers are flexible and allow the pastoralists to retain their freedom to sell to another trader or in another town. Such relationships are known from many economies where 'the goal of each actor is his or her economic self-interest, yet the maintenance of the relationship is valued over an immediate short run profit' (Plattner 1985: 136). These economic bonding patterns can be seen as 'instruments to work within the context of high risks' (Finan 1988: 708); they enhance both parties' success and give them 'protection from the vagaries of the market' (Mintz 2004: 262). All this applies to the caterpillar fungus trade in Golok. Especially for the pastoralists, it is important to have a shop where they can get preferential or at least 'honest' treatment. The latter is important given the bad reputation that the trade enjoys: as a domain populated by fraudulent traders waiting to exploit the naivety of suppliers.

The degree to which the traders rely on recurrent suppliers shows their embeddedness in the local society: the longer they lived in Golok and the broader experience they had, the more they relied on long-term contacts. Some traders, who had lived in Golok for a decade or longer, started out as caterpillar fungus diggers and only later turned to the trade. The contacts established during those early years explain the geographical specialization of some shops: traders who once dug caterpillar fungus in Rarja have suppliers from there, and those who dug in Domkhok buy the fungus from this township.

Apropos the volumes of caterpillar fungus that the traders buy, most informants declared stock of 200-300 or 300-400 *jin* per year. A few traders mentioned 500-600 *jin* and one man over 1000 *jin*. Not every trader could tell which parts of Golok he was buying from. This was particularly the case for small traders without an established position who depend on irregular suppliers and buy 'everything they can get', as they themselves said. Other traders indicated that caterpillar fungus from Domkhok formed 10-20 per cent of what they bought. That would mean that a trader who buys 200-400 *jin* per year, buys 20-80 *jin* from Domkhok. Traders who bought proportionally more were those who regularly cooperated with this township. This was the case with Mr. Liu described above: up to 70 per cent of his fungus came from Domkhok. Given that he bought 500-600 *jin* per year, he must have been buying between 350 and 420 *jin* from this township alone.

The questionnaire was informative not only about the traders, but also about their suppliers. The traders estimated that an average supplier sells between one and two *jin* of dry fungi. Women occupy a special place, but their contribution to the market is insignificant: 'Perhaps 20 per cent of all suppliers are women. But they bring little, only what they have found, sometimes 20-30 fungi, whereas men bring *yartsa* for the whole family.' Another group are monks, who constitute some 10 per cent of suppliers. They either sell the fungus they have dug themselves or do it on behalf of their relatives. The traders observed that the number of monks in the trade declined after 2008, but they still had some monk-suppliers who brought sometimes significant quantities of the fungus.[13] Finally, there are also the diggers from outside of Golok. They often sell their caterpillar fungus directly to the pastoralists on whose land they have worked or take it home to sell there. The traders do not like doing business with them, because – as they said – the 'outsiders' have high expectations and are not flexible: 'They invested big money to come to Golok and now they want it back. They can't wait with selling *chongcao* like the nomads do. So, they go back home hoping to get a better price there.'

Does Ethnicity Matter?

Most pastoralists say that ethnicity is unimportant to them and that they always just 'sell to whoever pays the best price'. In practice, however, they sell to the Huis and seldom to Tibetans or Hans. The fact that at the market in Dawu the Huis dominate explains why the pastoralists have little choice but to trade with them. At the same time, the Muslim traders are subject to negative stereotyping.[14] My informants said that they are cunning, deceitful, and exploit Tibetans' naivety and lack of experience. They can replace 'good *yartsa*' with a 'bad one' so skilfully that the owner does not notice it. They use unbalanced scales and do not hesitate to rob gullible sellers of their fungi. Such stories are a warning against the risks awaiting inexperienced

13 Scholars observed (and my informants confirmed) that some Buddhist monks dig caterpillar fungus (Boesi and Cardi 2009; Yeh and Lama 2013; Schneider 2013). I also heard that diggers dress as monks in order to pass the checkpoints during the digging season without being checked. Whenever I have travelled with monks, our car has never been flagged down.

14 The pastoralists called them *Hehe* or *Huerig* (from Chin. *Hui hui*) i.e. 'Huis', or *Kache*, which originally denoted Muslims of Kashmiri origin in Lhasa. Both names for them also covered Salars. Officials from the Industry and Trade Bureau also treated Salars and Huis together. Moreover, some Hui traders in Xining spoke about Salars as if they were part of their own *minzu*.

sellers. They create a picture of the caterpillar fungus trade as dangerous and attracting morally suspicious characters. But they also concur with the stereotypes known in other Tibetan regions, according to which Muslims abuse their Tibetan neighbours persistently and on many levels (Fischer 2005b, 2013). My informants supported this belief with two examples: armed conflicts between Tibetans and Huis in the past and present-day economic relations.

The contacts between Tibetans and Huis are burdened with difficult memories, especially from the period of the Republic of China when political power in Qinghai was in the hands of the Hui Ma warlord family. The Ma warlords launched a series of military campaigns into the Tibetan regions to extract taxes, explore mineral deposits, and bring the region under their control.[15] These campaigns were met by heavy resistance in Golok, the memory of which is still preserved by the pastoralists.[16] My informants believed that the Ma warlords were not politically or economically but religiously motivated: it was a jihad of the day, as one man claimed. Similar anti-Buddhist motives are attributed to Muslims today. Many people in Golok believe that the Huis are still conspiring against Tibetans, only their methods have changed. During my research, people spoke about Muslim cooks who, through the food served in their restaurants, try to force Tibetans to convert to Islam. They put the funeral ashes, bath water or urine of their imams into the meals; the idea is that the customers, upon eating such a meal, automatically 'turn Muslim'.[17] That such beliefs can flare up into open aggression was illustrated by the events in Gabde in 2007, where Tibetans attacked Muslims after a customer at a local eatery found something resembling a human tooth in his dish. Consequently, not only the eatery, but also the building site of a mosque was demolished. The army intervened and prison sentences followed.[18]

In daily life, such anti-Muslim sentiments manifest themselves in periodic boycotts of Muslim restaurants and in a certain stigma falling on their

15 For more about these conflicts, see Horlemann 2015; about the Ma family, see Lipman 1994; Lin 2007; and Forbes 1986.

16 These events are remembered on both sides. A Hui trader admitted that the Huis killed many Tibetans, but only because 'the Tibetans wanted to break off from China and the Huis helped the government to control them'. He was convinced that conflicts between Tibetans and Hui today are a result of economic competition.

17 Another widely circulated story spoke about Hui motorbikes who threw the ashes of their imams onto Tibetan pedestrians to elicit the same forced conversion effect.

18 Pirie mentions a fingernail (2013: 79). See RFA 2012b. These events perhaps explain why this part of Golok was relatively quiet during the 2008 protests. 'People still remembered 2007', as one man said.

customers.[19] These should be viewed in the context of the economic competition between these two groups. In the decades of economic liberalization, the Huis have successfully colonized many of the new economic niches in Tibetan urban centres, taking jobs in gastronomy and other services, and nearly monopolizing certain trade sectors, such as with caterpillar fungus. Tibetans, who sometimes see themselves as economically marginalized, say that the Muslims have built their economic prosperity at their expense. It is hard not to think that the Huis are a 'replacement enemy' and a focus for people's frustration. It is symptomatic that whereas the boycotts of Hui restaurants were discussed in Golok, there were no calls to boycott Han enterprises, despite the fact that it is the Han who are the main exclusionary force in the local economy and that they directly benefit from the Tibetans boycotting Muslim businesses. The boycotts did not impact other economic sectors where there are fewer alternatives and where the pastoralists depend on close cooperation with the Muslims, such as in the caterpillar fungus trade.

In terms of critical voices about caterpillar fungus traders, one hears mostly, but not exclusively about the Huis. Closer observation shows that the critique is not limited to one ethnic group or religious denomination but covers the whole occupational group. Huis are affected the most because of their position in the market, but the pastoralists were equally critical about many other people involved in the trade. Tibetan traders from Ngawa do not enjoy any more sympathy and the pastoralists tell the same stories about them as they do about the Huis.[20] Anyone who has achieved success in business seemed suspicious, because this success, as many people believe, is proportional to one's ability to deceive others. If a flair for trade is synonymous with a flair for fraud, it applies to all business people, regardless of ethnicity.

Whatever the stereotypes imply, they do not trouble relations with individual traders: those whom one is familiar with and who do not confirm the stereotype. When the spirit of commerce enters the stage, economic interests win over ethnic antipathies. The Huis and the Salars are the pastoralists'

19 These boycotts were challenging for many consumers. During one boycott, when I went to eat with a respected Tibetan, we made several attempts to enter a Hui restaurant where he wanted to eat before he finally summoned the courage to go in. 'I'm finished if people see me here', he said. About the boycott movement, see Fischer 2008.

20 Many of these traders come from farming areas and it is common in Golok to hear that farmers are dishonest and untrustworthy. It is interesting that one can hear among farmers or townspeople that pastoralists are 'more honest' than other Tibetans (see also Schrempf 2010-2011).

trade partners not only because they dominate in number, but because they pay better and buy more. As one pastoralist said, 'I'd really love to sell to Tibetans, but there are so few of them! There're several from Ngawa, but really so few who have money. So even though I think I should sell to Tibetans, in the end I sell to Muslims'. This statement should not suggest that Tibetans pay less, but that they have a different buying capacity: while Tibetan traders who I interviewed never declared purchasing anything above 100 *jin* of the fungus per year, the Huis bought two or three times more. The more the traders buy, the bigger profit their suppliers can earn and Tibetan traders do not offer them this chance. If the pastoralists really wanted to sell to Tibetans, they would need to sell their fungus in parts. Waiting for the trader to buy more, they would risk a drop in the price.

Muslim traders are more competitive and the pastoralists – whether they like it or not – have to trade with them. However, there can also be some sympathy between these two groups. Several Huis told me that they enjoy working with the pastoralists: 'They value good relations, not money. They wouldn't do business with someone with whom there are in bad terms. They aren't like the Han who care only about money.' Another trader stated that Tibetans and Huis have much in common, both economically and politically: politically because they cannot express themselves freely and economically because they work in the lower levels of the market, earning only a fraction of what the Han do.

There is another factor that encourages the pastoralists to trade with the Huis or rather discourages them from trading with Tibetans. Many pastoralists prefer to sell to persons from beyond their community and Huis are convenient partners because of their outsider status. They are a 'middlemen minority': even if they stay in Golok for a long time, they only 'sojourn' there (Bonacich 1973: 585).[21] They are part of the local economic scene but remain outside the local divisions and petty conflicts. Selling to a Tibetan implies selling to a defined Tibetan identity, from a certain *dewa* or *tsowa*, which the seller is on good or bad terms with. In such a situation, fairness is not expected readily. The Tibetan traders complained of the pastoralists' lack of 'ethnic solidarity' and said that if the latter sold to them, their position would be better. The pastoralists, on the other hand, suspected that unless they had good connections with a trader or his family, they would be cheated and the traders would seek a larger than usual profit at their

21 Even well-established Hui and Han traders only 'sojourn' in Golok: some of them stay there for many years and are fluent in Tibetan, but they still live in spartan conditions in a kind of temporary home, while their children live with their relatives in their original place of residence.

expense. By selling the fungus to outsiders, the pastoralists directed the gains from the trade out of their community. The Hui, Salar, or Han traders take their wealth away from Golok, making it invisible to the pastoralists. On the other hand, a situation where a Tibetan got rich 'at the expense' of other Tibetans was hard to bear for some of my informants who did not want to observe how a person, especially from another township, *dewa*, or *tsowa*, became wealthy from the fungus bought from them.

From Hand to Hand

To the untrained eye, caterpillar fungus markets give an impression of unbroken confusion. Groups of people exchange information under their breath, hold each other's hands in their sleeves, and discuss something vigorously. A bag with caterpillar fungus moves from hand to hand and a wad of banknotes travels in the opposite direction. It is not easy to make sense of what is happening. What causes this bewilderment is a visible 'overproduction' of traders. Is it possible that so many people can find employment in one product's market? Can they really all support themselves with this trade?

A useful tool for understanding the degree of interconnectedness of different actors in the caterpillar fungus trade is a commodity chain analysis. A commodity chain, a 'network of labour and production processes whose end result is a finished commodity' (Gereffi et al. 1994: 2), shows the path that goods travel from producers to consumers. In case of caterpillar fungus, this 'good' is a natural resource and the commodity chain has to be understood as a 'conduit through which commercialized natural resources [...] are ushered from the land, through their fabrication, to their final users, whether rural, urban or "international"' (Ribot 1998: 308). Commodity chain analysis allows for tracking the flow of a product from its source to its destination and documents the participation of different economic actors in its production, processing, marketing, and distribution. These actors may be individuals, groups, institutions, enterprises, or other bodies. Their role can be producing the raw material, transporting it to the processing unit, processing it to make it usable by the consumer or by other actors who will convert it into a more complex product, and finally bringing it to the retail market. All these processes may include several stages of work by various actors who build a real chain spanning geographically distant areas and long time periods.

Following caterpillar fungus from its producers on the Tibetan plateau to its consumers shows how financial benefits from this business are distributed and how they grow on the way from the plateau through different trade enterprises and industries to urban markets. However, commodity chains, especially those with a large number of interlinked actors, can be disaggregated into smaller units. These smaller units or sub-chains may differ in function and be responsible for raw material production, its transportation, processing, and so on. The analysis of these sub-chains allows a better insight into complex (often local, but not always) relations between the actors involved. A smaller sub-chain showing the trade in caterpillar fungus going on in Golok reveals the complex ways in which the fungus moves between different traders before leaving the prefecture. It facilitates the understanding of the complexity of this part of the commodity chain and explains why this market offers such a large employment space.

The journey of caterpillar fungus through the market starts with the moment when the pastoralists sell the fungus to the first trader. The furthest point when my informants ventured to sell it themselves was Xining, the provincial capital. This is the end of the sub-chain within which they functioned as sellers. From here, caterpillar fungus is transported to other parts of China, bought by pharmaceutical and other companies that process it or distribute it to retail shops. Here, it can also be bought by individual customers or sold abroad. The pastoralists enter this sub-chain at different points and with different parts of their products. Small quantities of fresh fungi are sold in the township, while larger ones are brought to the county town, Dawu, or Xining. The same applies to caterpillar fungus traders based in the townships: from their hands, the fungus travels to the county town, to Dawu and out of Golok, or, alternatively, it can take a short-cut and omit several stages. Township and county traders may deliver it directly to Xining or to their partner in another province. In-migrant diggers can sell the fungus at any of these points or in their place of residence. Finally, shops and pharmaceutical companies can send their middlemen to Golok to buy the fungus at any of these levels.[22]

22 The employees of several pharmaceutical companies in Xining told me that they buy directly from the pastoralists and have 'contracts' with them. It seemed to be a part of the new rhetoric to stress direct contact with the pastoralists; it guaranteed that the fungus was in its pure, unadulterated form and showed that the company lets the pastoralists maximize their gains. An employee of Qinghai Aweto International Trade Center said that this company is 'a mediator between the state and the nomads. One of [our] goals is to protect the ecological balance and regulate the prices, which also means that the nomads' income will be regulated'. The goal of the company was 'not to make profit, but achieve sustainability', he said.

In addition to the vertical movements that caterpillar fungus makes between traders of different sizes located at different levels of the network, some fungus moves horizontally or, rather, circularly. Observation of county and township markets shows that the fungus changes hands many times before it leaves the market. Trader A may sell it to Trader B, this one to Trader C, and so on, until one of them (Trader *n*) sells it out of the locale. This resembles the trade in matsutake, a culinary mushroom that witnessed an economic boom similar to that of caterpillar fungus.[23] Also in the case of matsutake, a lot of buying and selling goes on in the space of one village before the mushroom starts its journey to more distant markets (He 2010: 32). However, matsutake is consumed fresh and must reach consumers quickly. Caterpillar fungus, on the other hand, has long durability and a longer 'pending period' when it circulates in Golok. The local traders between whom these transactions take place might, but do not have to differ in the scale of their operations. Especially in small markets, traders have businesses of a similar size and they sell the fungus to each other. A common problem experienced by these small traders is being unable to accumulate large quantities of the fungus. Moreover, in order to buy new fungus they often have to sell some of the product that they had previously bought. Indeed, they sometimes sell it to each other with a plan to re-buy it later. Consequently, caterpillar fungus circulates widely between traders. As one Tibetan described it: 'I buy from you, you buy from me, and so *yartsa* circulates on the market without the end.'

These circular movements reveal the hidden market life of caterpillar fungus and explain why the market of this one product offers employment to so many people and gives them a chance to earn at least some income as a result of the difference between the price the fungus is bought for on one day and sold on another. They also explain why some traders told me that they seldom buy from pastoralists, even though they work in a very pastoral region. Finally, these circular movements account for the constant buying and selling. One trader, when asked in which season he has the least work, said: 'No season is lazy for *yartsa*! *Yartsa* circulates on the market all year round.' This is not only because the pastoralists sell their fungus at different points of the year, but also because the traders do the same.

23 See Yeh 2000; Arora 2008; He 2010; and Alai 1997. See also Tsing 2015.

5 Market Operations

The pastoralists have various possibilities for selling caterpillar fungus, which can be sold fresh or dried. Fresh caterpillar fungus is sold by piece, in small amounts: a few dozen fungi travel to the market wrapped in a plastic bag – the number depends on the locality, time, and luck of the digger. Fresh caterpillar fungus is sold in the same form as it was extracted from the ground: covered with a layer of soil, which keeps it moist and protects it from breaking. Selling it this way has advantages. It saves people the effort of cleaning and drying the fungus at home. This job will be done by traders and their assistants: groups of them, often women, sit in the shops and on the street, cleaning the fungus and guarding it as it is spread out in the sun to dry. It is also easy to find a buyer for small amounts of fresh fungi. Finally, even if selling the fungus fresh only brings in a small income, it is a fast income, coming almost immediately after the fungus is found.

A quick sale of fresh caterpillar fungus is believed to indicate that the seller needs money: 'Poor people sell *yartsa* fresh', as many pastoralists say. This association meant my informants were hesitant to admit that they did it too. But selling caterpillar fungus fresh does not have to signal poverty. Malinowski and De la Fuente, in their study of rural markets in Mexico, show that the villagers bring only small quantities of their products to the market: they sell precisely as much as they need to survive until the next market (1982: 186, Sulek 1986: 69). Selling caterpillar fungus fresh does not bring pastoralists big money but sufficient income for a time. It does not have to indicate lack of money in general, but lack of cash at home and could be a result of particular preferences in managing the household finances. Even well-off families sell some fungus fresh to avoid withdrawing money from the bank. The banks are located in the remote county town, while the market is on the pastoralist's doorstep, in the township. The market functions as an 'always ready, always available and amenable bank' for anyone who needs cash (Malinowski and De la Fuente 1982: 186). The sale of fresh fungus is thus subsistence oriented. It is typically followed by shopping: the sellers turn into buyers and spend their newly earned income on food and other goods. Because the money earned is not big, the shopping often exhausts it. Regardless of whether it is spent on food or entertainment, this money leaves the pastoralist's pocket soon after it enters it.

Every household in Domkhok sells some caterpillar fungus fresh. This is done not only for economic reasons. The markets are social events: this dimension is particularly important in a region where people live dispersed across a large territory and where little is happening in terms of public events. A crowd of onlookers and self-appointed advisors throngs the marketplaces, testifying to their social pull. People visit the market for the pleasure of getting a good bargain, of earning money, but also to meet others and exchange news. They also do it for the pleasure of 'showing off' and measuring themselves in a '*yartsa* competition'. People discuss who the most successful digger is and who has found the largest specimen. Particularly large and well-shaped fungi are a trophy and people boast about them in a way similar to what mushroom collectors or anglers do. These social factors are inseparable from the other motives that pull the pastoralists to the market and explain why people like to sell at least some caterpillar fungus fresh during the digging season.

If pastoralists do not sell their caterpillar fungus fresh, they have to invest time in processing it. After returning from the pastureland, they clean the fungus of the soil covering it. This is done with special care. If the larva breaks, it can be fixed with a toothpick, but damaged specimens are valued lower. Cleaned and dried fungus is sorted according to its size and quality and sold per weight. Because of its long shelf life and light weight (0.2-0.5 g per specimen) it is a very convenient commodity and serves as 'cash storage'.[1] Many people keep at least half a *jin* or several hundred specimens at home to sell when they have unexpected expenses or to use as a nest-egg for some investment. In every house there is a metal safe where the fungus is stored together with other valuables: it is not to protect it against physical deterioration, but against thieves who, people say, have been plying the highlands with increased intensity since the caterpillar fungus boom started.

Similar to the case with fresh caterpillar fungus, the pastoralists have a theory about who sells the fungus dry. While fresh fungus is said to be sold by people in need of money, dry material is sold by those who can wait longer for money: the gap between the time the fungus is dug and when it is sold indicates how dependent one is on the income from it. The traders in Dawu confirm this: people from more affluent townships sell more fungus dry. Selling the fungus dry is attractive for two reasons. First, dry fungus is more expensive. At first sight, 30-40 yuan paid for a fresh specimen is attractive: one sells three fungi and earns 100 yuan. But this is less than what one would earn for the same fungus sold dry and in a larger quantity. Additionally, the prices of dry fungus increase after the season. So, some people prefer to wait to sell their fungi until the price is higher.

1 Only in the top class, where a single specimen weighs more than 0.5 g. Low-class fungi weigh less than 0.2 g, but are seldom stored.

In comparison to the quick sale of fresh caterpillar fungus, which is done quite spontaneously and often at the nearest market, the sale of dry fungus is preceded by extensive reconnaissance. The pastoralists wander from shop to shop and from trader to trader and inquire about the prices. This helps explain why the traders say that the number of people visiting their shops is not proportional to the number of transactions they make: from a stream of potential sellers only a few people decide to actually sell. This procrastination can last weeks: the pastoralists call the traders every few days to figure out what the current trend is. They use every opportunity to collect information about prices. Knowing that I conducted interviews at the market in Xining, people asked me for the traders' telephone numbers. They did not hesitate to spend a day in a car driving to the city, hoping to get a better price there. Many pastoralists complained that they do not have contacts to help them sell their fungus directly to pharmaceutical companies in mainland China: finding such a 'shortcut' through the market was their dream and they believed it would increase their profits. Some ventured even further and dreamt about selling the fungus abroad.

In order to sell caterpillar fungus, the pastoralists travel to the township village, to the county town, or a bigger city. However, they can also sell it from home. Some traders tour the high-altitude valleys where the pastoralists' houses and tents are to buy the fungus directly there. These itinerant traders usually come from outside Golok. Buying directly in the highlands, they gain access to the fungus from a reliable source and in wholesale quantities, at the same time saving the costs of opening a shop in one of Golok's towns. They do exactly what the pastoralists would like to do, but in reverse: they pass over intermediaries through whose hands the fungus normally travels. This helps them increase their profit, but this profit is related to the amounts they buy rather than the price. Caterpillar fungus, if bought in the highlands, is often more expensive than in the town: 'The nomads are so bold! They can demand any price when we go to their homes', as several traders told me – not without some admiration.[2] In many of the bigger shops in Xining there were traders who regularly travelled to the pastoral prefectures. One of them told me more about his work. Every week in spring he went to Golok to buy caterpillar fungus. Back in the city, he hired cleaners and delivered clean and dried fungus to the shop. From one journey he could bring up to four *jin*. In winter, he travelled less, but bought more than in spring. There were times, he said, when he bought

2 This was confirmed by Tibetan mastiff sellers, who said that it is cheaper to buy dogs from the pastoralists coming to town than doing it directly in the highlands.

even twenty *jin* in one journey. Associated with the shop that he worked for, there were over 60 men who did this kind of work. This invites guesses as to how many people do such a trade in the whole of Xining, where caterpillar fungus shops are counted in the hundreds.

The roads leading to the high-altitude pastoral townships are far more frequented than one might imagine in such a remote region. In Domkhok, every valley is visited by traders from the city (Figure 11). The pastoralists are positive about selling caterpillar fungus to them. They believe that selling the fungus at home is safer because the traders 'don't play their usual swindles when they aren't on their own territory'. But this positive attitude does not mean that the pastoralists wait for the traders to knock at their door rather than go to the town themselves. Because of this, many traders leave the highlands disappointed, as with some of those I encountered:

> On the road to Soglung we were followed by another car. Lucky for us, as it was soon to turn out. When our engine died, none of the four men in our car knew what to do. But the Huis from the other car did: they rolled up the sleeves of their white shirts and got to work. After a few minutes our engine started. These men were *yartsa* traders. Their car was packed with bags of food and other goods. They knew every family in Soglung, stopped in front of every house, exchanged greetings and handed one gift-bag in at each. They followed us until we reached our home. This was also their destination. Soon after, we sat together in one room: the traders at the place of honour, with food and drinks in front of them. 'Don't take pictures', the host lady hissed to me. But the traders weren't bothered, they were curious. A European woman, wearing a Tibetan robe and not speaking Chinese, I seemed to them like something fallen from the sky. After taking their courteous cup of tea, the traders got to work again: a neighbour brought a bag of *yartsa* and they squatted in the corridor sorting and weighing the fungi. They had a long discussion, but it didn't seem that there was much to buy. 'We sold everything a few days ago', our host said. In the evening, the traders dined and watched television with us and other diggers living in our house. They slept in the car. Next morning, they washed and left for another valley. (FN 2009)

Quality and Prices

The quality of caterpillar fungus is assessed on the basis of several criteria: size of the fungi; relative proportions of the larva and the stroma; and colour.

Figure 11 Some traders go to pastoralists' homes

The highest valued is fungus with a large larva and a short stroma: the best fresh specimens have 5-6 cm long larva and 2-3 cm long stroma. In good quality specimens the larva's skin is firm and resilient: 'hard' or 'soft' are common adjectives used to describe it.[3] The longer the stroma relative to the larva, the lower the fungus's value and specimens in which the stroma dominates are worth the least. According to this criterion, caterpillar fungus is grouped in three classes. The top quality (*rab*) consists of the largest specimens that are also the heaviest: one *jin* contains about 1000 dry fungi.[4] The second, middle-quality (*drang*) designates a mixed sort, where bigger fungi are mixed with smaller ones. This class is less homogenous and lighter; one *jin* contains 1500-2000 fungi. The bottom-end class (*ta*) consists of small, partly hollow fungi with short larvae and long stromas. Traders rarely know how many specimens are in one *jin*: they count the fungi in the two upper classes, but rarely do so for the bottom one. The middle standard is the

3 The terms for referring to soft fungi are particularly interesting. They are described as *julug* or 'drooping intestine' and *wulug* or 'low rider *yartsa*' (a term used for a low-fastened Tibetan robe). Thanks to Huadan Zhaxi for this information.

4 Other regions produce bigger specimens: in Dzartod Township in Yushu the top class has 700 fungi per *jin*.

Figure 12 Caterpillar fungus: three quality classes

most common. The traders estimate that 70-80 per cent of all what they sell belong to this, while the remaining two classes comprise 5-10 or 15-20 per cent of the market.

The size, weight, and proportions between the larva and the stroma are related to the point in the fungus's biological lifecycle at which it was harvested. The largest specimens are those in the early phase of the lifecycle. They are usually found soon after the season begins. At that time, the stroma has only just started to grow. The larva's exoskeleton is firm and has the outer features of the former host still well visible. Such fungus is called *nga* or the 'early' one. Later, the stroma grows longer and a spore-producing tissue develops on top of it. The larva shrinks and turns into a hollow, empty shell, light and soft. Such fungus, with a shrunken larval part and a long-twisted stroma, is called *tsar* or 'final' one. The 'early' and 'final' fungus, as well as *bar* or 'middle' one, is shown in Figure 12.

The best quality caterpillar fungus is usually found in early spring. Later, the chances of finding it are smaller, but not non-existent. This connects to different timing for the digging season at different elevations. However, not all the fungus dug in the first three weeks of spring is of similar quality: the largest fungi are always set apart from the more ordinary ones. In some places, the top-quality stuff does not grow at all. Digging the fungus in early and late spring differs in terms of the effort one has to invest in proportion to its results. Digging one *jin* of top class fungus at the beginning of the season may take considerable effort. However, it is nothing compared to the work required to find the late-season fungi, of which one needs several thousand to fill one *jin*. Not only is the work immense but the income is small. Would the cheap, late-season fungus attract the diggers in June and July strongly enough? It probably would, but there is a chance that some better-quality specimens are still hiding in the grass. The fact that even in July people bring fresh fungi of the better two classes to the shops shows that this hope is not unfounded.

Different prices of caterpillar fungus at different stages of its lifecycle connect to its different medicinal value. The early fungus is believed to be the most potent. Later, when sporulation occurs, the medicinal value of the fungus declines and so does its price.[5] The price differences between such 'early' and 'late' fungus are spectacular: in spring 2010, the top quality specimens in Golok cost 70,000-80,000 yuan per *jin*; the middle quality 30,000-50,000 yuan; and the lowest 10,000 yuan. Table 1 shows these price differences not only between the three quality classes, but also *within* these classes. The latter are explained by different colours and other features linked to the place where the fungus comes from.

All top-quality caterpillar fungus in Golok contains around 1000 specimens in one *jin*, but this does not mean that one *jin* from two places fetches the same price. The differences between the best fungus from Domkhok and, for example, Huonkor Township in Darlag County can reach as much as 5000 yuan per *jin*. Darlag is known for good caterpillar fungus, but the traders say that the fungus from there does not have the right colour (light orange, a colour that Tibetans call *serkar*). Even bigger differences exist in the middle quality class: they span 10,000 yuan per jin (34,000 yuan in Domkhok and 24,000 in Darlag, May 2010). Similar differences are observed between Machen and Gabde: the fungus from Chemdry and Kochu townships, which are some of the best producers in Gabde, seldom reaches prices comparable to Machen.[6] But these differences are also visible in Machen. Table 2 shows prices from Xueshan, Domkhok, Dawu Zhuma, and Chamahe. The first three townships are the caterpillar fungus 'first

5 In spite of this, the traders in Xining said that this late fungus is bought by pharmaceutical companies.

6 Chemdry produces one the largest amounts of caterpillar fungus measured in *jin* per area unit. As the traders said, in Domkhok or Xueshan 'the living standard is already so high that people dig less, while in Chemdry it is lower, so people dig more'.

Table 1 Caterpillar fungus quality classes and prices, Dawu town, May 2010

Shop No.	Caterpillar Fungus						The Fungus' Approximate Origin
	Top Quality		Mixed Quality		Low Quality		
	Price (yuan/*jin*)	Specimens per *jin*	Price (yuan/*jin*)	Specimens per *jin*	Price (yuan/*jin*)	Specimens per *jin*	
1	70,000-80,000	1,000	30,000-40,000	up to 2,000	10,000	3,000	70% Domkhok
2	70,000	1,000	32,000	2,000	6,000-7,000	5,000?	10% Domkhok
3	60,000	1,000	30,000	2,000	6,000-7,000	unknown	30% Domkhok and Xueshan
4	70,000	1,000	40,000-50,000	2,000	10,000	unknown	–
5	70,000	1,000	30,000	2,000	6,000-7,000	unknown	mostly Gable
6	70,000	1,000	–	2,000	7,000	unknown	40% Domkhok
7	60,000	1,000	28,000	2,000	7,000	unknown	mostly Rarja
8	80,000	1,000	40,000	2,000	12,000	unknown	–

league', but the fungi from Domkhok is bigger than that from the other two townships, and that from Xueshan has better coloration. What the fungus from one township lacks in terms of colour is compensated for by its size and vice versa and, ultimately, all three townships get a similar price for the top-class fungus. The differences are more evident in the middle class: the fungus from Domkhok and Dawu Zhuma costs 2000 yuan less than that from Xueshan. Chamahe does not produce the top class at all and its middle-class fungus costs only 62 per cent of what is paid for the same class from Xueshan.

Table 2 Caterpillar fungus prices in four townships of Machen County, July 2010

Township Name	Caterpillar Fungus Price (yuan/*jin*)	
	Top Quality	Middle Quality
Xueshan	60,000	32,000
Domkhok	60,000	30,000
Dawu Zhuma	60,000	30,000
Chamahe	data not available	20,000

These local variations between caterpillar fungus from different parts of Golok manifest themselves in greater and lesser demand for the fungus theoretically from the same quality class. Diggers from outside of Machen can have difficulties selling their fungus, especially in Dawu where the traders are 'spoiled' by the availability of the fungus from the best three townships. The following excerpt about a pastoralist from Jigdril County illustrates these difficulties:

> Gombo is from Menthang, but he couldn't sell his fungus there. The township market was too small for the amount he had to sell. He wanted to go to Ngawa, but heard that traders in Dawu pay better, so he came here. His surprise was great when no one wanted to take a look at his several *jin*. The traders asked, 'Where is the *yartsa* from?' This was enough for them to tell that they weren't interested. Gombo was desperate. He carried his backpack full of *yartsa* with him all the time: he was afraid that if he left it in the guesthouse it would be stolen. When he started losing hope, his cousin performed a *bsang* ritual to the family's mountain god. The next day, everything changed as if some magic spell was lifted. The traders bought all the fungus on the spot, even though before this Gombo had spent two weeks looking for a buyer. (FN 2009)

While even the best caterpillar fungus from outside Machen does not always find buyers in Dawu, that from Machen is famous well beyond Golok. The best known is that from Xueshan. The further the fungus travels, the more difficult it is to tell which part of Golok it comes from and much of Golok fungus is 'branded' as coming from Xueshan – as I observed at the market in Xining:

> Almost every seller tells me that his *yartsa* is from Xueshan. There is no way for someone uninformed to tell the difference and I expressed my suspicion that it is not really possible that their fungus is from there. The traders finally admit that their fungus is from another part of Golok, but the 'Xueshan quality' is so famous and the name ('Snow Mountain') so attractive that they cannot resist using it. It's not a big lie, they say. After all, their *yartsa* is very good and it also comes from Golok. (FD 2010)

Bargaining

The range of prices paid for different quality classes of caterpillar fungus explains why there is a broad space for price negotiations. The size of the specimens plays the main role, but other factors, such as colour, can push the fungus up or down the price scale. A further field for negotiations opens when the fungus is difficult to classify according to its water content and the level of processing. Most fungi are sold either fresh and not cleaned or cleaned and dried. When cleaned but not completely dried fungus is sold, long discussions ensue about how to assess its value.

Caterpillar fungus trade is done with a large amount of bargaining. It involves a system of hand signs, which has previously been documented in Tibet. William Rockhill wrote in the nineteenth century:

> The buyer and seller takes each other's right hand, well covered by the long sleeves, and having agreed upon the unit to count, rupee, ounce of silver or brick of tea, the vendor takes hold of a certain number of the other's fingers, thus telling him the price at which he offers his goods. The other makes his offer in the same manner; and so, with many knowing winks, shakes of the head, and remarks to bystanders, the trade goes on till finally they come to an agreement. (Rockhill 1891: 252)

This description could easily come from Golok during the years of the caterpillar fungus boom, where the trading parties convey their bids not by showing the numbers but by *feeling* them: they hold or press each other's

Figure 13 Bargaining in a shop in Koja

fingers to communicate the sum they propose.[7] Tibetan robes, with their knee-length sleeves, offer good cover for the hands, but anything else can be used for this purpose: a jacket, a hat, a basket, or other accessories found in the shop or at the market (Figure 13).

The hand sign system consists of numbers from one to nine. Haggling about the price, people discuss sums counting in tens if not hundreds of thousands of yuan. When one *jin* of the fungus is on the scales, it is clear that the talk goes on around five-digit numbers and zero is not needed. If 33,000 is proposed, the sign for three is pressed twice. Number three is shown with the three middle fingers extended and the little finger and the thumb closed: a person bidding this price holds the other person's three middle fingers (Figure 14). To propose 70,000, only the sign for seven is shown. It resembles the letter L, with thumb and index finger stretched and the other fingers closed: to bid this price, one holds the other person's thumb and index finger.

7 In Mongolia, a similar manner of bargaining, called *khantsui naimaa*, *khantsui dotor naimaa khiikh*, or *khantsuilakh*, was used in the livestock trade. Thanks to Ulrike Gonzales and Jan Rogala for this information. In Tibet, it is used with livestock and for expensive modern goods, such as cars or motorbikes.

Figure 14 Hand signs used in bargaining; in each case, the hand on the left is proposing a price

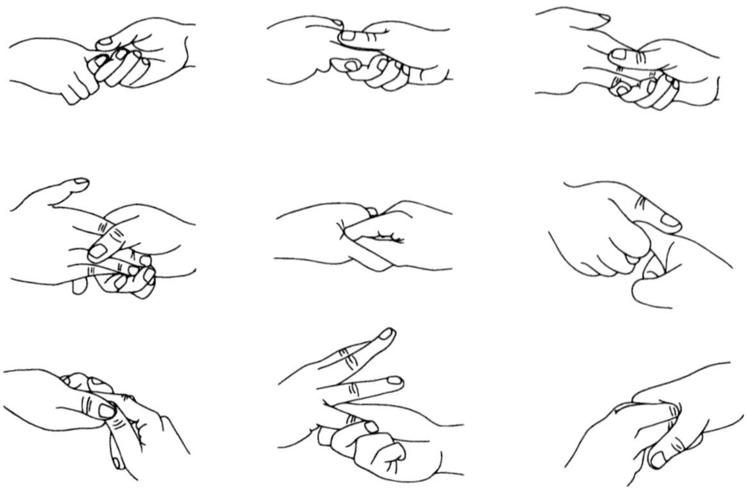

Haggling with hand signs has advantages. It allows a degree of confidentiality in a situation where the deal is struck in a public space and in the presence of dozens of people. Some of them might also have fungus for sale and wait their turn to approach the trader. Others are traders who, waiting for customers, watch their colleagues at work. Regardless of how dense the crowd is and how much people strain their necks, they cannot see all the details of the negotiations.[8] Very rarely the price is said aloud, usually when it concerns small purchases. The hand signs are also helpful when linguistic communication is impaired. As a Hui trader told me: 'It's very good to use them in places like Qinghai, where there're so many ethnic groups who don't speak each other's language.' Although in Golok many traders speak Tibetan, some ambiguities can always creep into the transactions. The hand signs give both parties a common platform where they feel secure and help them avoid misunderstandings and complaints about the financial aspects of the deal.

Bargaining, as Clifford Geertz observed in the context of Indonesian markets, can be seen a 'reflex of the fact that the absence of complex bookkeeping and long-run cost or budgetary accounting makes it difficult for either the buyer or the seller to calculate very exactly what, in any particular case, the "reasonable" price is' (1963: 32-33). This makes bargaining typical

8 The link between the hand signs and the secrecy of the trade gives the former a certain aura of mystery. Though the signs are known to almost everyone, few people dared demonstrate them in front of a camera.

Figure 15 Emotions run high: Aren't these scales cheating?

of societies with a low level of literacy and poor information flow, where 'specific comparative and historical data are simply not available' and where buyers and sellers operate in a 'gap of ignorance between a price obviously too high and one obviously too low' (ibid.). However, in Golok, the pastoralists are well informed about the range of prices paid, even in more distant markets. Here it is rather the seasonal fluctuations of prices and the large diversity of quality classes that make bargaining important. It is also a traditional way of settling a trade deal of considerable value, when prices are not fixed and where transactions are burdened with a risk of loss.

The following scenes show price negotiations between pastoralists and traders. They illustrate several previously discussed points. The pastoralists sell the fungus sorted in different quality classes and arrive at the shop well prepared. They have inquired about the prices in other shops, know how much they want to get, and either decide to wait to sell their fungus until the price goes up or look for a better price elsewhere. These scenes also show the rhetoric used by the traders, who stress the risks associated with the trade and question the pastoralists' ability to evaluate the quality of the fungus. Warnings against the risks awaiting unexperienced suppliers convey a message: this is a risky business and one should stay with a reliable partner even if he does not propose the best price (Figure 15).

Scene 1

In the flat of a Hui trader. Having laid their motorbike helmets on the bed, two men from Gabde stand at the table. Robes rolled around their hips, thick pullovers, crow-black hair. They unpack a box full of caterpillar fungus. The trader sifts it through his fingers. The pastoralists discuss the price with hand signs. They use the same method to communicate with the trader. When bargaining with the trader, they cover their hands with a straw hat. When they discuss the price with each other they do it in their sleeves. The conversation is in Tibetan.

Trader: Will you exchange for this [price]? (The trader shows the price.)
Pastoralist 1: No.
T: But you've said this much.
P1: I didn't say it, some others did. Let's do this. (He shows the price.)
T: No, we don't have such a price. We can't do business this way.
(P1 exchanges a look with his companion. The other rubs his forehead.)
T: I have these scales, *good* ones. (He points at shiny metal scales.) But others have *bad* scales.
(Silence.)
T: Tell me how much others are giving. (The trader tries to break the silence.)
(P1 shows him the price.)
T: Hmm, that man gave a high price!
P1: Yes! 30,000!
T: Oh! This person must be someone who wants to *steal* it.
P1: Yeah… I thought he might want to *steal* it. So, I didn't dare sell.
T: So please sell it to others.
P1: Come on, give a reasonable price. I don't dare showing it to others.
T: The others even if they offer a big price, when they bring you somewhere, they might *steal* it. Bolo [the name of the other pastoralist] knows the price. We *know* each other.
P1: Yes, that's why we came.
T: Ok. I will tell the others that there's *yartsa* for that price.
(Silence. The pastoralists are discontented.)
T: Is there any way?
P1: No way. *Dukpaekur*, the price really went down.[9] (He starts packing away the fungus.)

9 *Dukpaekur* is a popular Tibetan swearword which means 'I'll bear the sin' or 'I'll swear to bear the sin'.

Pastoralist 2: Yeah, the price went down.
T: Is this ok? (The trader bids a new price.)
P1: No.
T: Then how much is ok?
(The pastoralists discuss the price between themselves.)
P1: Tell us a nice price.
T: But the *yartsa* isn't nice.
P1: Say a nice price.
T: You say a nice price.
P1: Let's do this. (He shows price with his hands). *Dukpaekur*!
T: No way.
P1: *Dukpaekur*! That's a cheap price.
T: I don't have money today. If I don't have money, it's all for nothing.
(Silence.)
T: If I pay this much would you sell? (The trader tries one more time.)
P1: No, no, no.

In the whole conversation a price was said aloud only once and this was a price offered by a previous trader who was not present. After the pastoralists left the flat, the trader revealed that they wanted 27,000 yuan per *jin*. He offered them 4000 yuan less.

Scene 2

A tall man with a brown, pock-marked face enters the shop. From a bag hanging from his neck he pulls out an empty can of lychee lemonade with a few fungi inside. They have a long fruiting body and a shrunken larval part. This fungus is not worth much, but the man has more fungus to sell. The bigger its value, the closer to his body he keeps it, and the best fungus he keeps in a shoebox inside his robe. The conversation is in Chinese.

Trader: How much do you want?
Pastoralist: Five [yuan per piece].
T: Three.
P: Fine. (He takes the money and produces a bag with more *yartsa*. It looks better than the first one, but there are only 55 specimens.)
T: Fifteen [per piece].
(The pastoralist pulls a calculator from his shirt and checks how much it would be).

P: Fine.
(The trader gets the money from the safe.)
T: Do you have anything more?
P: Yeah. (He pulls the shoebox from the folds of his robe.)
T: Can't you just show everything at once?
(In the box there is a white cloth bag with the owner's name and caterpillar fungus inside.)
T: Let's see… (The trader places the fungus on some electronic scales). 130 [grams].
P: No way.
T: Look by yourself. (The trader points at the numbers on the display.)
P: It's not *balanced*.
T: No? Maybe you just don't know these modern scales?
P: Nonsense.
T: Want to try the old ones? (The trader shows him some big metal scales that look more like a shop decoration.)
P: Go ahead. (The trader weighs the fungus.)
T: So? 130! (Triumph is heard in his voice.)
P: Fine.
T: How much do you want?
P: 10,000.
T: What?! (The trader laughs) 5000.
(Now the pastoralist laughs.)
T: Look, it's not all good. (He points at some smaller pieces.)
P: In every good load of *yartsa* there are some bad ones. How can they be identical?
T: Really? Look at this! (The trader takes a plastic bag of roughly one *jin* of the fungus from under his desk). I paid 70,000 for this. Do you mean that your *yartsa* is just as good?
P: Mine is better.
T: Pardon?!
P: The colour is better and the size bigger.
(The trader is laughing.)
T: I'll give you 5300.
P: No chance.
T: 5400.
P: Come on!
T: Look, I pay you several thousand, it's a good price.
P: Maybe it's good for you but not for me. (He puts the box back into his robe and starts to leave.)

T: What's your last word?
P: 10,000.
(The pastoralist is already outside the door.)
T: 7000!
P: If not 10,000, then there is nothing to talk about. With other *yartsa* I just helped some people. As long as I have done that, I don't care about mine. I can wait.

Market Fluctuations

The price of caterpillar fungus is subject to seasonal and annual fluctuations. The price at a given moment is not representative of the whole year or even digging season. What can be said with certainty is only how much fungus (of a particular quality) cost at a particular moment or what was the range of prices within a longer period. Both the pastoralists and traders name the instability of prices as one of their biggest problems: risk is inscribed in this trade more than in any other. Prices change not only within one year, but within one season. Traders say that they 'fall overnight'. What will the price be tomorrow? 'How could I tell? Today it's 30,000 yuan. But a while ago it was 40,000. What will it be tomorrow? I'd also like to know!'

The caterpillar fungus market experiences two peaks a year. The first is observed in May: when fresh fungus starts appearing at the market, the prices reach their upper limit. The scenarios that follow depend on the amount of caterpillar fungus in a given year. 'The scarcer the fungus, the better the price' is a gross simplification, but a certain balance between what the land produces and what the traders can buy shapes the situation. What starts as a 'good' season can easily turn into a 'bad' one, when the supply outweighs the demand.

After the initial peak, the prices decline. This happens sometime in the middle of the digging season: the size of the harvests impacts whether this decline is more or less spectacular. In 2010, prices in Dawu stayed stable until 16 June and then started dropping. Table 3 shows the prices before and after that date. The smallest decline was observed in the top class: the traders estimated it as 10,000-20,000 yuan, that is, 12-25 per cent of the original value. A bigger decline was observed in the other two lower classes: 25-35 per cent in the middle and as much as 40-75 per cent in the lowest class. This change appears even more dramatic when one realizes that in late June and July the diggers find mostly the fungi of these two classes whose prices declined the most.

Table 3 Changes in caterpillar fungus prices, Dawu town, May-July 2010

Quality Class	Caterpillar Fungus Price (yuan/*jin*)			Caterpillar Fungus Price (yuan/piece)	
	Before June 16th 2010 ("Top" Price)	July 7th 2010 "Top" Price	July 7th 2010 "Normal" Price	Before June 16th 2010	July 7th 2010
Top	80,000	70,000	60,000	–	–
Middle	40,000	30,000	25,000	25	4
Low	12,000	7,000	3,000	–	–

In the second half of the digging season, the market seizes up. People still bring caterpillar fungus for sale, but the traders can no longer sell on what they have bought before and cannot buy more. This is visible in Dawu, but even more so at the county and township markets where the traders' purchasing power is smaller. At the same time, the radius of the pastoralists' travels grows. Groups of pastoralists walk between the shops in Dawu looking for a buyer and often leave for Xining, hoping to sell their fungus there. Much of the caterpillar fungus that they do not manage to sell in this period waits for winter when the market improves again.

Between December and February, the market revives and groups of pastoralists carrying bags, backpacks, or cardboard boxes with caterpillar fungus appear on the streets again. Winter prices surpass those from spring. In December 2007, a Tibetan trader in Gabde recalled that the same fungus that he had just bought for 78,000 yuan was worth 10,000 less in spring of the same year. This winter surge of prices explains why 'people who aren't in dire need of money wait until Losar to sell', as he said. Losar, or Tibetan New Year, is celebrated between January and March. Sumptuous family feasts cost money and the winter rise of caterpillar fungus prices could not have better timing.

The pastoralists say that the caterpillar fungus trade is a risky enterprise. Nothing demonstrates this better than the events of 2008, when the market in Golok faced a serious decline. In 2007, the top-quality fungus cost 110,000 yuan per *jin* and the pastoralists believed that the prices would keep growing. Many people expected that the Summer Olympic Games in Beijing would help advertise caterpillar fungus around the world and the prices would rise. Expectations were high, but instead of rising, the prices fell. In early 2008, they were still at the level of 80,000-110,000 and 40,000-50,000 yuan/*jin* for the top- and middle-quality fungus, but in late May or early June they started declining. My informants recalled how from 80,000 yuan the

Table 4 Fluctuations of caterpillar fungus prices 2007-2010, Dawu town

Shop No.	Time Period	Caterpillar Fungus Price (yuan/*jin*)		
		Top Quality	Mixed Quality	Low Quality
	2007 spring			
1		70,000-80,000	40,000	10,000
2		110,000	–	–
3		80,000-100,000	40,000-50,000	20,000
4		100,000	80,000	–
5		70,000	37,000-38,000	6,000-7,000
	2007/2008 winter			
1		110,000	50,000	20,000
2		110,000	47,000	–
3		80,000-100,000	40-50,000	20,000
4		80,000-90,000	40,000	4,000-5,000
5		110,000	40,000-50,000	10,000 and more
	2008 spring			
1		50,000 and more	20,000	–
2		70,000	15,000-17,000	3,000-4,000
3		40,000	10,000-20,000	5,000-6,000
4		–	–	–
5		50,000	20,000	–
	2008/2009 winter			
1		60,000	20,000	–
2		80,000	27,000	6,000
3		60,000	30,000-40,000	6,000-7,000
4		–	–	–
5		60,000 (80,000?)	24,000-25,000	10,000
	2009 spring			
1		65,000	30,000	15-16,000
2		50,000	20,000	6,000-7,000
3		40,000	25,000-26,0000	6,000-7,000
4		40,000	20,000	4,000-5,000
5		40,000-50,000	17,000-20,000	4,000
	2009/2010 winter			
1		80,000 and more	47,000-48,000	20,000
2		50,000	40,000-50,000	6,000-7,000
3		70,000	40,000	10,000
4		40,000-50,000	20,000-30,000	4,000-5,000
5		70,000 and more	36,000	6,000

price fell to 77,000, then to 72,000 and 67,000, to reach 60,000 in December 2008. The traders estimated the scale of this decline as 35-55 per cent (top class), 60-75 per cent (middle class), and 85 per cent (lowest class) of the initial value. Table 4 shows this decline and a subsequent recovery. Between January and April 2009, the prices gradually increased, and from May they went up, as the traders noted, almost daily. Regardless of the crisis, the seasonal rhythm remained unchanged and in winter the prices were still higher than in spring.

The 2008 crisis cooled people's enthusiasm for storing their financial assets in caterpillar fungus. Mistrust or scepticism about the trade was widely voiced, but many people still kept some fungus at home. Only the quantities were smaller. Few pastoralists wanted to tell how much they had lost in 2008, but many admitted that they had indeed kept their fungus to sell in 2008 and even bought more from neighbours and relatives. Jigmed Dorji had 10 *jin* to sell during the Olympic year and expected manifold gains. He did not want to say how much he lost. Similar to many other pastoralists, he connected the decline of the caterpillar fungus market with the political situation in many Tibetan regions, which faced dramatic political protests in the wake of the Olympic Games. Many informants believed that this was a centrally orchestrated punishment for their political insubordination: 'Someone on the top decided that we won't get good money for our *yartsa* anymore. They turned off the tap with money.' Other people speculated that 'Chinese businessmen in the cities pushed the price down' and that a media campaign was used to discourage people from buying caterpillar fungus. 'The television said that *yartsa* was worth not more than carrots', as someone told me. Other pastoralists opined that the fall in caterpillar fungus prices was linked to the global economic crisis. In any case, these two dimensions – economic and political – come together in the narratives of the market decline, as in the case of a young man who was one of few open enough to reveal the scale of his personal financial loss:

> I not only kept my own *yartsa* but borrowed money to buy more. I'd heard that *yartsa* will be exported during the Olympic Games and the price would rise. As a result, I lost nearly 250,000 yuan. In 2007, one *yartsa* was twenty-five yuan, and in 2008 it was only ten. We lost fifteen yuan per piece! But it wasn't only about *yartsa*! In 2007, butter was seventeen to twenty yuan per jin, and in 2008 it was only ten. *Chura* was twenty-five, and is now fourteen.[10] *Khulu* was ten to twelve, and now even if you sell

10 *Chura* or *chura kompa*, yak-milk cheese dried in small pieces in the sun.

it for one yuan nobody wants it.[11] Many people hoped that the Olympic Games would boost the economy. But things went bad for us. I'm not saying it because I'm a biased Tibetan, but to tell the truth many army trucks arrived here that spring. The army provoked us and conflict developed. So, both economically and culturally 2008 was a very bad year. I think it was the worst year since 1958.[12]

11 Soft undercoat hair collected from yaks' bellies and hind legs, used for making blankets, ropes, and weaving tent cloth.

12 1958 is the year that symbolizes the repression of the anti-Communist rebellion in Golok and the neighbouring regions, and the beginning of radical changes introduced by the new authorities. This year is used by my informants as a caesura between the 'old' and 'new' society, a symbol of difficulties, injustice, and misery.

6 The Law in Action

As long as the pastoralists were organized in the people's communes, the trade in caterpillar fungus was controlled by the state. This started changing in the 1980s. The communes were closed down and the economic reforms freed the trade from the constraints of the command economy. The pastoralists were allocated land and livestock and started selling their products, including caterpillar fungus, on the private market. At the same time, people across China embraced their newly regained mobility and sought income opportunities outside their place of residence. The growth in the popularity and price of caterpillar fungus triggered seasonal migrations of diggers to pastoral regions where it grows. Tensions over the land intensified and so there emerged a need to create a system governing access to the land and caterpillar fungus resources. How was it to be organized, if anyone else, apart from the pastoralists, wanted to dig the fungus? Should this be possible free of charge or upon some sort of payment? Who should decide about the payment and the use of that income? The following decades witnessed a series of attempts to regulate access to the caterpillar fungus-producing land: new policies replaced old ones that turned out to be contested or ineffective.

During the late 1980s and 1990s, diggers coming to Domkhok paid a Grassland Tax (Tib. *rtsacha*) to the township where they wanted to gather caterpillar fungus. The size of this tax was decided by the herders committee (Tib. *drok u khang*), the smallest administrative unit in the township, and it depended on the caterpillar fungus production capacity of a given area. Between 1988 and 2000, this tax in Domkhok rose from 500 to 1000 yuan per person. The diggers could dig 'wherever they wished' within the area they paid for, as the local officials informed me. Individual households' rights over their land were of secondary importance. The major part of the Grassland Tax went to the herders committee and the rest to the county legislature.[1] In Domkhok, this money was spent on road and house construction, fencing pastures, connecting electricity to the township, and building a bridge over the Domchu.[2] A man who worked in the local administration during that

1 Goldstein says that Xueshan paid 5 per cent to the county (1996: 19). Domkhok paid 30 per cent. The herders' committees also collected a tax in caterpillar fungus paid by in-migrants for the *waimao*: in Domkhok it was two *shang* per person (see also Costello 2008: 94).

2 In Xueshan Township it was spent on similar purposes: building roads, subsidizing fences and houses for the pastoralists (Goldstein 1996: 18).

period recounted: 'This money went on developing the local society. It was neither eaten up by the leaders, nor divided between people as cash.'

These two decades were crucial for the development of local infrastructure. As many pastoralists said, Domkhok did not have 'proper roads' before and was isolated from the rest of the prefecture. Thanks to the high productivity of its land and people's interest in caterpillar fungus, Domkhok gained a degree of economic self-sufficiency and could finance important infrastructure projects. The prerogative of accepting diggers rested with the herders committees and it was they who managed the income. Although money from the Grassland Tax was not distributed among the pastoral households on whose land the diggers worked, the infrastructure projects financed with it benefited the whole population.[3] Most of the pastoralists I interviewed recognized these benefits and believed that they were, at least in part, possible thanks to the Grassland Tax funds.

The Open Door Starts to Close

The situation started changing in the 2000s, when Golok Prefecture announced a series of steps aimed at bringing caterpillar fungus exploitation under a more direct control. This change came in a period of intensified legislation in China aimed at regulating land use. In 2002, the Beijing government passed a revised version of the Grassland Law, which regulated use of the country's grasslands and addressed the problems of environmental protection by restricting and prohibiting many practices identified as having a negative impact on the grassland environment. These included exploiting mineral resources, quarrying soil or stone, reclaiming land for agriculture, overgrazing, using pesticides, and even driving motor vehicles. A separate article informed that it is prohibited to 'dig plants [...] or engage in other activities to the detriment of grassland vegetation' (Art. 49) (GLPRC 2002). At around the same time, a series of other regulations was passed that addressed the exploitation of particular plant and fungus species. These included liquorice and ephedra, which were popular in Inner Mongolia, but also caterpillar fungus, which in 1999 was declared the 'second-class state-protected species' (SFAMA 1999, Mu and Zhu 2005, Wang and Yao 2011: 43). This status did not imply that caterpillar fungus stood under a threat of

3 Horlemann remarks that it was the pastoralists who suffered most from the destruction of their land and so should receive money from the Grassland Tax (2002: 262).

extinction but showed that the state recognized its economic importance.[4] From now on, digging the fungus and trade in it required official permission:

> Those intending to sell or purchase wild plants under second-class state protection shall apply for approval from the Department of Wild Plants administration under the government of the province, autonomous region, or municipality directly under the central government or the agency authorized thereby. (RPRCPWP 2001, Art. 18)

Caterpillar fungus, being covered by this 'second-class state protection', was classified between species covered by 'first-class state protection' – trade in which was forbidden and collecting allowed only for scientific purposes – and the species protected by lower administrative levels (provinces and autonomous regions). This shows the special place that the fungus has among other species: although it grows in few provinces and these could manage access to it by themselves, the central authorities decided that any operations involving this commodity have to be controlled by the state.

The changes in caterpillar fungus management policy should be placed not only in the context of intensive legislation aimed at controlling exploitation of medicinal plant and fungal species, but also in the context of debates about desertification, which during that period was recognized as one of burning problems in China. In the official discourse, desertification was explained by anthropogenic reasons – among others, overcultivation, overgrazing, and exploitation of natural resources.[5] Farmers were blamed for reclaiming the grassland for agriculture and pastoralists for having herds that were too large. A separate place belonged to digging medicinal plant and fungus species. Thousands of diggers trampling the grassland were blamed for destroying the soil structure and contributing to environmental problems. This is how the situation in Martod, the most ecologically degraded county in Golok, was explained.[6]

4 Some pastoralists stated that there is less fungus now than in the 1980s. This was for them a result of the boom: 'People dig *yartsa* year after year. It must be dwindling.' Increased competition among the diggers can also explain why people perceive caterpillar fungus to be scarcer than before. On the sustainability of caterpillar fungus, see Weckerle et al. 2010, Cannon et al. 2009, and Winkler 2013.

5 There is extensive literature on this topic, with many authors who support the official view as well as those who argue against it; see Harris 2010; Williams 1997; and Ho 2000.

6 The pastoralists recalled how quickly the situation in Martod deteriorated. One man recounted how during the Cultural Revolution people buried religious statues under a sand dune to save them from destruction. When they went to dig them out in the 1980s, the statues were already exposed where the structure of the dune had been eroded.

While overgrazing and demographic growth were considered the main causes of desertification in Martod, digging caterpillar fungus was said to have significantly contributed to the process (Du 2012: 117). Digging caterpillar fungus was thus identified as dangerous to the environment and needed to be controlled.

The new strategy for managing Golok's caterpillar fungus resources implied eliminating something that the policymakers called 'wild digging'. As some county officials mentioned to me, during the earlier period when the townships were responsible for accepting the diggers, 'everyone could dig everywhere' and this posed a threat to both the sustainability of this species and to the wider environment. Limiting the number of diggers was thought to be necessary to bring the situation under control. This was not only expected to safeguard the sustainability of caterpillar fungus, but also to bring many other positive effects. It was hoped that it would 'turn back the tendency of environmental degradation and let damaged grassland recover and get back its natural functions', but also to improve the water retention of the soil, prevent erosion, and even natural disasters (Li et al. 2010: 33).

The new approach implied that digging the fungus was to be done in a rational, scientific, planned, and organized way. Such attributes were used to juxtapose the new approach against the earlier one, which was called irrational, wasteful, and predatory. The previous management system was portrayed as lacking an organized character: it was not a system, but chaos, as one official said. From now on, the county government had to impose limits on the volumes of fungus dug and ensure that they were not exceeded. The townships had to prepare annual collecting plans and submit them to the Agriculture and Animal Husbandry Bureaus in the county for approval. The county bureaus, in their turn, had to consult the prefecture. They also had to decide how many diggers would be allowed entry as well as to delineate areas where digging was forbidden and others where a rotating system was to be used.

The new policy granted the pastoralists usufruct rights over caterpillar fungus resources located on their land. It protected them against in-migrants who, as the policymakers pictured it, wanted to deprive them of the income from caterpillar fungus and destroy their land. Paradoxically, the government called for the introduction of limits on the volumes of caterpillar fungus dug, but at the same time claimed that the new system was characterized by the principle of 'self-management by the people' (Li et al. 2010: 33). The role of the prefecture and county was to 'provide collaboration schemes and monitor work', but it was the pastoralists who were supposed to manage their own land. Thus, at least theoretically, the new policy vested them with the responsibility to protect their land and its resources. The pastoralists were to be the 'protectors of the environment', as it was framed in the language of the new policy.

A consequence of this new approach was the decision that only the pastoralists would be permitted to dig caterpillar fungus in Golok. This marked the beginning of the closed-door policy: diggers from outside Golok – 'outsiders' as they were called in the legal Newspeak – were no longer welcome.[7] However, announcing such a decision was easier than enforcing it. On the one hand, the size and character of Golok's terrain made enforcing it difficult. On the other, certain channels of entry remained open.

Non-Golok residents could apply to the Animal Husbandry Bureaus for a so-called Medicinal Grass Certificate, allowing them entry into the prefecture.[8] These licences were issued to organized groups of diggers sent to Golok on agreement made between the counties: one county applied to another, in the same or another prefecture, for permission to send their residents to dig caterpillar fungus. This process did *not* require the consultation or consent of the pastoralists on whose land the diggers worked.[9] Thus, although the new policy was supposed to protect the pastoralists' rights, the licence system excluded them from both participation in the decision-making and from any financial benefits arising from the fact that someone was digging caterpillar fungus on their land. The money derived from selling the licences was, according to the officials I interviewed, spent on different purposes, including programmes for cleaning the grassland when the season was over. Winkler mentions such programmes too but observes that the diggers' campsites after the season 'were marked by empty and broken beer bottles, discarded instant noodle containers, and plastic bags', which suggests that these cleaning programmes have not been successfully implemented (2008: 297-298). The pastoralists in Domkhok said that these programmes were a dead letter. They were also sceptical as to whether, given the scale of the phenomenon and the size of their land, they could ever be realistic anyway.

7 Chin. *wailai renyuan*, lit. 'personnel' or 'persons [coming] from the outside'.

8 This bilingual document (Tib. *mentsi fanyikh*) contained the holder's name, address, and ID number, validity period, and the place where the person is authorized to work. Its price in Machen was 1000-5000 yuan depending on the area. Winkler says that in 2006 the licences in Golok were the most expensive in Tibet; in other regions they cost 300-1500 yuan (2008: 297).

9 Costello reports that 'local and outside entrepreneurs [...] organize groups of diggers themselves and smuggle them into the prefecture or conspire with local officials to share their fees' (2008: 94). Horlemann mentions over 100,000 diggers entering Golok during this period (2002: 263).

During the same period when the Animal Husbandry Bureaus sold these licences, another way of entering Golok emerged. According to the new policy, only residents of pastoral townships could dig caterpillar fungus there. In order to prove this status, one needed a household registration or *hukou* that listed one's name among other persons living under a given address. Both pastoralists and officials admitted that it then became common to add new family members to the *hukou* booklets, which swarmed with newly discovered cousins and in-laws from other parts of Qinghai, sometimes even non-Tibetans. These manipulations of the household registration showed close cooperation between pastoralists and diggers. It had been years since the first diggers started coming to Golok and many had well-established contacts there. When the authorities restricted access to the prefecture, many diggers made private agreements with the pastoralists and leased the land for digging directly from them. The year 2002, when the restrictions on entry to Golok were introduced, is remembered by the pastoralists as the first year when the diggers came directly to them and paid them cash for digging fungus on their land.

After several years of selling the Medicinal Grass Certificates, the licence system was abandoned. In 2007, when my research started, it was already part of history. But why was the system abandoned? Several officials admitted that it was a failed experiment – a 'total failure', as one official called it. It failed not only because people forged documents, but also – and this factor the officials did not mention – because the licences were contested by the pastoralists, whose dissatisfaction erupted into open protests and violent conflicts.

The best-known example of such a conflict took place in Yushu, a pastoral prefecture neighbouring Golok in Qinghai. In May 2005, Tibetans in Yushu, as the media reported, accused the authorities of imposing a 'tax' on caterpillar fungus diggers and misappropriating this money.[10] Information from Dzartod County, where it all happened, helps to understand the background of the conflict. The Yushu Prefecture authorities sold licences to Tibetans from Nangchen (another county in Yushu), allowing them to dig in Dzartod. But the pastoralists in Dzartod did not recognize the legal power of these licences and blocked the diggers from entering the grassland. This led to a serious standoff (there were casualties), which was only resolved with military intervention. The profile of this conflict appears clear: the prefecture sold the licences without consulting the pastoralists and sharing any financial benefits with them. The pastoralists protested this decision as

10 The conflict was reported on by ENS (2005) as well as Lama (2007) and Gruschke (2011d).

ignoring their interests and the diggers tried to execute the right that they had acquired in buying the licences. In the same period, violent clashes also took place in other parts of Sichuan (WTNN 2007), Qinghai (Gruschke 2011c), and the TAR (TCHRD 2007, Winkler 2008a). Accusations of officials sending diggers to the pastoralists' land and 'pocketing' the money were heard in Golok as well. Costello reports that the 'locals have been increasingly vocal in protesting against the large numbers of diggers' (2008: 92) and Winkler notes that they 'bitterly complained about the licencing approach, which not only deprived local herders of one of their most lucrative resources, but also left them stuck with degrading holes in their pastures' (2005: 80). Bad publicity forced the authorities to take steps in order to prevent similar situations in the future. A year after the events in Dzartod, Yushu stopped selling the licences and Golok followed (Lama 2007: 58, Sonam Lindub 2007: 58).[11]

Banishing the Licences: The Door Closes

After the transition period, when large numbers of diggers were still allowed into Golok, the time had now come to close the door more tightly. The second half of the 2000s brought radicalization in the official stand and a complete ban on digging of the fungus by residents of other prefectures:

> In order to strengthen management of [*yartsa gumbu*] resources in Guoluo TAP, and in order to protect the natural environment, pastures, and plant cover in the Three Rivers Area, and to ensure social stability, the Guoluo TAP People's Government has decided to continue implementing its policy of banning the collecting of [*yartsa gumbu*]. [...] Whoever enters our prefecture by whatever means for the purpose of digging shall without exception have their illegally acquired gains confiscated, they shall be forcibly returned to their places of residence, and they shall be responsible for all damages caused.[12]

11 Chinese participants of the International Conference on *Ophiocordyceps sinensis* convened by the Ministry of Agriculture in Xining in 2010 claimed that the licences were still used. A book accompanying the conference stated: 'Local governments consolidated administration on collecting and procurement, and strictly followed the permit system' (Ma 2010: 125). Photos showed diggers wearing the permits pinned to their jackets and policemen climbing the mountain slopes (!) to check them.

12 The notice appeared at the Golok TAP website with the date 14 February 2008. See also Winkler 2008d.

From then on, official channels for entering Golok ceased to exist. Checkpoints were set up on the roads to ensure that nobody could sneak in. If there were people determined to enter Golok in spite of the law, they had to be prepared to bear the consequences.

The details of this new policy were laid down in 'The executive programme of protective and administrative work of the Chancery of the People's Government of Maqin County in 2010 on Chinese caterpillar fungus resources' (hereafter 'The Executive Programme', see Appendix). Passed in 2010 by the Machen County government, this fifteen-page document contains the rules regulating the digging of caterpillar fungus and related activities that were in force during my research as well as measures for how to ensure that these rules are implemented. The document leaves no doubt: it is prohibited for in-migrant diggers to enter Machen during the caterpillar fungus season – there is a 'strict prohibition on digging caterpillar fungus by outsiders' (Appendix: 3-4). In fact, local residents of Golok employed in other sectors than the pastoral economy and having no legal connection to the pastureland cannot dig either:

> It is strictly forbidden that whoever from outside this county and any city dwellers from the prefecture and the county, cadres and staff workers (including relatives as well as children), pensioners, unemployed persons (laid off, awaiting employment, unemployed), people with minimum-livelihood husbandry roles, self-employed industrial goods producers and traders, students from caterpillar fungus non-producing areas and other people be allowed by any means to collect the fungus. (Appendix: 4-5)

An exception was made for 'ecological migrants', that is, pastoralists who had been relocated to the town through the sedentarization programmes: they could dig the fungus in other townships than their original residence, but only 'in a planned and organized way' (Appendix: 4).

In order to ensure that the new policy was properly implemented, the government departments in the townships, county, and prefecture had to follow a path of close cooperation, as the document says. They should realize that 'efforts to ban the collection of caterpillar fungus is complex and protracted work of a strongly political nature' (Appendix: 7). The document prescribes a list of methods helpful in this work: informative and educational campaigns, various kinds of controls, and, finally, punitive measures to be applied when other methods fail.

Education and information campaigns were to be organized both in Golok and beyond its borders. In the townships, officials were ordered to visit pastoralists' houses to instruct them about the law and 'cause the propaganda contents to penetrate into people's hearts, increase the responsibility and knowledge of the environment among the populace' (Appendix: 8). The counties should organize educational campaigns on a larger scale too: with the help of printed materials (posters, banners, magazines in both Tibetan and Chinese), radio and television, through loudspeakers mounted on vehicles, and even via text messages they should reach as large a number of people as possible.[13] Similar campaigns should be organized outside Golok. The Executive Programme says that the prefecture should strengthen cooperation with neighbouring areas and calls for their authorities' support in disseminating information about the law restricting digging caterpillar fungus in Golok.[14]

To supervise the enforcement of the policy and ensure that no diggers violated the ban, the checkpoints or so-called Propaganda Stations for Environmental Protection were to be set up across the prefecture. Their role was to control traffic, instruct people in the importance of protecting the environment, identify cases of violation of the law, and 'persuade outsiders to return to their original residence' (Appendix: 11). The checkpoints were to employ officials 'from security, traffic, land resources, environmental protection, water resources, agriculture, animal husbandry and so forth' as well as 'representatives of the masses', whose common goal was to 'mutually and closely coordinate, cooperate, supervise one another, earnestly fulfil their working obligations' of 'guarding the gate, taking care of the people, managing the affairs [properly], safeguarding stability, stimulating an increase of income' (Appendix: 8-9). The document specifies that the county is responsible for checkpoints on the main roads, and the townships for those located on local roads leading into the areas under their jurisdiction. Even herders committees were encouraged to create their own checkpoints. Further to that:

13 These campaigns reached their peak in April, which was named All District Environmental Protection Propaganda Month (Chin. *quan zhou sheng tai huan jing bao hu xuan chuan yue*), and May. A man working for the Golok Television station, which broadcasted spots calling on people to abide by the law, compared their efficacy to 'shooting arrows into empty space': when the station put out these spots, people were already digging.

14 It mentions Xinghai (Tib. Tsigorthang, a county in Hainan TAP, Qinghai), Ngawa Prefecture, and Gansu Province. Large numbers of diggers from all of them arrive in Golok each year.

> All the townships and herders' committees have to organize inspection teams [consisting] of basic cadres, party members and people's militia members to patrol the grasslands in their areas and make sure that no outsiders are left in any house and any farm. (Appendix: 11)

Setting up checkpoints on the roads and dispatching inspection teams to the grasslands are only a small part of the immense work of the state control apparatus that was set in motion each spring. Increased traffic checks were put in place to combat the smuggling of diggers into Golok as well as control of flat rentals in town to make sure that no accommodation would be rented to such individuals. If, in spite of all this, some diggers did manage to reach the grasslands, they would be, as the document says, 'cleared up' and punished (Appendix: 14). The document does not specify how this operation should look, but it mentions the use of security personnel and armed police, showing that the use of force is authorized. The evicted diggers are to bear the costs of this operation and would be punished not only for digging caterpillar fungus, but also for any other offences committed during their stay on the grassland, such as lighting fires, poaching, littering, and destroying vegetation (Appendix: 14-15). Not only diggers should fear punishment. Heavy fines 'amounting up to a tenfold value of the illegal income' (Appendix: 13) also await any pastoralists who help diggers to enter the prefecture. The pastoralists risk more than just a heavy fine:

> Cases of contractors using mountains, gullies, banks or rentals to individual proprietors and soliciting labourers to collect caterpillar fungus [...] will be fined politically and economically; those who violate the abovementioned regulations will also get their right of user revoked. (Appendix: 13-14)

The 'right of user' applies to land lease contracts, that is, the official form of land lease in China, where rural residents do not formally own the land, but lease it – on a long-term basis – from the state. These contracts, introduced in the 1980s, allow the pastoralists to use the land and its resources and draw income from it, but also oblige them to manage the land in a way that does not cause its destruction. Any serious violation of this rule could be used by the authorities as a reason or excuse to cancel this contract. Finally, punishment also awaits other individuals who help diggers to enter Golok or who benefit from such activities. This applies to the personnel of the Propaganda Stations who, if they 'engage in favouritism or commit irregularities, [...] will be punished administratively in accordance with the law and [...] handed over to judiciary institutions for criminal investigation' (Appendix: 14).

The regulations listed in The Executive Programme reflect the social reality: they show how many activities, in the law-makers' perception and/or experience, have emerged with the caterpillar fungus boom or are related to it financially. The document frames them as offences against the law. In this sense, the law reveals much about the unlawful: it penalizes a wide range of activities and warns that even if seemingly benign they will be legally prosecuted. However, the document not only lists legal regulations and methods of enforcing them, but also reveals something of the thinking behind these rules. It shows the structure of argumentation and explains that protecting caterpillar fungus is important not only due to environmental concerns, but also for economic, social, and political reasons:

> Enhancement of protection of the ecological environment [should be] beneficial to stimulation of structural economic adjustment and transformation into the growth model, bringing about faster and better development; [should be] beneficial to raising the level of knowledge of ecology within the whole society, stimulating creation of new, socialist pasturelands; [should be] beneficial to safeguarding long-term profits of broad masses of pastoralists, providing a good livelihood and room for development for future generations. (Appendix: 3)

Controlling the use of caterpillar fungus resources leads to an increase in the pastoralists' income and to enhanced economic growth. And this, as the document says, guarantees social stability and contributes to the creation of a so-called harmonious socialist society. Giving the pastoralists priority rights over caterpillar fungus resources is also expected to boost the people's satisfaction and prevent social unrest. By producing satisfied citizens and enhancing harmony between the people and the environment, the policy contributes to the building of a so-called 'ecological civilization'. This is a concept publicized by president Hu Jintao who, in 2007, announced that China would 'build a resource-conserving and environment-friendly society that [...] harmonizes economic growth with the population, resources and the environment' (2007: 5). Following this ideal, Machen County declared its determination to

> scientifically plan and manage the caterpillar fungus resources in conformity with the law, enhance the protection of the ecological

> environment of the grasslands, stimulate harmonious development between humans and nature, create a harmonious socialist society so that the concept of 'the ecological civilization' penetrates [everybody's] conscience and makes this region ceaselessly advance along the road of civilized development. (Appendix: 3)

A Checkpoint

Together with the restrictions on access to Golok, Propaganda Stations started to appear in the prefecture. They were erected at Golok's borders and on the roads between the counties, but also on smaller roads leading into the high-altitude pastoral townships. Put in place just before the season commenced and removed a few weeks later, the checkpoints were short-lived, but reappeared year after year, becoming firm points on the local road map. Their position and size revealed which destinations are most strategic for both the diggers and the state forces in charge of controlling them.

The Sumdo checkpoint (Figure 16) sits at such a strategic point: at a junction on the road leading from Dawu to Domkhok and Xueshan, two townships famous for their caterpillar fungus.[15] I can still remember the emotions accompanying me when I crossed Sumdo for the first time. I was sitting (or, as I felt, hiding) in Jigmed Dorji's car, together with Tsering Drölma's younger sister. The spring holidays had started in Dawu and the students were going home to help their families with digging. Wrapped in a Tibetan robe and wedged between sacks of rice, pasta, and steaming loaves of bread, I tried to organize my mind to offer a reasonable explanation as to why I was there: anything at least half-convincing for the policemen at the checkpoint that we were going to cross. Stories of people turned back from the road did not make me optimistic. However, our car was never checked. We stopped at the checkpoint and spent some time in the car waiting for Jigmed Dorji, who disappeared for a while. A short time later, we continued the journey. Returning the same way several days later, I already felt more daring and, seeing a crowd of people and vehicles, I shot a series of photographs which for me, at that time, felt almost like war photography. A man in a police uniform looked into the car but did not ask

15 *Sumdo* denotes a junction of three roads. Here it is used as a place name. In fact, some parts of Domkhok lie before Sumdo and certain valleys can be accessed without crossing the checkpoint. Still, the main part of the township, and the more fertile one, lies beyond this point.

Figure 16 Sumdo checkpoint

any questions. I continued my journey undisturbed. I crossed Sumdo many times subsequently and never encountered any problems. If I managed to pass through, how many other people did? Were other cars checked more carefully?

The Sumdo checkpoint employs twenty people, who take turns working and sleep in tents pitched at the roadside. Such a large team can be a surprise on a road leading to two sparsely populated pastoral townships. Outside the caterpillar fungus digging season, little happens at Sumdo. The air is still. Sometimes, a car or motorcycle passes by. But during the digging season, this place is barely recognizable. A cloud of dust hangs in the air. A line of cars waits in front of a barrier blocking the road. Drivers honk their horns nervously. A young motorcyclist tries to squeeze between the cars, avoiding the controls, but is stopped by the police. Drivers and passengers queue in front of the police tents. The twenty people working at Sumdo have much to do. Drivers arrive even in the night when the road, at least theoretically, is closed. The checkpoint regulations at the roadside inform:

> Environmental Protection Propaganda Station for Domkhok and Gangri Townships.
> Attention!

1. All persons driving cars and motorcycles into the caterpillar fungus production source region have to unconditionally undergo control.
2. No cars (machines used in agriculture, motorcycles, and others) may enter the caterpillar fungus production source region between 9 pm and 8 am.
3. Persons riding motorbikes have to take off their helmets and face coverings during control.

The Propaganda Stations in places such as Sumdo control the traffic and check the reasons behind anyone attempting to enter Golok during the caterpillar fungus season. Someone from outside Golok must prove that they are coming for work or family reasons. However, the exact procedures and strictness of control differ between the checkpoints. Identity cards and household registrations are checked on bigger roads where the controls take time, especially when one enters Golok from the north, via Rarja, or from the south, via Jigdril and Bamma. On public buses stopped at these roads, the police look for passengers hiding under the seats or in the luggage hold. All journeys end with a delay. In 2007, a bus on which I travelled from Jigdril to Dawu was stopped at four checkpoints. It took thirteen hours to cover around 400 kilometres. Even between Rarja and Dawu, over a distance of 68 kilometres, there were four checkpoints and the journey took four hours. After such controls, the buses continued their journey with half the number of passengers. The other half had to look for a transport back home or for another idea about how to enter Golok while avoiding controls.

The deeper one goes into the highlands, the less formal the controls become. So, it was at Sumdo. Its personnel were recruited from different institutions in the county and the township: Agriculture and Animal Husbandry, Forestry and Traffic Bureaus. Most of the checkpoint officials were from Machen and many hailed from Domkhok and Xueshan. Employing local people was thought to guarantee a quick and uncomplicated control process. The officials knew everyone living in the area that they were responsible for: 'There're so few people living here, we're like one family. We know nearly everyone by name', one man told me. In the pastoral townships inhabited entirely by Tibetans, the presence of non-Tibetans cannot go unnoticed. But when Tibetans arrived at the checkpoint, the situation was also simple. The personnel quickly recognized who was from Machen, or from another county, or a different part of Tibet. One glimpse at a face usually sufficed; sometimes a few questions followed. Another official remarked: 'For us Tibetans, if we see people's face and clothes, and hear the language, it's easy to tell who lives here and who's cheating.'

It seems to be easy for the checkpoint personnel to recognize who among people arriving at Sumdo lives in Domkhok and who has come to dig caterpillar fungus. However, not everyone was convinced about the efficacy of the checkpoint's work. A young Tibetan said:

> The government closed Golok and put checkpoints everywhere. The goal is to protect the environment. So how is it possible that so many people pass these checkpoints? The secret is that the drivers have connections there. People who have money can pass. Those who don't have money can't pass. You've got money, you go. You haven't got any money, you don't go.

Such statements were common. People commented that the work at the Propaganda Stations guarantees good profits. Asked if the checkpoints are effective, one pastoralist laughed: 'They're effective, especially for their staff!' It is possible that the accusations of bribery or partiality in sieving those who can from those who cannot enter the township are voiced by people who were negatively affected by the checkpoints' work. To discover exactly what happens at a checkpoint, I had to find a person working there who would tell the story from the inside. This task was not easy. Although many of my informants admitted that they had friends or relatives working at Sumdo, none of them wanted to talk. Finally, after relentlessly trying for months, I succeeded. A local environmental activist arranged a meeting for me with a county official who worked at this checkpoint. The following excerpt is from that interview:

> ES: What happens if there are any diggers caught at the checkpoint?
> Mr. X: We educate them. We tell them that they should protect the environment, that digging *yartsa* has a bad impact on it and so on. If they don't listen, we fine them. If they try to run away or fight, we can arrest them.
> ES: And what if you catch someone coming from the mountains with a bag of *yartsa*?
> Mr. X: Generally speaking, to enter is more difficult than to leave. If you're caught on your way back, you'll be fined, but not too heavily.
> ES: How big can the fine be?
> Mr. X: Between 500 and 5000 yuan. If you've got friends and are nice, then maybe 2000. It depends on the relations you have. If you've good relations, you pay as little as 500. If you don't have good relations, you pay as much as 5000. This is paid according to people's face.
> ES: So, if I'm from Domkhok and have five diggers in my car, what should I do to pass?

> Mr. X: Even if you're from Domkhok and have diggers in your car, you'll be educated first. To educate people and make them think about the environment is the most important thing now. In more serious cases you'll be fined. It doesn't matter whether you're from Domkhok or not. But if you have friends, you'll have green lights.

This official estimated that 500-600 diggers arrive at Sumdo each day during the first two weeks of the digging season. How many of them are turned back? 'Very few. Only those who have no money at all or are really impertinent.' These words were confirmed by other informants. A member of an environmental association in Machen agreed that the checkpoints are only difficult to pass for people who are particularly resistant to cooperate:

> Not even 10 per cent of them is sent back. Before, it used to be more, maybe 20-30 per cent. The police were stricter. This year, the police are relaxed and let people in. Unless the diggers start fighting, because then the police really have to arrest them, punish, and send them away.

In Domkhok, a township inhabited by less than 2000 persons, the population surges dramatically during the caterpillar fungus season. The man working at Sumdo estimated that the population increases by ten times. This would mean that the township is visited by over 15,000 diggers. Other people gave only slightly more moderate numbers: a township official said that 10,000 diggers come there in spring each year and a member of a local environmental NGO estimated it as 12,000 people. Not all of these diggers arrive via Sumdo, many chose other ways:

> ES: Do diggers try to enter Domkhok in other ways too?
> Mr. X: Of course. They cross mountain passes and rivers. Between Xinghai and Golok, for example, the border is on Chebchu. They swim and are here.
> ES: Do more people cross the checkpoints or use these other routes?
> Mr. X: In the last few years more people have come through the mountains and rivers, although one can pass the checkpoint by paying a very small fine. But there're some people, for example from Hualong, who want to spend as little as possible, so they climb the mountains and swim the rivers just in order not to pay.
> ES: What if they are caught and sent back?
> Mr. X: If someone is determined to get through, he will. Even if stopped and sent back, he will find another way. There is no way of stopping it.

The Pastoralists' Opinions

The new policy outlined in The Executive Programme was met with mixed feelings from the pastoralists. On the one hand, it claimed to defend their rights over the caterpillar fungus growing on their land, and from what has been said about the social climate surrounding the licence system, this change was supposed to at least partly answer their wishes. On the other hand, banning entry to Golok to in-migrant diggers was not exactly what people wanted. The government decided that pastoralists should manage their land by themselves, but at the same time limited their freedom to manage it *as they wished*. This was the main point of critique: my informants argued that under the cover of protecting their rights, the authorities had actually limited them.[16] If their land produced more caterpillar fungus than a family living on this land could dig, then not allowing others to dig it would be a waste, as people said. And if the pastoral family was small or for some reason unable to dig, this policy was even more unjust: it let the fungus rot away and left people without income. The best solution, my informants believed, would be to let the diggers in but make them pay a fee directly to the pastoralists – in other words, to officially sanction the existing practice. This view was shared by both the pastoralists as well as members of the local administration. A township official commented:

> In 2002, the *yartsa* digging was [officially] stopped. From then on only local people have been allowed to dig. I think this is a wrong decision. Here, we only have the ecological migrants from Martod and Chamahe, the Goloks, and the Wranakhs, they dig the fungus, but they don't have to pay the fee. I think it would be good to bring some people from the outside to collect the fees from them.

Many informants were convinced that the Golok authorities had a hidden agenda when they introduced the ban on in-migrant diggers. When the licence system was used, the pastoralists argued, it served both those officials who issued the licences as well as those who organized groups of diggers. The decision to close Golok to in-migrant diggers must serve the

16 Some pastoralists said that Golok acted too autonomously and that its law is inconsistent with that in other regions. A similar point is made by Li et al., who write that 'Golok Prefecture behaves as if it was Golok Republic' (2010: 33). Inconsistencies between laws passed on different legislative levels are explained by Mu and Zhu as being due to conflicts of interests: the prefectures are concerned with their economy and pass laws contradicting the state ones or never execute those latter (2005: 9).

officials' interests, too. Fines imposed on people violating the ban create the opportunity for officials to earn money on the side, people said. The Executive Programme stated that the fines paid at the checkpoints must be transferred to the county, but my informants suspected that this is not always the case.

More importantly, the pastoralists doubted whether the new policy really served its goals and helped the environment. They were not at all convinced about the negative environmental effects of the caterpillar fungus digging. There were other more destructive activities, they argued, which should be controlled but are not. The most important of them is mining.

The pastoralists' knowledge about mining is not purely abstract. One of the reasons this topic was so widely discussed was due to a mine in Derni valley, south-east of the Domkhok Township seat. The mine, which produces a range of non-ferrous minerals, such as copper, cobalt, sulphur, silver, and gold, was vehemently criticized not only by pastoralists, but also by Tibetan officials in the township and the county. They questioned the mining company's right to exploit mineral resources on their land. For them, it was an example of a particular lack of sensitivity or arrogance: the mine is located in the vicinity of Amnye Wayin, a sacred mountain of the Metsang *dewa*. My informants stressed that gold and other riches hidden in the mountain are the property of the *zhibdag* and should not be exploited. Raising the discussion to a more political level, pastoralists argued that the natural resources of their land do not belong to the Han, but to Tibetans. In their eyes, the Han have no moral right to their land at all. Finally, the pastoralists spoke about cases of environmental pollution, toxic discharges, as well as human and animal diseases. As two men from Domkhok told me:

> Amnye Derni is rich in gold and everything else you can imagine. It's the treasury of Amnye Machen. Emptying it causes floods, storms, and diseases. People petitioned the government to close the mine, but the government sold it to a private company instead. They can do anything they want and don't contribute to the local economy.

> The effects of mining are obvious! Look at the weather! In winter it snows all the time and in summer there're these heavy rains. Besides, the mine is discharging some black liquid which flows into the Machu and Guchu. When the yaks drink it, they lose weight, become lean and ill. When you butcher them, you see that their internal organs are deformed. We're forbidden from entering the mine, but I went there once to take pictures. I started having very strong migraines afterwards.

Almost every conversation about the environmental effects of digging caterpillar fungus ended with talking about the mine. Digging the fungus does not destroy the land, so my informants said. Besides, caterpillar fungus is a renewable resource. The government's tolerance for the mining was for my informants a sign that the authorities do not care about the environment but use environmentalist rhetoric to introduce new forms of discrimination against the pastoralists. The policy forbidding them to lease their land to diggers was seen by my informants as such discrimination. My informants called for the government to fight real crimes instead of 'harassing innocent people'. It was a paradox to them that digging caterpillar fungus should be so strictly controlled, while the mine goes on operating undisturbed:

> The government officials say we should protect the environment, but they do exactly the opposite. They allow mining! We have some associations here which say we should protect the environment for the future. But if the mining goes on, there won't be any future. We ourselves will face hunger. In comparison to mining, digging *yartsa* is harmless. *Yartsa* grows on the surface of the earth. Comparing it to a human body, mining is like pulling internal organs from a living organism and digging *yartsa* is like plucking a hair from one's beard.

The Law and Control

By issuing laws regulating the use of the caterpillar fungus resources, the authorities set the lines of what is legal and what is not. State institutions of different levels were charged with responsibility to control whether this law is implemented. The pastoralists often mentioned the danger of 'control being sent upon' them. It was unclear to me whether this was more of an imaginary danger or if it really could happen that a group of officials suddenly materialized on someone's land. Was it an excuse not to share information with me, a reflection of a more serious concern, or a proof that such controls take place?

The county institutions do send their officials to the townships to inspect the state of the grassland and the implementation of caterpillar fungus-related law. But my informants were not concerned about these types of control. They received information about the coming controls well in advance: 'People lease their land secretly, but the leaders help us and tell us when the control is coming', as one pastoralist said. The officials sent on such field trips do not carry out their duties very strictly: 'They normally go to the checkpoints. They have no way to check the valleys, anyway. People dig in places which are hard to reach even by horse. They're not so stupid as to dig next to the road', another man stated. Several officials from the Forestry and Grassland Bureaus mentioned their busy work schedules during spring. However, the efficacy of this work was, in their own eyes, rather questionable:

> Mr. Y: We do sometimes go on the field trips. If we see some diggers, we tell them to leave.
> ES: What if they don't listen?
> Mr. Y: Then they don't leave. We can't arrest anybody.
> ES: Do you have a gun?
> Mr. Y: (laughing) Of course not.
> ES: And if it gets dangerous...?
> Mr. Y: It's my responsibility for it not to get dangerous. Anyway, I know everybody there. I know that people have diggers. Besides, what should I do if I 'catch' my own uncle?

Officials sent on such field trips often have their 'uncles' and other relatives in the townships they control. They have to choose between work duties and loyalty to their community. For the man quoted above, work in the county administration was not a fulfilment of his dreams, but he failed to pass his university exams and could not pursue another career. Several other officials whom I interviewed told me that the job they had was what they wanted, but they were not the least bit enthusiastic about controlling their own township and informing their superiors about who was violating the law. It seemed that their superiors were not always interested in knowing the truth either. Some officials admitted that to make sure that county officials turn a blind eye to what is happening on the grassland the pastoralists pay a special 'fee':

> It doesn't matter whether you have fifty or ten diggers, the government charges you a fee, let's say 1000 or 3000 yuan. Nobody complains about it. People are very keen on paying it. But this is paid under the table, not officially.

These routine controls, planned well in advance, did not worry the pastoralists very much. There was something else they feared: these were the sudden controls, real interventions, dispatched fast, often at night and without any warning. Backed by police forces, these interventions had little of the leniency of the standard controls. They took place, my informants said, when the government received information from inside the township that 'something very wrong' was happening in a particular household and when it had to react, sometimes on the orders of a higher institutional body.

In order to get hold of such information, The Executive Programme called on county institutions to build a network of contacts in the townships. The township authorities installed confidential letterboxes, which the residents could use to report the legal abuses of their neighbours. To encourage people to cooperate, the document proposed that the informants would be rewarded with 30-50 per cent of the fine paid by the person they reported on (Appendix: 11). Thus, whereas pastoralists used their contacts in the town to get information about the coming controls, the authorities sought information in the township about illegal activities taking place there. A retired official explained:

> The Communist Party and the people stand back to back. The Party put a box in the township, where you can throw in letters, and this way they check if there're outsiders digging here. And people check if the officials are coming on their field trips. You see how they're back to back, don't you? Do they trust each other? No, they don't. There's a messenger between them. It's like in 1958. I was twelve years old back then. When the messenger arrived calling 'Hey, the Chinese are coming!', we hid. Now they don't say the 'Chinese'. They say 'fieldworkers' (Chin. *xiazhang*). When the messenger calls 'The fieldworkers are coming!', people hide.

What happens when the message comes too late and the diggers cannot hide? The same man continued:

> Terrible things happen. Imagine that these diggers borrowed money to pay the digging fee and come here. But they're chased away and their tents, food, and clothes are burned. For me as a Party member saying this is like slandering my own country, but it's really like it was when the Japanese arrived. The people, the *dadui* leaders, and the prefecture and county officials all take part in it. Some people joke that Golok implements three eliminations: burn all, loot all, and kill all – like the Japanese did.[17] Once,

17 The Chinese term he used was *san guang: shao guang, qiang guang, sha guang*. It refers to the 'policy of three eliminations' (Chin. *san guang zhengce*) implemented by the Japanese troops in China during the Second Sino-Japanese War. It implied killing civilians, destroying their dwellings, burning or carrying off their food reserves or work tools (Fairbank and Goldman 2006: 320).

> when officials and policemen came here for a field trip, they clashed with the diggers. The diggers said: 'We paid the fee!' and didn't want to leave. The officials ordered them to go. But they didn't. And a fight broke out. A man was shot in the leg. They didn't shoot him from a distance. It happened during a fight.

Legality and Licitness

Analysing the practices of leasing land to in-migrant diggers from the perspective of the state leads to seeing this in terms of binary categories of legality and illegality. In fact, it is better to consider the practice in light of two sets of categories: legal (versus illegal) and licit (versus illicit). This distinction was proposed by Itty Abraham and Willem van Schendel for the analysis of illegal flows or flows of commodities, persons, and ideas outlawed by the state. In many fields, the categories of illegal and legal appear to be insufficient: 'When we shift our nomenclature to the distinction between "licit" and "illicit", we refer less to the letter of the law than to social perceptions of activities defined as criminal' (Abraham and Van Schendel 2005: 18). Such activities are often conditioned by the interaction between formal political authority and non-formal social authority. These two are not symmetrical: social practices do not have to adhere to the law and the legislation does not have to reflect social practices. 'Legal restrictions often come up against socially sanctioned practices, and while this may have the effect of driving these practices into the sphere of formal criminality, it does not eliminate them nor does it necessarily force them into hiding' (Abraham and Van Schendel 2005: 19). Practices deemed illegal by the state can be seen as legitimate (hence: licit) by the citizens and what is legal for the state can be contested by the population as illegitimate or illicit.

The categories of legality/illegality and licitness/illicitness and their overlapping and/or conflicting character are evident in the different status of different groups of diggers in Domkhok (Table 5). The first group works legally: these people live in Golok and dig in another township, paying to the pastoralists for working on their land. The second group are people from outside Golok: they come to Golok illegally but pay the pastoralists a fee for leasing their land. The third group comes from outside Golok but does not pay the fee. These three groups situate themselves at different points on the scale of legality or licitness and their status is differently defined by the state and by the pastoralists. The first group is legal as long as the diggers pay the fee. The second one is licit from the perspective of the pastoralists, but illegal according to the state. The third group is illegal for the state and illicit for the pastoralists. The latter two groups come from outside Golok: the officials estimated that at least 80 per cent of diggers coming to Domkhok belong to these two categories.

Table 5 Legality and licitness of caterpillar fungus diggers

	Residents of Golok TAP		Non-residents of Golok TAP	
	Caterpillar fungus digger's fee		Caterpillar fungus digger's fee	
	paid	**not paid**[18]	**paid**	**not paid**
Pastoralists	licit	–	licit	illicit
State law	legal	–	illegal	illegal

For the pastoralists, it was the third group that caused the biggest nuisance. These diggers (my informants called them 'thieves') were a subject of continuous complaints. The most notorious were Tibetans from Ngawa, who entered Domkhok in organized groups.[19] My informants did not blame the checkpoint officials for not stopping these 'thieves' because 'what can you do if suddenly a group of 180 men arrives on motorcycles at night?!', as one man said, indicating the scale of this phenomenon. Sometimes, threatened by a group of armed pastoralists, such diggers move one valley farther away, but when nerves are tense, violent conflicts happen.[20] An elderly man commented:

> People from Ngawa used to come here in the past to dig without paying and they don't want to change their habits now. They ask, 'Why should we pay at all?' They say, 'We've got no money, we're all Tibetans, please, please [he shows the thumb up, the sign for begging]. Please let us stay.' If you are soft-hearted, they'll go to dig and disappear without paying

18 I never heard of people from Golok coming to Domkhok without paying the fee. If such cases happened, they would be illicit for the pastoralists and illegal according to the law.

19 'Thieves' and 'Ngawa' are used almost interchangeably. When I worked with an assistant from Ngawa, he tried to hide his identity, but language differences disclosed it. 'I come from where all the thieves come from' was a joke he made that sometimes helped, but we also heard very clear statements such as: 'We don't like Ngawa people here.'

20 Many diggers come also from Rebkong and Hualong. Ironically, these are again Tibetans. In 2007, I shared a flat with women from Rebkong who planned to sneak into Domkhok without paying the fee. They had few belongings: one set of clothes, some blankets, *rtsampa*, and instant noodles. They were not very worried about their safety, but much more concerned that if they were caught the pastoralists would take their fungus away.

> you anything. But there're also those who come and don't ask if they can stay. With these people, if you tell them to leave and they don't go, there's only one solution: to fight.

The large area and rugged character of Golok's terrain make it challenging for state institutions to control the influx of diggers, but it is also difficult for the pastoralists to keep their land under constant watch. Just as 'outsiders' enter the township by evading the law, so illicit diggers manage to escape the attention of the pastoralists. Once they are caught, they become a target of the pastoralists' anger. Those who venture into the highlands alone or in a small group find themselves in a particularly dramatic situation. Being both illegal and illicit, having neither the state's nor the pastoralists' support, these people are lost in an alien environment and left to their fate. The following story was told by a *khandroma* from Dawu, a female diviner whose help is often sought by in-migrant diggers:[21]

> Recently diggers from Rebkong called me to do a divination. They wanted to know when to set off and whether they'd succeed. I made *mo* and told them to set off at a certain time and go in a certain direction, and this helped.[22] Some people ask about the checkpoints and policemen, and whether they will be caught by the pastoralists. The people from Rebkong called when they had lost their way in fog. I told them to keep walking. The man said they couldn't see anything, hadn't eaten for the whole night, were hungry, and could walk no farther. He said that pastoralists had beaten them up badly, so I wanted to help them.

Golok is an arena of competition and conflict between different understandings of the legality or licitness of the economic activities related to caterpillar

21 *Khandroma* (Skt. *ḍākinī*) can be translated as 'the one who goes through the sky' or a 'sky-dancer'. This name denotes female celestial beings or deities, but also the women considered to be their human embodiments who act as the 'divine consort' of a male tantric practitioner and/or themselves are religious practitioners. In Golok, this name is applied to female religious specialists of various kinds. The *khandroma* quoted here was not affiliated with any monastery. She was one of the three most popular diviners in Dawu. People asked her advice in many life matters, ranging from health and marriage issues to questions whether to put a horse in a race or to buy a car.

22 *Mo* are divination practices performed in situations of life difficulties, illness, loss of property, but also before any serious undertaking to see whether the situation is auspicious or how to make it so. Many different methods of divination are used in Tibet. The *khandroma* used divining dice; other diviners employed a divining mirror or prayer beads. On *mo*, see Ekvall 1964b: 258-259.

fungus. The law outlined in The Executive Programme conflicts with the pastoralists' understanding of what should be allowed, or what is licit and what is not. These distinctions help us understand what the social factors are that limit the efficacy of the law and why people disobey it. One of these factors is the low level of the internalization of the law and its poor compliance with the social norms of the local society. Other factors include inconsistencies in the law, partiality or corruptness of state institutions responsible for implementing it, and the lack of legitimacy or recognition of the law-making government among the citizens.

The pastoralists in Golok consider themselves the rightful owners of the land they inhabit, even if – formally speaking – they are not its legal owners, but only long-term users who 'lease' the land from the state. However, they believe that they have the moral right to manage their land as they wish and see the county law regulating the use of the caterpillar fungus resources as harmful to their interests. They do not perceive themselves as beneficiaries of the law and their acceptance of it is small. They have neither internalized, nor do they identify with this law. The case of mining further undermines the pastoralists' trust in the government's concern for the environment. My informants questioned the sincerity of the environmentalist rhetoric used by the government to explain why the law was introduced. Finally, the obvious lack of efficacy of the law and its partial implementation does not help either: my informants contended that if the state institutions do not take the law seriously then why should they? If officials are selective in implementing the law, people can be selective in following it as well.

More generally, pastoralists disobeying the law can be explained by the contested status of the law-making institutions in China. Fernanda Pirie, in her analysis of the Tibetan pastoralists' preference to resort to their old mediation and dispute settlement strategies from prior to the PRC, remarked that the pastoralists deny the legitimacy to the state's criminal justice system, and see it as ineffective and unjust (2013: 77-78). Similarly, the state institutions that passed the caterpillar fungus-related laws are observed by the pastoralists with distrust and suspicion. My informants regularly accused county and prefecture officials of looking after their own interests and disregarding the needs of the population. This was a common criticism, even though on the level of daily practices a degree of cooperation between the officials and the pastoralists was visible. Some county officials openly told me that certain regulations are not implemented because they do not serve people's interests. Still, the pastoralists spoke about the prefecture and county governments as driven by self-interest rather than serving the wider society.

The legitimacy of the Chinese rule in Tibet is, as Pirie notes, questioned by many Tibetans and this influences their attitude towards the law and explains why 'there is a good deal of resistance to the more unpopular policies [...] which includes noncompliance and foot-dragging' (2013: 84). This lack of recognition for the state's legitimacy fuels something that can be called a 'sentiment for illegal conduct': non-compliance with the law can become a matter of principle and a form of resistance to the government whose legitimacy is questioned (Podgorecki 1962: 178). In Golok, the legitimacy of the government (of whichever level) is not questioned by the pastoralists on a permanent basis, but rather in the context of particularly unpopular policies that run counter to people's interests. The issue of the government's lack of legitimacy is not raised when the authorities' decisions agree with people's expectations. Rather, it comes up when these decisions, as in the case of caterpillar fungus, diverge from what people expect and it is used by them as one more reason not to obey the law.

7 Money

The pastoralists in Golok earn from the caterpillar fungus economy in two main ways: from digging and selling the fungus and from leasing their land for digging. The income from these two sources differs in many aspects. Digging caterpillar fungus brings money that the pastoralists, symbolically speaking, earn with their own hands. This income depends on the time and effort they invest in work, on the weather, and on the size of the family or availability of the labour force. If the family does not have enough workers or if the season is bad, they will dig fewer fungi and earn less. But to earn this money, they not only have to find the fungus, but also to sell it successfully and this can be complicated. Selling caterpillar fungus requires good reconnaissance of the market, many visits to town, and often luck. The size of the income is also not guaranteed, since prices change fast: one can earn, but also lose out if the market declines. This income is susceptible to changes and is *not* guaranteed. It is often delayed as well, either because pastoralists wait for the prices to rise, or because the market jams up and the traders cannot purchase more fungus.

The income from the land leases is different. A lump sum from the fees that the diggers pay for leasing the land goes into the pastoralists' pockets even before the season starts and regardless of any external conditions. For this money, pastoralists help diggers cross the checkpoints and offer them some protection during their stay in the mountains. Thus, the income from land leases comes principally from facilitating other people's work. It is perhaps a high-risk income, since the law forbids leasing of land to in-migrant diggers, but it is also a low-effort one and the pastoralists themselves say that this is 'quick money' – attractive and easy to earn. Most importantly, this money is *guaranteed*. It forms the basis of the pastoralists' family budget: regardless of what happens, they can count on it in planning family expenses.

The differences between these two types of income are clear when one explores how much a pastoral household earns in a year. Tsering Drölma and Tendor, the couple I dug caterpillar fungus with, leased their land to 36 diggers. Each of these diggers paid a digging fee of 11,000 yuan, which gave Tsering Drölma and Tendor a 396,000 yuan income. The situation is more complicated when it comes to selling caterpillar fungus. Tendor could not easily tell how much they earned from it because this income is not regular. The family usually digs around three *jin* of caterpillar fungus; this is the material that they clean and dry at home and subsequently sell by weight. However, they actually dig more: some fungus they sell fresh without processing it at home. Neither Tendor, nor Tsering Drölma could tell me how much they sold fresh: they do it quite spontaneously, in small amounts, and tend to spend the income immediately. They wait until winter to sell dry fungus: with sufficient income from the digging fees, they are not in a hurry and can wait for the price to increase. During my survey, Tendor said that they currently had at least one *jin* of top-quality fungus and two *jin* of medium-quality stuff. He hoped to get 80,000 yuan per *jin* for the top and 50,000 yuan for the medium quality, but he was ready to sell it for 10,000 yuan less. If he succeeded in getting the desired price, the family would earn around 180,000 yuan. This is less than a half of what they earned from digging fees. The number of diggers that Tsering Drölma and Tendor had was by no means the highest in the township. In other families, the difference between these two types of income was even bigger.

Money Talk

I met many difficulties when collecting information about household incomes. The irregularity of the income from selling caterpillar fungus, which is seasonal and prone to changes, makes it difficult for people to estimate how much they earn in total. The tendency to exaggerate expenses compared to income is another problem. My informants often claimed that they earned less than they spent: they wanted to show their discontent with the economic situation and, at the same time, their ability to cope with it. Declaring income from activities located at the edge of legality does not come easily either, even if this is only supplementary income. But in Golok, it constitutes the *main* income and forms the basis of household finances – and this makes declaring it even more difficult.

At the beginning of the research, my informants claimed that they did not dig caterpillar fungus at all or underreported the size of their harvests. They complained that their land was of poor quality and that their health made them unfit for such work. Those who admitted to digging said they got only several hundred or at most 1000 fungi per year per household. One thousand specimens dug in 40-50 days gives twenty fungi per day. My fieldnotes from Soglung show how unrealistically low these numbers are:

> In the evening Tseso sits with her *English ABC* textbook on her lap. A heap of fresh *yartsa* is spread out on it. Other people sit with a piece of cardboard or a plastic bag. The toothbrushes in everyone's hands swiftly remove the traces of soil from the *yartsa*. People count and compare how many fungi they have found. *Ama* [Tsering Drölma's mother] found 43 pieces today. (FN 14 May 2007)

> Norbu called from the mountains (you could hear the wind in the telephone). He found 67 *yartsa* yesterday. Domkhok enjoys a good reputation for its *yartsa*. If Norbu's family is so big, and everyone finds even 30-40 pieces a day (and Tseso apparently gets around 100), their land must be really fertile. (FN 2 June 2007)

> I meet Tseso on the street. She found 1200 *yartsa* during three weeks of the school holidays. One may wonder how many she would find if the holidays were longer. (FN 9 June 2007)

Similar numbers were repeated in other valleys. In some places in Domkhok, lucky diggers found 200 or 300 fungi per day. But even in those valleys where people complained about the quality of their land, a haul of 50-60 fungi in a day was common. The questionnaire that I conducted in Golok Tibetan High School directly after the spring holidays, which the students spend at home helping their families dig, also showed how productive Golok is. Not surprisingly, the most successful diggers were students from Domkhok, Dawu, and Xueshan. A student from Domkhok found 120 fungi in one day and 1300 during the entire holidays. Her family collected 6000 fungi in the first three weeks of the season. Two students from Dawu found 203 and 300 fungi in a day respectively, and 1553 and 3000 during the holidays as a whole. A student from Xueshan declared that he had found 365 fungi in a single day and 2500 over the course of the holidays. His family, he said, had found 10,000 fungi. The values from other counties than Machen were lower, but still attractive: 73-103 per day in Darlag; 50-80 in Jigdril; 25-206 in Gabde, and 26-130 in Bamma.

The quantities of fungi that the students dug were a common topic of discussion after returning to school. The numbers they declared for the whole family were certainly not accurate, especially in light of the fact that as they returned to school their families were still out digging. But when the students spoke about their own work, they often gave precise-seeming numbers. This suggests that the students take notes about how much they find. Do pastoralists really keep a close account of their caterpillar fungus harvests? Are there any written records that would help to assess the actual productivity of the Domkhok land? Such records do exist. As I observed in the early stages of my research:

> In the evening, people sit in the kitchen. They watch television, clean the fungi and spread them on the stove to dry. They count the fungi carefully, note the numbers down and store the notes on the shelves covering the back wall. (FN 15 May 2007)

These household records are made for the pastoralists' own use and list reasonably accurately the numbers of fungi that people dig every day. The private status of these records implies a potentially exact character for the data, but it also makes them difficult to access. Although I saw how carefully people counted their fungi and recorded the numbers, as soon as I wanted to inspect these records they suddenly became 'non-existent'. My informants swore that they never took any notes or openly refused to show them to me. Then a breakthrough came unexpectedly:

> It rains the whole day. We kill time watching television and chatting with our host lady. She shows me her photos and explains the details of women's hair ornaments, which are another identity marker of the Metsang. In the same drawer, there are other papers. I ask if she has any notes about *yartsa*. After some hesitation, she pulls out a dog-eared school notebook. Among telephone numbers, names of debtors and debts to pay, sums spent or earned, there are long columns of numbers, running page after page. Each column has a name at the top and stands for one person: one for her, one for her husband, and one for their divorced son who recently returned home. This is what I was looking for: numbers of *yartsa* that the family members found each day. (FD 28 June 2009)

These household records contain information about the length of the harvest, numbers of the fungi people dig, and the productivity of the land. They show that the digging season lasts up to seven weeks, longer than is generally assumed: the longest record had entries covering 47 days. But they do not indicate the date or the amount of time people spent digging each day. They fail to note when the pastoralists took a break for religious celebrations or if they came home empty handed. It is possible that low numbers were a result of bad weather or other circumstances that kept people at home. But high numbers show how productive Domkhok can be: they leave no doubt that one digger can find 1000 or more caterpillar fungus specimens in a day. According to one of these sets of records, someone collected 1914 specimens: more than double what some informants declared for their whole household! Values of around a hundred fungi are recorded even towards the end of the season when the productivity of the land declines: one record lists 68 fungi on the 39th day of the season, four days before the record ends. Although it is difficult to estimate the quality of these late-season fungi, the quantity itself is remarkable (Figure 17).

Figure 17 Caterpillar fungus records

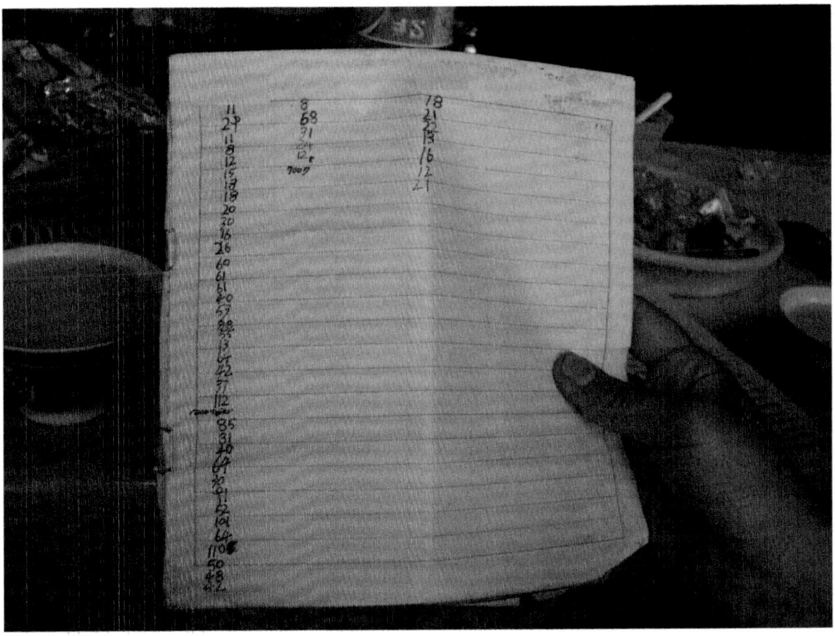

These caterpillar fungus records are made soon after people get in from work. However, they still report numbers lower than what people really dig. In the records I studied, all household members were listed separately, but all were in *one* book. Their work results were visible to anyone who could access it. But many pastoralists had a two-pocket system: most of the fungi they put into the 'common pocket', but some they put into another, 'private' one. At home, they disclosed the contents of the common pocket: this fungus went to the family budget. The contents of the other pocket remained private. Thus, people secured extra income for themselves, which they could use as they so wished, including in ways their parents or spouses disapproved of. This two-pocket system is especially important for persons with limited access to cash, such as women and teenagers. Men, who sell caterpillar fungus on the market, can always spend some of the proceeds without their family knowing about it. Other family members do not have this opportunity. Because caterpillar fungus is small and light, it is also easy to hide and keep for private expenses.

Talking about the amounts of caterpillar fungus that pastoralists dig met with difficulties, but an even heavier veil of silence covered the practice of leasing the land to in-migrant diggers. Here again, I needed time to break through the barrier of mistrust. Initially, my informants assured me that they did not lease their land, either because they did not need the extra income or because their land was inconveniently located: too far from the road and hence difficult to reach for the diggers or too close to the road and so too easy to control for the police or other state institutions. However, even short-term observation revealed that it is not only local pastoralists who dig caterpillar fungus in Domkhok. During my first days in Soglung, Tsering Drölma showed me a group of people climbing some mountains opposite the house. 'Chinese', she said.[1] Worried that intruders were trespassing on her land, I asked what we could do about it. She waved her hand. 'It's ok', she said. The view of dark-clad silhouettes moving in the distance accompanied us during our work and I soon learned that these were not intruders.

The longer I stayed in Domkhok, the more open people became. Gradually, my informants started to admit that they did have some in-migrant diggers working on their land. The number of diggers they declared grew with time. Sometimes, the answer 'We have two diggers, but not more' changed to 'We usually have twenty, but not more', inviting guesses as to what the answer would be had I asked this question again. In the end, most families declared that they had between several and 60 diggers, often stressing that it was their conscious decision not to accept more.

In smaller valleys, neighbours decided together how many diggers to accept in order to secure a good income without putting themselves at too great a risk. Soglung is a large valley and there was no common agreement between all residents. Many families decided individually, others tried to do it in small groups. Tsering Drölma and Tendor share their land with Jigmed Dorji, Tsering Drölma's uncle. The two households decided together to accept twelve diggers on their winter pastures and 60 on their summer land. But Tendor claimed that they could take double this number:

> There're always more and more people who want to come. If we don't say 'it's enough', they would trample us to death. So, we said: 'Enough'. We don't need more. We earn enough, they earn enough, and we don't risk them destroying the land and the police raiding us.

Tendor estimated that Soglung is visited by several hundred diggers, perhaps even a thousand. This is an average of 40 diggers per family, a number that many informants openly admitted to. However, a few pastoralists suggested

1 *Rjami*, as she called them, is used not only for the Han, but sometimes for other non-Tibetans (see Schrempf 2010-2011: 327). I was also sometimes called *Rjamo*, but I was usually referred to as *Mekogi* ('American') or *pshirjid* ('foreigner').

that 'people have more, but they won't say it'. One man stated that his neighbour had '300-400 diggers' on 6000 *mu* of land. Is it possible that one household really had several hundred diggers? In a widely commented case reported to me by both the county officials and the pastoralists, a family had leased their land out to 360 diggers. The consequences were severe: the family's land lease contract was revoked. Two years later they were still appealing to the prefecture authorities to get the land back. In another case:

> The wife leased the land to one group of diggers and the husband to another. These two groups clashed and the government sent the People's Armed Police to pacify them.[2] There were more than 200 people on the spot. If it wasn't that many, the government wouldn't take the pastoralists' land away.

The authorities take action only when compelled to do so, my informants said, and people feel safe as long as they do not challenge the limits of the state's tolerance: 'Ten or fifteen diggers on 3000-4000 *mu* is fine. The officials wouldn't bother to spend a day in the car to drive the diggers away. But if there're one or two hundred, they have to do something.' According to this, fifteen diggers is too insignificant a number to cause alert, 30 or 40 should not cause problems either, but anything above 100 is risky. Some of the expressions that the pastoralists used were also indicative of the scale of this phenomenon. Talking about ten to twenty diggers, they said it was 'not too many', 50-60 was 'normal', and 'plenty' meant over a hundred. If the largest group of diggers that people reported to me counted over 300 people, how high were the numbers in other cases? By suggesting that the limit of the authorities' tolerance was around 100 diggers, the informants implied that any lower numbers represented a quasi-tolerated norm. On the scale of the numbers of diggers working in Domkhok, there must be many values missing between what was declared by the pastoralists as the average numbers they accept and the 200-300 that made the government intervene.

That leasing land to 'outsiders' is common practice becomes clear when one realizes that it is not why people lease the land that needs explanation. Rather, it is the rare cases where someone does not lease out their land that mean a deviation from usual social practice. As a local environmental activist remarked:

2 The People's Armed Police or Chinese People's Armed Police Force is a Chinese paramilitary organization primarily responsible for internal security, law enforcement, as well as providing support to the People's Liberation Army. It follows the rules and regulations of the PLA and enjoys the same treatment as PLA troops.

> There're two types of families. Some lease their land and some don't. If they don't, it's because they have enough people in the family [to harvest all the available fungus themselves]. If they have six people and each gets one *jin* of *yartsa*, times six, times 50,000 yuan, this gives 300,000 yuan income. Many families prefer to make money themselves. Besides, the outsiders don't care about the land and so destroy it.
> *ES: How many families in Domkhok do not lease out their land?*
> 20 or 30 per cent. Not more.

The reasons my informants gave to explain why some people refrain from leasing their land varied. If a family had little land or if their land was of poor quality, they would keep it for their own use. The fear of the consequences of violating the law could play a role as well. However, people said that it did so only in those families whose members had previously had conflict with the law and feared being under surveillance. Another reason was a belief that digging caterpillar fungus has a negative impact on the environment, but my informants were only moderately convinced about it. In order to extract a fungus from the ground, one has to dig a hole some 10 cm in diameter. After removing the fungus from the lump of soil, one should place the soil back into the hole. If this is done properly, no real damage to the grassland should be caused, as Tsering Drölma explained. Whether the in-migrant diggers do it this way is an open question. Many pastoralists complained that they do not – either because the diggers do not know that they should, or because 'they don't care about the land, because it's not theirs', as some pastoralists asserted.

All these reasons had an impact on pastoralists' decisions about leasing out land, but they generally made people limit the number of diggers they took rather than refrain from leasing their land altogether. There is one more factor that played a role: pressure from the local community. This can work both for and against the land leases, but whatever the decision of the majority, going against this is not easy. Living in a valley where all other residents accept in-migrant diggers and deciding not to follow this practice required courage. It put one at risk of becoming a social and economic outsider and – more importantly – a potential suspect, as the neighbours might consider you to be a secret informer reporting to the police on their activities. In Soglung, only one family did not lease their land. 'It's not good for the land. This is more important than money', the man of the family said. It is perhaps no coincidence that the family only had their summer pastures in Soglung and the rest of their land was in another valley. Thus, whereas their neighbours in Soglung continued leasing their land, this family could stop doing so because they were part of the community only for a short part of the year.

To lease land, the diggers pay a fee (Tib. *rtsala*), which gives them the right to work on the pastoralists' land, live in a tent on the grassland, or in the pastoralists' house, and allows them to count on some degree of protection: the pastoralists often fetch their diggers from the town and arrange their safe passage through the checkpoint.[3] This digging fee is either decided upon by the households individually, or among neighbours – if they prefer having a single, collective fee policy to avoid competition and conflict. The fee rises and falls in line with caterpillar fungus prices. At the beginning of my research, in 2007, the fee in Soglung was 10,000 yuan per digger. In 2008, when the caterpillar fungus market declined, it fell to 6000 yuan and then grew to 8000 yuan in 2009, before reaching 11,000 yuan in 2010. But the fees in Soglung were not the highest. In 2010, the fee in some valleys in Domkhok reached 15,000 yuan and in neighbouring Xueshan Township the diggers paid even as much as 20,000 yuan. However high the fee was, though, there was no shortage of people ready to pay it. A high fee was an investment in future success and it carried a promise of attractive income. Lower fees carried a risk that this money would not be returned: in less fertile valleys, some pastoralists accepted so many diggers in order to compensate for the low income from the fees that the competition between diggers made their work impossible.

In most cases, pastoralists leased their land to the same group of diggers every year. The majority of them had been coming there for many years, sometimes since 2002. New diggers were accepted on the recommendation of someone who had worked in Domkhok before: pastoralists unwillingly accepted complete strangers. This preference for long-term cooperation was also shown by the fact that the diggers often booked the land in advance. By the end of spring, most of it had been booked for the next year: the diggers must have been satisfied if they wanted to return to the same place. Many had done it for so many years that they were well integrated with their host family, lived in their house, and took over some of the household tasks:

> During the day the house is quiet, empty, and cold. Only in the evening do people start arriving. They are not pastoralists. They're *yartsa* diggers coming from the mountains. A group of women from Tungde cooks rice noodles for dinner. One woman next to the stove stirs them. Men sit on

3 This word is used in a variety of situations and means any payment for leasing land or for grazing.

> small stools or squat at the floor sipping their noodles. Everybody watches the adventures of the Monkey King and the 'Old House', another soap opera on the Tibetan television channel. After taking their meal people clean their *yartsa*. All these people (I counted 30) stay in this and the neighbouring house. In the morning they drink tea and go to work. (FN 2009)

Long-term arrangements between pastoralists and diggers benefit both sides. The diggers know that there are no surprises in the unwritten contract; they can estimate their gains and know that their hosts will not accept too many diggers, thus making their work difficult. They can be confident that their investment will be returned. That this is not always the case is clear when one observes life in Dawu at the end of the spring, when the town fills with bankrupt diggers drowning their sorrows in alcohol. Their stories are always similar: lacking experience and good contacts in Golok, they leased land from someone whom they did not know and in a place about which they had no knowledge. Some of these stories end dramatically, like that of a young Tibetan from Kamdze. Together with some friends, he leased land from a pastoral family in Machen. They paid for it in advance but could not find any fungus. Believing that they were being cheated, the young digger demanded that the family return their fees. The pastoralists refused and argued that the diggers themselves were responsible for their failure. The argument escalated into a fight. The young digger rode off on a pastoralist's motorbike, taking it – as he told me – as rightful compensation for the money he lost. However, he was caught and taken captive by the pastoralists. They accused him of breaking the motorbike and threatened to kill him if he did not pay for the damage. The digger's friends had to go to the town to borrow money. Only when the demanded 6000 yuan was paid did the man regain his freedom.[4]

What helps the pastoralists defend their position in such conflicts with diggers is the Buddhist law of karma (Tib. *le*), or the power of religious merit accumulated during one's past lives. Good eyes and a fit body as well as access to fertile land are necessary for success in digging, but karma helps explain why among people who work on the same land, some fail to find any fungi whereas others find many. The pastoralists rarely spoke to me about karma in the context of their own successes or failures, but they often used the concept when discussing others. If karma was the main determinant of a digger's success, then this success could not be guaranteed by anyone, let alone by the pastoral family on whose land the digger works. When I

4 This story was related in detail in Sulek 2008.

Figure 18 In a pastoralist house: in-migrant diggers make notes on their harvests

asked an elderly man, what happens when the diggers working on his land pay their fee but do not find enough fungi said: 'It's their problem. Maybe they don't deserve it.' Hearing such arguments, it is hard not to think that the concept of karma is used instrumentally by some pastoralists to avoid responsibility and is to the diggers' disadvantage.

Similar expressions of slight disrespect for the diggers were heard in other conversations about the digging fees. My informants said that they did not accept diggers who have no money to pay the fee in advance. 'They all bring cash. If they don't have cash they don't come to dig', one man said. The logic was simple: 'If they can't pay, they can't dig. If they have no cash, there's no way.' A pastoralist boasted that there is always a queue of people ready to replace those who cannot pay. Such statements show the pastoralists as being in a position of power: they dictate the terms of cooperation and leave the diggers only enough space to decide whether to accept these terms. These statements create an image of a saturated employment market and a township where the land is in high demand and competition fierce.

Although my informants wanted to show themselves as having the upper hand in contacts with the diggers, the relations between these two groups are more complex than that. The diggers are not as powerless as the pastoralists would like them to be. As one man explained:

> The neighbours normally turn a blind eye to what is happening on other people's land unless they have had a conflict and want to take revenge. But the diggers can also call the police. Perhaps they had a disagreement with the nomads about the lease or about payment and want to take revenge for bad treatment.

The pastoralists are also more empathic than their declarations show. If they know the diggers well and have had good experiences with them in the past, they give them a certain amount of trust and, for instance, agree to receive payment of the fee in instalments. However, the usual insistence on upfront payment is not only a sign of limited trust, but also a security measure: it sometimes happens that diggers who only pay part of the fee before the season run away without paying the rest. Many pastoralists were keen to prevent such situations. Tendor explained his method. The diggers face a problem regarding how to store their caterpillar fungus: they cannot go to the mountains carrying with them everything that they have found before. In Tsering Drölma and Tendor's house they deposit their fungus in a metal safe. The fungus leaves the safe only when the final part of the fee is paid. This final part, as Tendor said, can also be paid in caterpillar fungus itself.

The Mystery of Spring

The pastoralists' unwillingness to talk about how much caterpillar fungus they dig and the income they earn from it is understandable if one looks at the official discourse in China, which blames pastoralists for engaging in activities that lead to land degradation. The prefecture officials in Golok complained that the pastoralists have too much livestock, dig too much caterpillar fungus, and 'destroy their land'. Some Tibetan NGOs active in the field of environmental protection share this opinion: their activists in Golok told me that the pastoralists have 'a very limited understanding of the environment', are short-sighted, and interested only in immediate gain.[5]

The pastoralists in Golok are targeted by state and non-governmental campaigns and education programmes aimed at changing their attitude to the environment and teaching them a 'scientific' or 'rational' way of managing it. Educational sessions organized in the townships as well as

5 This contrasts with the image of 'green Tibetans' as nature friendly and possessing a unique knowledge of how to manage the environment in a sustainable way (see Huber 1991; Huber and Pedersen 1997).

television programmes and magazines published by the local NGOs spread a message that the environment should be protected and that the pastoralists do not show enough effort in doing so.[6] The topic of caterpillar fungus featured there, too. A magazine published by one NGO in Machen claimed that digging the fungus and other medicinal plants contributed (to the tune of 31.8 per cent, or so they argued) to the desertification of Martod County (MGGT [n.d.]: 11). My informants found such accusations to be unjustified, but they preferred to play down the scale of their intervention in the environment. By claiming that they only find a few fungi and do not lease their land, they hoped to avoid criticism and show themselves as 'rational' and 'environmentally friendly'. However, there are also other reasons why pastoralists are not keen to discuss their income from caterpillar fungus. These relate to the spatial and temporal restrictions on digging, which forbid doing it in certain places at certain times of year.

Together with Tsering Drölma and Tendor's family, we set off to the mountains every morning. Only when it snowed or rained heavily did we stay home with our noses glued to the window, waiting for the sky to clear up. Tsering Drölma said that she goes digging 'any time she can'. In practical terms, this meant when the weather allowed it, or when there was enough fungus to dig. If there was little, Tendor said, 'people stay home for a day or two until there's more [*yartsa*] growing again'. The neighbours in Soglung also said that they stop digging when the weather makes the work impossible and when low quantities of fungi make it not worth the effort to climb the mountains. However, they admitted that digging caterpillar fungus on certain days is inauspicious, too. These are days of religious observance (Tib. *düchen*), which fall on the 8th, 15th, and 30th of each Tibetan month. The most important *düchen* days are those during the fourth Tibetan month, called *saga dawa*, when three pivotal events in Buddha's biography are observed: his birth; enlightenment; and entering Nirvana. During this month, many people abstain from eating meat, drinking, or gambling and they distribute alms. Less motivated people focus on the three *düchen* days. Any karmic effects of positive actions taken on these days are believed to multiply. However, negative actions do as well. As one monk warned me, 'If you do something good, it will grow a million times. But if you do something bad, it'll also grow million times bigger.' The s*aga dawa* month falls, depending on the year, between April and June and overlaps with the caterpillar fungus season. Can one dig caterpillar fungus on *düchen* days

6 Golok television broadcasted, for example, a video clip showing a group of Tibetans digging caterpillar fungus on a lush meadow. In the next scene, the meadow had turned into a desert.

during the *saga dawa* month? My informants said it was inadvisable. Some of them argued that on these days one should strive for religious merit rather than economic gain and others explained that digging the fungus is a non-virtuous action (Tib. *digpa*) and therefore should be avoided in particular on days of religious observance.

Whereas it is clear that caterpillar fungus is dug for economic gain, there was no consensus among my informants about the ethical value of this work. The Tibetan name of caterpillar fungus, *yartsa gumbu* or 'summer grass winter worm', captures two points in this organism's lifecycle and an idea of the metamorphosis taking place between them. But is this metamorphosis complete in spring when the digging season starts? If it was complete and caterpillar fungus was a 'grass', digging it would not create any ethical problems. However, if caterpillar fungus retained some of its 'worm life', 'tearing it out' of the ground would be a violent act, killing the buried creature. In the context of Buddhism, this would be a *digpa* – an action leading to an accumulating of negative karma, which has an adverse impact on the person's future rebirths. In the Tibetan mindset, widely shared in Golok, killing large animals that provide a large amount of meat is more tolerable than killing small ones. Breeding animals only for meat is ethically problematic and killing for profit, especially small animals, and to satisfy the whims of the palate is particularly distasteful. If caterpillar fungus *was* a 'worm', killing it would share the last two features: it is done for profit and certainly not to satisfy hunger.

There was no consensus in Golok on how to classify caterpillar fungus during spring and no definite answer about the 'ethical weight' of digging it up. Opinions varied between seeing it as an innocent activity not entailing any guilt and as a *digpa*. Between these two extremes there stretched a field of uncertainty built of doubts, guesses, and speculation:

> There's no *digpa* in digging *yartsa*, since *yartsa* is dead, and not alive. It's more like digging *droma*. Nobody says that digging *droma* is a *digpa*, right? (Former monk, 27 years old)

> Of course, it's a *digpa*, because it's somebody's life! Maybe of a person who did something bad in his former life? (Tantric practitioner in his mid-40s)

> I went with my wife to Nagchu.[7] I couldn't see any *yartsa*, but she showed me one. I covered it with a piece of dung so that it could live. If it's a *digpa*, why should I commit it? (Trader, over 60 years old)

7 A county in a prefecture of the same name in the TAR.

> Maybe it's a *digpa* or maybe not. People say different things. But when I dig *yartsa*, I say *om mani padme hum* for each of them. (Pastoralist, 27 years old)

The opinions expressed by my informants show a desire to define digging caterpillar fungus in ethical terms and reveal that this activity falls into a grey zone. Many people classified it as *potentially* non-virtuous. This denotes an action the status of which is 'pending' and conveys a sense that something definite will be announced about it at some point, perhaps its condemnation. Until this happens, pastoralists are left to make their own judgement. The lack of a unified stance on the side of the Buddhist clergy was mentioned by some people who supported their view that digging the fungus is ethically neutral with the fact that it has not been officially condemned by their religious leaders. Several monks said that it is precisely because caterpillar fungus is such an intricate organism that they refrain from pronouncement: 'It's hard to say anything definite about it without making a mistake.' But its economic importance matters as well: 'Condemning something economically so vital is not easy without risking being criticized oneself', as one ex-monk stated.

In spite of these discrepancies, there seemed to be agreement that one should not dig caterpillar fungus on *düchen* days. Most pastoralists stated that they stay home on *düchen*. However, observations contradict this. Kunga Lama recalled his meeting with a woman in Yushu, which took place on a *düchen* day. The woman claimed that people do not dig on such days. A moment later, her relatives arrived at the tent with freshly dug fungi (Lama 2007: 79). In Golok, elderly pastoralists complained that society was preoccupied with one thing – caterpillar fungus: 'Here everybody goes digging! Every single day! Even on *düchen*!' or 'My daughter would dig even at night if she only could. Of course, she digs on *düchen*! She wants to earn money!' The majority of informants in the survey declared that at least half of their valley's residents dig caterpillar fungus on *düchen* days. It was difficult for them to admit that they also do so, but they openly spoke about methods for evading neighbours' attention: some families dig on summer pastures, farther from home; others send only their children to work. Some families openly admitted that they dig on *düchen* days, like the family of Karkho, Tsering Drölma's neighbour. Karkho suffered from glaucoma and could not dig. Only his wife, son, and daughter-in-law went digging. When the daughter-in-law had a baby, she also stayed home. Karkho said they were forced to sacrifice *düchen* for the sake of earning money: 'Last year we didn't dig every day, but this year we go even on *düchen*. I know that this isn't good, but only two people from our house can dig now. This is very few. I don't think we can afford to sit and be idle.' The family made the decision collectively, but in many cases, it was individual choice, not always approved of by relatives, as the complaints expressed by the elder generation show.

Sacred Mountains

Spatial restrictions on digging caterpillar fungus concern doing it in places inhabited by mountain deities, in Domkhok especially on Amnye Wayin, a mountain associated with a *zhibdag* of the same name. The Tibetan word *zhibdag* (lit. 'the owner of the base') denotes *dei loci* or territorial deities who inhabit the landscape and are an important element of pastoralists' religious life.[8] Amnye Wayin, who is the main *zhibdag* in Domkhok, belongs to a group of deities related to Amnye Machen, the famous sacred mountain in Golok. Amnye Wayin is one of its four gatekeepers.[9] In comparison to Amnye Machen, whose fame extends beyond the region and draws pilgrims from far away, Amnye Wayin has a more local importance. It is a 'deity of the local territory' who, as Samten Karmay explains, is at the core of the local community's regional identification, but whose cult does not extend beyond the locality's borders (1998a: 433). Amnye Wayin does not attract pilgrims, but he is nonetheless the main reference point for local pastoralists.[10] He is a subject of community rituals performed at important points of their migration cycle. His support is sought for group and individual enterprises and is crucial for community and individual well-being. As an elderly monk remarked:

8 Concerning *zhibdag* and *yulha* ('god of the locale'), see Samten Karmay 1998a and Huber 1999: 22-25.

9 The others are: Amnye Dzargen in the west, Dunchiong Dzamar in the south, and Weri Dukmtso Lake in the north. Each of them presides over a territory which it protects and where it is worshipped. Amnye Dzargen is worshipped in Dawu Zhuma, Dunchiong Dzamar in Dawu, and Weri Dukmtso in Dawu Zhuma in Machen, and Mtsonag in Martod County.

10 The pastoralists explain the lower status of Amnye Wayin with its lower level of religious realization: while Amnye Machen has attained *bodhisattva* status, Amnye Wayin is still on his way to it. Amnye Wayin was converted to Buddhism by Tsongkhapa Lobzang Drakpa (1357-1419), the founder of the Gelug school, and since then he has reached only the eighth stage of religious realization. Until he attains the tenth stage or the enlightenment, he will remain a *jigtenpi lha*, or a 'god of this world', whose powers are limited, but who is useful to people in their daily affairs (Samuel 1993: 166).

> Amnye Wayin isn't good on big things. But he can help if you need something here and now. He can remove obstacles or, on the contrary, put them in your way, making you unable to achieve your goals. If you have business to do, he can help you find the right person. If you lost livestock and are searching for it in the west, he will make you change your mind and go to the east, in the right direction. If you have lost a horse, he can hide it away from the thieves. And if you left your house empty, he can scare the thieves away.

Amnye Wayin is rich in natural resources, as the name, deriving from Mongolian *bayan* or 'rich', suggests. The pastoralists call him a 'treasure owner' (Tib. *terbdag*) and say that he guards enormous riches hidden in the ground. They also tell of rich vegetation covering the mountain and excellent caterpillar fungus growing there. This makes Amnye Wayin a tempting goal for people seeking a shortcut to wealth. However, Amnye Wayin's mountain slopes are the *zhibdag*'s sacred precincts and trespassing on them is risky.

The pastoralists claim that Amnye Wayin is fond of interacting with people, but he is also short-tempered and quick to take offence. This manifests itself when someone breaks the rules of a particular *savoir vivre* that binds people as users of the land. Among the blacklisted activities are hunting, fishing, logging trees, and destroying vegetation, disrupting the surface of the earth and polluting the land, air, and waters, either with material rubbish, or even smells – for example of scorched meat. Digging caterpillar fungus brings together several activities from this list. The diggers disrupt the surface of the earth but remaining in the mountains for a longer time they are also likely to commit other offences. This applies especially to those diggers who stay in tents and not in the pastoralists' homes. Living in a tent, they need fuel (so they destroy vegetation) and food (so they cook and fish) and produce garbage. 'They are dirty and make the land dirty', as some pastoralists said with visible disgust. Considering how many negative aspects there are associated with digging caterpillar fungus, it should be no surprise that Amnye Wayin has little tolerance for it, especially if people venture where they should not and dig on his mountain slopes.

Once angered, Amnye Wayin can severely punish trespassers. A typical punishment is a lightning strike. Accounts of accidents when caterpillar fungus diggers have been struck by lightning from a clear blue sky show that the danger is real:

> A woman from Rebkong was killed there recently. She went to Amnye Wayin to dig *yartsa* and was killed by lightning. People told her not to go,

> but she said that *yartsa* is so good there and she went. She was killed the same afternoon. There was another person with her, but that one turned back. And the woman didn't [turn back] and died.

Such narratives, commonly heard in Domkhok, have a didactic value and are a warning against violating the *zhibdag*'s laws. They are constructed around an interaction between the pastoralists and in-migrant diggers: the latter are the trespassers, while the former try to dissuade them from doing it. However, the diggers disrespect the danger and get punished, sometimes barely escaping with their lives. But is it only the 'outsiders' who trespass on the *zhibdag*'s land? The answer is no, even if material confirming it is scanty. Kunga Lama analysed the case of Lama Norlha, a sacred mountain and a female *zhibdag* in Yushu, who attracted so many diggers that the pastoralists had to organize patrols to guard the mountain. Some pastoralists were caught among the trespassers, too (Lama 2007: 86). While in Domkhok persons from outside the township were blamed, some local residents also took the risk of digging on Amnye Wayin. During my research, several accidents happened in which people were killed by lightning high on a mountain range. My informants interpreted these accidents as punishment for crossing a line beyond which digging the fungus is forbidden. In one case, a relative of the victim admitted that the woman left home to climb Amnye Wayin and dig caterpillar fungus. This is also where her body was found.

An additional problem associated with digging caterpillar fungus, especially by non-local diggers, can be called a 'patriotic' one. The mountain or territorial deities such as *zhibdag* play an important role in building a sense of local identity, and also political one – as was demonstrated by Samten Karmay (1998a and b). They preside over a large territory and the community inhabiting it, and sometimes are even envisioned as the ancestors of local people. The *zhibdag*s play an active role in the life of local inhabitants and can even support them in military conflicts. Elderly pastoralists in Domkhok said that Amnye Wayin did so at least twice in the course of the twentieth century. Once, during a conflict with a neighbouring *tsowa*, he descended in the shape of a raging red bull upon the aggressors, who fled in panic. He also demonstrated his fury in 1958 when pastoralists were attacked by the Communist troops. This shows that Amnye Wayin has personal, ethnic, and even political sympathies mirroring those of the local pastoralists. He prefers pastoralists to farmers and Tibetans to other nationalities, especially Hui and Han. He is more likely to accept people from those *tsowa* who the local people are on good terms with than those with whom they are in conflict. Thus, the problem consists not only of the fact that Amnye Wayin's land is trespassed upon, but *by whom*. The very presence of non-local diggers disturbs him and it is easy for them to take one step too far. They have to be careful, as even actions positively valued when performed by locals can have the opposite effect if done by others. My informants recalled how a group of Tibetans from Rebkong performed *bsang* offerings to Amnye Wayin, perhaps without bad intentions.[11] They were warned that only local inhabitants could do so but did not listen. Soon after a storm carried their tent away and they barely escaped alive.

Amnye Wayin turns his anger not only against trespassers, but also, at times, the local community. Some pastoralists believed that they were in greater danger than in-migrant diggers, because the latter – especially if they are not Tibetans – are beyond the power of the *zhibdag* and often go unpunished.[12] On the other hand, the *zhibdag* can hold the local pastoralists responsible for not protecting his land. Who exactly is to suffer punishment, however, was unclear. Some pastoralists argued that it is only those people who actively assist the trespassers and others believed that the *zhibdag* applies a principle of collective responsibility: regardless of whether there has been direct assistance, passive consent, or simply negligence, the punishment remains the same. If lightning strike is a punishment that Amnye Wayin uses against individual diggers, he deployed other methods with a wider impact against the local community as a whole. My informants interpreted droughts, floods, and animal plagues as such. One of the fields in which the *zhibdag*'s anger can seriously affect the pastoralists' livelihood relates to the 'essence of the land' (Tib. *sabchud*). This term denotes the nutritional and resilience potential of the land, which conditions its ability to nourish livestock, and resist plagues of pests and processes such as desertification.[13] Its strength relates to the quality of the natural resources of the land, including medicinal plants and caterpillar fungus. Depleting them is believed to weaken *sabchud*. Two elderly pastoralists explained:

> I don't agree with all these Chinese and farmers coming here to dig. I only support the local folk. Otherwise it's disrespect to our *sabdag* and *yulbdag*.[14]

11 *Bsang* or *bsang chöd* is a fumigation ritual performed with juniper twigs by lay members of the community, in Golok only by men. See Karmay 1998a and 1998c.

12 Huber also observes something similar about hunters who poached on mountains belonging to other communities because 'any wrath of the gods was much more likely to fall on the local community than on them' (2004: 143).

13 On *bchud*, see Gerke 2012 and Da Col 2012.

14 *Yulbdag* refers to *yulha*. *Sabdag* are yet another sort of being associated with more localized natural features, such as rocky outcrops, stones, and small areas of land.

> These people hunt game in the woods, kill birds, and put out traps to catch deer. Digging *yartsa* weakens *sabchud* and brings trouble upon people and livestock. People may fall ill and bad things may happen to livestock. I don't agree with that. But since everyone else agrees, I can't stop it.

> The land must be losing its *sabchud* when people dig out millions of yuan from the ground. Gold is dug by the government and *yartsa* by Chinese, Tibetans, and farmers. It's good for them, but bad for the land. All people here are nomads and depend on livestock. When the land loses its *sabchud*, livestock give less milk and get weaker. This bad weather [it was a rainy summer] is perhaps also connected to *yartsa*.

Dangerous Money

It seemed curious that, when asked about when they do not dig the fungus, my informants initially mentioned bad weather and poor grassland productivity and only later spoke about proscriptions against digging on days of religious observance. It could mean that the latter are so deeply ingrained in pastoralists' minds that mentioning them did not appear to be necessary. But priority given to the weather or grassland productivity might also be a sign that religious proscriptions remain in the normative space, while the former two factors inform people's actions in a more direct sense: the snow actually blocks people's way into the mountains and low quantities of caterpillar fungus growing in the mountains really makes them think twice before setting off on a digging trip. The reluctance to talk about digging on *düchen* days shows that this normative space does inform people's opinions, even if it does not translate into their actions. The norms shared in any community cannot be considered 'direct indications or even descriptions of activities' taken by its members, but they can influence the way they talk about themselves (Holy and Stuchlik 1983: 22).

This discrepancy between words and actions was explained by some religious practitioners as related to the low levels of Buddhist practice in Golok. Jigmed Dorji's brother, a *ngagpa*, said:

> Buddhism came to Golok later than to the other parts of Tibet. So, people have little awareness of what is good and what bad. Take me for example. For most of my life I used to dig *yartsa* on *düchen* [days] and even on holy mountains. Only slowly did I understand that digging *yartsa* on

> *düchen* spoils your karma and that digging on holy mountains angers the mountain gods.

Jigmed Dorji's brother was one of many religious practitioners who spoke about Golok as a *terra nova* of Buddhism. As a pastoral region, Golok had fewer well-established monastic centres than other regions and was 'far behind the rest of Tibet', as he remarked. Poor knowledge of the Tibetan calendar contributes to people's ignoring *düchen* days, as well: 'To be a good Buddhist, you should read the calendar first', one monk said. But in Golok two calendars are used, the public solar one and the Tibetan lunisolar one. Consequently, according to the monk:

> People get confused. Some who dig on *düchen* [days] don't know what date it is. Others know, but they go digging anyway because they are poor and need money. And there're also those people who have enough money and know what date it is, but nevertheless go digging.

Does the money earned in such circumstances weigh heavier in the pastoralists' pocket than other income they earn? Is it possible that people do not want to talk about this money because in earning it they violate the spatial and temporal restrictions on digging caterpillar fungus?

Classical theorists envisioned money as anonymous and impersonal, a fungible, qualitatively neutral, and infinitely divisible homogenous medium of market exchange. For Georg Simmel, money is 'free from any quality and exclusively determined by quantity' and its essence lies in its 'unconditional interchangeability, the internal uniformity that makes each piece exchangeable for another' (1978: 279, 444). Simmel calls the notion that money could be qualitatively stained mere 'sentimentality' (1978: 441). His ideas echo Karl Marx's approach in which money lacks social or cultural shades and has only one, quantitative dimension: it is a 'radical leveller' that levels qualitative distinctions between things and even between intangible non-commodity objects and makes them ordinary commodities with only one dimension – the quantitative (Marx 1995: 87).

Contrary to what these classical theorists believed, money is neither culturally neutral, nor socially anonymous. It can be personalized, earmarked, morally and ritually ranked, and is qualitatively heterogeneous.[15] Not all

15 There is a large body of literature on money as incorporated into different systems of value, inscribed with meanings, 'earmarked', tabooed, or sacralized, e.g. Bloch and Parry 1989; Gregory 1997; Guyer 2004; Maurer 2006; Peebles 2012; Zelizer 1997.

money is equal and not all sources of money are equally good in symbolical or ethical terms. On the contrary, there are hierarchies of money, some of which have higher and some lower status. 'Bitter money' earned from selling land, tobacco, gold, and roosters (Shipton 1989: 28), 'money of shit' from cleaning latrines, selling beer in town, and domestic service (Hutchinson 1996: 84), and 'evil money', which sugarcane plantation workers earn thanks to, arguably, their contacts with the devil (Taussig 1980: 45) – these are all categories of money that are considered by their users as stained or symbolically impure. The reasons some money is considered qualitatively 'worse' differ. Sometimes it is a particular activity that 'pollutes' certain income (such as the Nuers' 'money of shit') and sometimes it is the sale of particular goods that should not be sold (e.g. land) that is the problem (as in the 'bitter money' of the Luo from Kenya). In many societies, mining produces money that is deemed 'heavy' or 'polluted' (Mongolia) because it comes from exploiting resources that do not *fully* belong to people, but to other, sacred beings inhabiting the land and because digging into the ground angers these entities and so should be avoided for religious reasons (High 2017: 61).

Is the money earned from caterpillar fungus similar to the categories listed above? Does it have a different status, qualitatively lower than other types of income available in Golok? I did not record any verbal expressions confirming it as such. My informants did not talk about this money as symbolically less valuable, ethically problematic, or polluted. Moreover, I did not observe any special status in terms of the pastoralists' expenses or the purposes for which caterpillar fungus money is used. Many anthropologists have noted that some categories of money are considered unsuitable for certain purposes. 'Bitter money' cannot be spent on religious offerings, to buy livestock, or used as bride wealth, because it is 'barren, sterilizing, and ultimately destructive of family welfare' (Shipton 1989: 76). The 'evil money' of sugarcane plantation workers cannot be spent on any long-lasting investments. It is pointless to invest it in land or livestock because it will make the land sterile and cause the animals to die (Taussig 1980: 13-14). Money from such ambivalent or negatively valued sources is often spent on quick expenses rather than substantial investments. In Mongolia, the 'polluted' money of goldminers is preferably used only for immediate consumption (High 2013: 682). The same applies to 'hot money' from goldmining and lotteries in Indonesia, which is spent 'as quickly as it was earned' (Znoj 1998: 199). Even the Nenets of the Yamal Peninsula, who get rich on the reindeer antler trade – an activity that at least some of them see as ambivalent – spend their earnings on entertainment or luxury shopping and cover their daily needs with other income (Stammler 2005: 307).[16] The money from caterpillar fungus differs from the above categories for one important reason. The pastoralists in Golok do not earn side-line income from the caterpillar fungus economy. On the contrary, almost all their cash originates from this source. It forms the basis of their economic functioning. The pastoralists spend it on a wide range of purposes: food and daily needs, entertainment and luxuries, but also education, infrastructural investments, and religious offerings.

The money from caterpillar fungus may not show obvious similarities with the categories mentioned above. Yet, some comparison can be made. Similar to these examples, money from caterpillar fungus is considered potentially dangerous. This fact emerged during my interviews with religious specialists, astrologists, and diviners, who are an important professional group in Golok. People seek their advice in many life matters, i.e. before taking any serious decision about marriage, studies, medical treatment, financial investments, or a long journey, to check whether the situation is auspicious or how could one make it so. They turn to diviners for help when facing unforeseen difficulties, such as diseases, accidents, or unexpected death in the family. The diviners I interviewed interpreted such unpredicted diseases and tragic accidents as a side-effect of the caterpillar fungus boom and a result of the accumulation of wealth earned from it. A *khandroma* from Dawu said:

> If a family earns 50,000, 60,000, or 100,000 yuan from *yartsa*, it will definitely face problems. Someone in the family will pass away or the livestock will die. I don't know why. Maybe among the *yartsa* there is something that belongs to the *zhibdag* and people take it away?

The pastoralists have no choice but to dig caterpillar fungus ('what else should they live on?', as the *khandroma* asked), but they should be prepared to shoulder the consequences. The sums she mentioned as creating risk for their owners are not high: every family who took part in my survey earned much more than that. This would mean that everybody is in danger. This danger relates closely to the idea that caterpillar fungus money does not belong to the people but is stolen from the land. It will need to be paid back. The question remains: how and when? Caterpillar fungus is the *zhibdag*'s property, as another man stated – suggesting that it is Amnye Wayin who is the ultimate owner of the natural resources in Domkhok. However, it is

16 Stammler observes that the Nenets do not value money from this trade less than from another source, even though some admitted that 'it might be bad for their herd to cut *panty* [the antlers]' (2005: 307).

not only the amount of money that people earn from caterpillar fungus that matters, but also the speed at which it is earned. A popular astrologist in Dawu mentioned previously unknown diseases and misfortunes befalling people who got rich 'too fast'. The risk grows proportionally to the material wealth accumulated and the speed of earning it. It is like with 'evil money': people who earn too much too fast can expect to die prematurely and in pain (Taussig 1980: 13-14). It was no coincidence, the astrologist remarked, that most of his customers seeking help in tragic family accidents come from Domkhok and Xueshan, the wealthiest townships in Machen.

Suddenly, caterpillar fungus money turns out to be associated with risk. The cash itself, in the sense of every banknote passing through people's hands, does not carry any particular danger, but the fact that people earn large fortunes from it is problematic. As the *khandroma* emphasized, these fortunes do not have to come from digging caterpillar fungus on sacred mountains or during *düchen* days: regardless of how such sums are earned, they can easily turn into the owners' misfortune. The opinion that all caterpillar fungus belongs to the *zhibdag* was not commonly (or at least not openly) expressed. If it described reality thus, the pastoralists would have to either stop digging the fungus or live in constant fear of punishment. While it is clear that people do not stop digging in spite of the risk it creates, it is worth asking what measures they take to minimize this risk and mitigate the consequences of this work. To prevent misfortunes, people seek divination about where and when they can safely go digging. A Buddhist monk who offered divination admitted that he includes in his services *düpa* and *shunkor* amulets, which help people avert the *zhibdag*'s anger and offer some protection during the work. Finally, when the money is already in the pastoralists' pocket, the owner should be generous in spending it on religious causes and offerings to the *zhibdag* to pacify him. Otherwise, as the *khandroma* noted, 'the money will not stick to you or, on the contrary, it will attract bad luck and make you suffer'.

The association between money earned from caterpillar fungus and potential problems should not be a surprise. In a book based on his travels through Golok Serthar and Dzachuka in the 1950s, Namkhai Norbu informs that the pastoralists 'often say that the *yartsa gumbu* and the *karmog* are the treasures of the earth spirits [...]. If anyone should pick them, his community will be struck by virulent epidemics that will spread all over the region' (1997: 68).[17] He notes that this ban on digging medicinal roots has

17 *Dzayung* is another name for *pimo*. *Karmog* is most probably *karmong*, Lat. *Taraxacum tibetanum* Hand-Mazz, a type of dandelion.

not always been observed and people have often violated it. If caught, they were savagely beaten, but they took the risk as they had 'little belief in the stories about those tubers and [were] far more interested in the tea, silk, cotton and ornaments they [could] procure bartering them with Chinese traders' (Norbu 1997: 68). This shows that also before the caterpillar fungus boom, the economic incentive occasionally won out over people's fear of the mountain deities and human reprisals. In the early years of the boom, the pastoralists in Golok seldom dug themselves, explaining it with a traditional taboo, as Horlemann reports (2002: 262). During my research, though, they dug it on a massive scale and benefitted from the caterpillar fungus economy in many other ways. They were also much less outspoken about the negative side of this activity or its former forbidden status. Even if beliefs about the potential negative effects of caterpillar fungus digging did not prevent them from engaging in this work, it is possible that they were nonetheless reflected in beliefs about money earned from it. This would mean that the conviction proscribing digging for caterpillar fungus moved to another level and was reformulated as a belief that in some way one has to pay a price for getting rich through such means. This belief would be another reason why the pastoralists prefer not to talk about their caterpillar fungus harvests and the income from it, as it is a kind of morally stained or at least problematic wealth.

8 Pastoral Life and the Market

Tsering Drölma and Tendor were born in the 1980s and are too young to remember the people's communes. However, Tsering Drölma's uncle, Jigmed Dorji, remembers this period well. The family then lived on the other side of the Dom River and was much less affluent than they are now. When the communes closed, the land was allocated to individual households. In 1983, Jigmed Dorji's parents (Tsering Drölma's grandparents) received a land lease contract stating that they were the legal users (though not formal owners) of new land in Soglung Valley. The contract had thirty years' validity but was later extended to fifty years. When Jigmed Dorji's parents died, the contract was transferred to him and his sister, Tsering Drölma's mother. In addition to the land, the family received a share of livestock from the commune herds. In the production team they were part of, every person received twenty-one sheep and nine yaks. Other teams were wealthier and people got thirty sheep and fifteen yaks, or even as many as forty sheep and twenty-three yaks. Luckily, Jigmed Dorji's parents had some animals that were assigned to them by the commune for their private use.[1] As a result, the nine-person family entered the new economic era with 189 sheep and eighty-one yaks from the commune herds as well as twelve yaks, twenty-one sheep and two horses which they had already owned.

Two decades later, when I conducted my research, the two families – Tsering Drölma's and Jigmed Dorji's – who lived on the land covered by the above contract were quite prosperous, even in their own eyes. During the survey, Tsering Drölma and Tendor declared that they (a family of three) had 123 yaks. In front of their house stretched a yak corral (*zogra*) where the *dris* and calves stay at night – one could see them through the kitchen window and hear their grunting (Figure 19). All the *dris* had names: Shiomo, Wobse, Keyig, Rogzylma, Horog, Thurog, Thiknag and so on. *Dri* is a general name for female yaks of reproductive age, but the group splits into several sub-categories. Pregnant females and those that just calved are called *thima*. Those that calved the year before and thus whose calves are yearlings are *yaryma*. Finally, the *dris* that have not calved for several years are referred to as 'dry' or *drirkam*. In a good year, Tsering Drölma gets up to thirty calves, in other years it could be just half this number. The calves are born in May, when on cold nights the herd still wakes up covered with a blanket of snow. Yet, the mortality rate of young animals is low: during the year of

1 Such animals were called *khazhu*.

Figure 19 Yak corral in front of the house: the women are collecting yak dung for fuel

my survey Tsering Drölma lost only one calf – it was killed by a male yak on the grassland.

Tsering Drölma and Tendor declared that they had forty male yaks: Karcha, Ngobok, Gyara, Rogra and the rest. All but one were castrated. Most people in Soglung castrate the male yaks, believing that a bull yak is too much of a burden, being semi-wild and difficult to control. Tsering Drölma had one bull yak (*norma*) kept in order to cover the *dris*; the neighbours would rent it from her. Four yaks in the herd could be used for riding and as pack animals (*nalo*), but they were seldom used in this role. Male yaks stay on higher pastures and are not brought down for the night. Once, when a bull yak came down to the summer encampment during a moonless night, panic broke out: dogs jumped wildly on their chains, waking up all the people in their tents. Apart from such unplanned incidents, male yaks are seldom seen in the vicinity of the house.

Tsering Drölma and Tendor did not have any sheep, unlike Tsering Drölma's uncle who lived in the neighbouring house. Until recently, all families in Soglung had sheep, but now Jigmed Dorji's family was one of just three out of fifteen who continued breeding them (Figure 20). In his sheep pen (*lugra*), as Jigmed Dorji said, there were nearly 250 sheep, among

Figure 20 Unusual sight: sheep taken for grazing

them some 120 ewes (*mamo*) and fifty males or *habzang*, including ten rams (*chom*) and forty wethers (*hoton*). In a good year as many as a hundred lambs could be born, but they are born in winter and many do not survive their first weeks of life. The pastoralists have to be ready to take care of the newly born animals, drying them and taking them home: they are kept in boxes made of plasters of yak dung and sheep's wool (*drekang*). Yet, for many lambs help comes too late; petite lambskins hanging on the sheep pen are a sign that life in Golok is not easy.

Apart from the livestock, which 'worked' in the household and had an economic role to fulfil, there was another group of animals in Tsering Drölma and Jigmed Dorji's herds, animals that have a special 'resident' status. These are yaks, sheep and horses for which *tsethar* ritual has been performed or that had been offered to mountain deities. *Tsethar* (lit. 'freed-life' or 'life liberation') is a Buddhist ritual in which certain domestic animals are granted life until their natural death. They cannot be killed, given away or sold if there is any risk that the next owner might kill them or sell them to a slaughterhouse. Jigmed Dorji would free at least one animal during each Tibetan New Year. Tsering Drölma and Tendor did the same, believing that it helps the family stay healthy and the livestock to flourish. The ritual is also performed in case of diseases, to prolong someone's life, avert misfortunes, or safeguard a good rebirth.[2] Theoretically, *tsethar* should be performed with sheep rather than yaks and with male animals rather than females. Females bring offspring and are important for the continuity of the herd, whereas males are bred mostly for meat. As a monk from Golok explained:

> *Tsethar* is done with sheep, because they don't live so long. If you do it with yaks, you have to look after them for twenty to thirty years. And when they're old, they get stubborn, don't want to stay with the herd but roam everywhere. Sheep are kind hearted and stay in the flock, and you don't need to tether them at night. If there're buyers coming, it's more convenient to sell the sheep and the purpose of raising sheep is exactly this: to sell or eat them. *Dris* give milk, so they live until they get old. Riding yaks can also have a longer life. And what is the use of sheep? You can't use them to carry loads. You don't milk them, either. Their destiny is to be killed or eaten. Ewes lamb, so they also might live longer. So, people usually free rams, because it's clear that if not freed, they'll be killed or sold [to be killed].

This rationale diverges from the pastoralists' practice and in Domkhok people liberate ewes just as often as rams: Tsering Drölma had eighteen female and ten male *tsethar* sheep, and her uncle fifteen males and eleven females.[3] Yaks, including females, are liberated too. As Tsering Drölma explained: 'We milk our *dris* for so many years that we want to give them some rest afterwards.' The same logic applies to particularly fecund *dris*: those that calve every year have their *tsethar* status almost guaranteed.

Another group of 'resident' animals are those offered to the mountain deities, the *zhibdags* and *yulhas*. The iconography of these deities shows them riding yaks, horses or other mounts, brandishing swords and holding other war attributes. The *zhibdags* have very clear preferences. Amnye

2 For discussion of the reasons for performing *tsethar*, see Holler 2002: 213ff and Gerke 2011: 174ff. Jigmed Dorji performs the ritual by himself, as do his neighbours. He starts with a symbolic purification of the animal, recites the mantras of Tsepame and Dzambala, asks for a long life for his livestock and family and for the flourishing of the family's wealth, and takes an oath not to kill the animal. The animal is decorated with colourful cloth and tassels and receives markings on its fleece and ears as a sign of its new status.

3 Some monks criticize this behavior. One monk complained: 'Because of the little contact people in Domkhok had with monasteries, they have limited knowledge of religion. In the rest of Golok, people know religion better. For example, a *lhayak* has to be healthy. It can't be injured or handicapped. If it has a broken horn, is blind in one eye or has a *nartag*, it can't be offered. When I visit people, they show me their *lhayak* and I see a yak with a big *nartag*. This is against tradition.' *Nartags* are ownership markings cut into animal's ears.

Wayin, for example, likes horses and yaks of a cream (*ngangan*) or 'fox cub' (*wabryk*) colour. The pastoralists offer him animals fitting these preferences as well as others whose unusual traits, colour or behaviour suggest that they would be selected by him. Tsering Drölma told a story of young *dri*, a beautiful animal that led the whole herd: wherever she went, the herd followed. However, the *dri* was barren. All this, including the animal's infertility, was taken as a sign that it was chosen by the *zhibdag*. Yaks and horses offered to the *zhibdag* are considered his property.[4] They usually stay with the herd but can also roam the land freely and may be ridden only by their former owner or persons from his close family.[5] They are offered to the *zhibdag* either during *bsang* rituals held by the whole community or by individual families seeking the deity's assistance. People believe that the *zhibdag* is 'obliged to look after the donor with gratitude, becomes his friend, protects him, bestows rain, wealth, and increasing herds on him, and averts misfortune and disease' (Holler 2002: 219). The pastoralists in Domkhok also hope that Amnye Wayin will reciprocate their gifts, look after them and help in any difficulties. This help can be very effective: a family whose land lease contract was revoked as a punishment for leasing their land to caterpillar fungus diggers received it back thanks to – as they said – Amnye Wayin's help. They performed a *bsang* ritual and offered the deity numerous animals. This helped settle the matter, in addition to formal appeals which the family made at the county institutions.

Pastoral Products and the Pastoral Calendar

Today, many livestock products are available in the shops: butter, cheese, meat and even yak dung can be purchased in town. However, the pastoralists still rely largely on their own products. The yaks supply them with milk, meat, hair, hides and dung. They provide large amounts of meat: the biggest male yaks can yield 500 *jin* of meat or more, average males 160-170 *jin* and barren females 180 *jin*. Female yaks are milked: a single *dri* gives around one *jin* of milk if milked once a day during the height of the milking season

4 Such horses are called *lharta* and yaks *lhayak* and *lhadrima*. The most exceptional ones are remembered for a long time: Jigmed Dorji could still tell stories about *lhayaks* his parents had in the 1980s.

5 Breaking this rule can be dangerous. It happened that a man rode a *lharta* to a cousin who did not know of the horse's status, hopped on its back and rode off to the mountains. The horse got ill and died. 'When such animal dies, it needs to be replaced quickly. Otherwise the *zhibdag* gets offended and things might go wrong', the man explained.

between late April and September. Milk is processed into dairy products, mostly those with long-term storage options: butter and dry *chura* cheese. The yaks produce two kinds of fibre: the coarse, long outer hair (*rtsipa*) and the dense woolly undercoat growing on the animal's belly, flanks and thighs (*khulu*). Yak hair is used for making tent cloth, blankets, ropes and bags. An average yak gives about six *shang* of *rtsipa*, bigger ones up to eight *shang*. As for *khulu*, a medium yak gives 7-8 *shang*, smaller ones 5-6 *shang* and the biggest ones up to one *jin*. Yak hides (*kowa*) provide durable material for sacks for grain, girdles, leather straps and other products. Finally, the dung is used as fuel. The yaks are also transport and pack animals, which can carry up to seventy kg loads.

The uses for sheep are less diverse. The pastoralists in Golok do not milk ewes.[6] Fleeces are used for felt bedding, horse blankets and saddle pads. An average sheep in Golok yields one *jin* and several *shang* of fleece, smaller animals several *shang* and the biggest males as much as 2.5 *jin*. Sheepskins are used for making winter robes, worn with the fleece inside, and other garments. Products of sheep slaughter are used for immediate consumption or making sausages and other products. An average sheep gives thirty *jin* of meat, a big ram more than double this amount. Finally, both sheep and yak stomachs are used as containers.[7]

The calendar of work in pastoral households is geared to fit the time for harvesting and processing pastoral resources. The most intensive periods are late spring and summer, when much of this work takes place, and late autumn, when the pastoralists prepare for winter.

In spring and summer, the biggest workload is connected to milking and processing milk. These activities define the rhythm of daily routines in the tent which during the day turns into a site of milk processing. In the morning, Tsering Drölma gets up early and – still half in darkness – untethers the calves.[8] She lets them suckle, drags the calves away, tethers them again and starts milking. During other seasons, she gets just enough milk to serve

6 Some informants said that people in Golok never milked sheep. Others argued that they did so before but stopped because ewes gave little milk. This change, they said, took place in the nineteenth century. See also Costello 2008: 81.

7 The stomachs are washed, dried and soaked for one day in hot water. In such bags, called *joby*, butter is stored, and blood and animals' entrails are frozen.

8 Milking is a women's job. In several houses where I stayed, men helped with milking, but they would not openly admit to it. This help was necessary because of their wives' health, but these men were also young, socially engaged and relatively gender conscious. I also met a 64-year-old man who lived alone and milked his *dris* by himself. The livestock thrived under his hands, so the neighbours said.

with the tea, but in summertime the milking takes a good hour of work. When this is done, the herd leaves to graze, and Tsering Drölma moves on to other work, processing milk being one of the tasks. She spends hours heating milk, setting it to make yoghurt, kneading butter and crumbling cheese. She carries endless buckets of milk and wide aluminium pots, in which the milk is boiled. The work is entirely manual: the only novelty in terms of kitchen equipment is a milk separator. Every day, around noon, Tsering Drölma pulls it from the corner of the tent and, sitting in the sunshine, separates milk into cream and skimmed milk. The sheepskin bags used for it in the past are not common in Domkhok anymore, but Tsering Drölma dreams of buying an electric milk separator that was advertised on television. When ready, butter is packed into containers made of sheep stomachs and, although it slowly goes rancid, it remains edible for several months. Skimmed milk or *dara* is used for yoghurt and *chura* (cheese). This cheese, in rock-hard crumbs, is another staple that keeps its quality for months and is eaten with butter, *rtsampa* and tea in what is called *chaka*.[9]

The amount of time that women spend milking and processing milk depends on the size of their herds, but also on the milking frequency. Until recently, during summer women milked *dris* before the herd left to graze, at 5-6 am, and when it returned home, at 6-7 pm. In recent years, they have stopped milking in the evening. The main reason Tsering Drölma gave referred to the declining productivity of the herd: 'I feel pity for the calves. They get hungry and don't grow properly. This is important because the *dris* give less milk than before.' According to the pastoralists, this change took place during the last two decades. However, they could not estimate the scale of the decline. The owner of a dairy company in Dawu could not explain it either. He said that it seems there was less milk in Golok now, but suspected that it could be a matter of subjective judgement: 'It's hard to tell. Maybe people simply milk less or maybe the *dris* really used to give half a *jin* of milk more?' Regardless of whether this decline has really been observed, my informants in Domkhok had started to consider milking *dri* twice as a sign of human greediness: 'What we get is enough for us', people often said. 'Everybody milks once a day. People say that if you milk twice, the calves grow weak', the wife of Jigmed Dorji remarked. Reducing

9 *Chaka* is eaten mostly in the morning. To make it, one mixes *rtsampa*, cheese, butter and sugar in a bowl, then adds tea, drinks up some of the tea and kneads the rest into a big dumpling. The box with *rtsampa*, cheese and butter always stands in a visible place in the house. The order of adding ingredients (dry ingredients first and then tea, or the opposite approach) is believed to differentiate the Wranakhs from the Goloks.

milking frequency changed the pastoralists' daily routines. When the *dris* were milked twice a day, they had to be brought in from the pastures early in the evening. Hence, they also had to leave earlier to graze. When they are milked once, they can leave later, which makes the whole day in the household more relaxed. This change has impacted the women's workload: they now have one hour less milking work, and do not have to boil the milk to prepare it for the next day, when it is mixed with the milk from the morning's milking and processed further. But this change has not only affected women: when the *dris* were milked in the evening, the calves had to be kept separate for at least part of the day and this required extra workforce.

Spring and summer are the time for other important work that requires cooperation between neighbours. This includes preparing and processing hair, wool and meat. Sourcing fibre starts in April or May with *khulu* or the yak's woolly undercoat. The yaks start shedding hair during this period and the whole process would not be complicated if not for the fact that the animals have to be caught, hobbled and pulled to the ground before work can start. This is repeated in August when *rtsipa* or outer hair is sheared: a yak is brought down with all its legs hobbled, and only then can the men get their shears to cut the *rtsipa* – as close to the skin as possible so that the fibre is of maximum length. July and August are also the time for shearing sheep. Although sheep are easier to handle, the neighbours help each other with this work too. But since most households have sold off their flocks, and keep only *tsethar* sheep, shearing is no longer as time consuming as it used to be. Cooperation is also necessary when livestock, especially yaks, are slaughtered. The pastoralists sometimes slaughter a yak or a sheep in summer. Meat from this summer slaughter is used for immediate consumption. The households take turns in performing this task and share the meat. Sharing meat is a means of payment for the neighbours' help, but also a necessity, since summer temperatures do not allow meat storage. Larger numbers of animals are slaughtered in November or December, just before winter starts and when the low temperatures allow the storing of large amounts of meat for the coming months.

The pastoralists in Golok kill their livestock by sealing the animal's mouth and nostrils and leaving it to suffocate. The act of killing is done by the animal's owner, but the preparations and everything that happens later require the cooperation of several men who help with hobbling the animal and skinning and butchering it later. Yak hides and sheepskins are either left for tanning or sold untanned: a scene of pastoralists sitting on the pavement next to fresh yak hides is common on the outskirts of

Dawu. Yaks' internal organs and blood are stored and used for cooking.[10] The butchered carcasses are left to freeze and dry. In the cold, dry air the meat from the autumn slaughter maintains its quality until the end of spring. By then, these reserves are finished: when I stayed with Tsering Drölma in May, during the caterpillar fungus season, she apologized for the shortage of meat and sent Tendor shopping when she wanted to cook something better.

Slaughtering livestock is crucial for the household's survival as meat is the main high-protein food in the region. However, Buddhist ethics condemn killing and say that intentional harming of sentient beings leads to the accumulation of negative karma and inauspicious rebirth. Slaughtering livestock is thus not an activity that people take pride in or are willing to discuss in detail.[11] A few people claim that they do not kill their yaks at all but eat only those killed by predators or that 'fell off a cliff'. This topos is known from other Tibetan Buddhist societies where consuming animals that have died in such accidents is considered a 'smaller evil' (Childs 2004: 127). Most of my informants laughed at this idea as an obvious lie, but they showed sympathy for the animals – as this young woman did:

> Women should not be in the place where yaks are slaughtered. A yak, seeing a woman, could think that she is kind hearted and that she will save its life. One should not wear red, either, because a yak might think that it's a monk and hope that its life will be spared. Before the yak gets killed, we pour water with a holy medicine into its mouth, and chant *mani* and *opame*, and turn a prayer wheel on its head. And when it's dead, we offer butter lamps for it.[12]

10 Blood of live animals was used in the pastoralists' diet, as well. It was consumed, for example, after being cooked with butter and salt in a dish called *chiagyi*. My informants explained that blood should be drawn from a yak's tail or legs in summer or autumn when the animals are in the best shape. Ekvall (1963: 147) and Thargyal (2007: 89) noted that the pastoralists consumed it during periods of meat shortage. One of my informants recalled that they did it during the Great Famine.

11 I encountered a similar problem when I collected data about hunting. Elderly men shared their knowledge as long as we spoke about hunting in general and not about their own experiences. As one monk said, 'In the traditional society, not all Tibetans were religious, same as today. Some people acted against religion and hunted animals. Tibetans believed that hunters' families would die out or become beggars. People called a hunting rifle an "evil rifle" (*dygwu*) and hunter's hands were known as "evil hands" (*dyglag*) and they avoided being touched by them.'

12 *Mani* refers to the mantra *om mani padme hum*, and *opame* to the *mantra* of *Opame* (Skt Amitabha), the Buddha of Infinite Light.

Before the slaughter takes place, each family needs to make a careful decision about which animals to kill. The households surveyed slaughtered male yaks and barren females not younger than four or five years. This stood in contrast to the animals they sold to slaughterhouses: these were younger, two- to three-year-old ones, which the pastoralists never kill for their own consumption. My informants stressed that they slaughter as few animals as possible: only as many as is really necessary. Such rhetoric is understandable in a Buddhist society, but it is also connected to campaigns led by some Buddhist monasteries which highlight the animals' suffering and the negative karma accrued from killing animals or selling them for slaughter. Posters on walls in the town thematized the cruelty of the animals' death and criticized meat consumption. A leaflet I found in many homes in 2007 showed a yak, a pig and a rabbit who appealed: 'Dear friends, please do not eat our meat! We are killed because of it, and we are very afraid of dying.' Another leaflet showed a yak head surrounded with dumplings and prayer beads, all sprinkled with blood. Again, the text was styled as if it was the animals pleading: 'When will you stop killing? Meat and blood are food of the mountain wolves! Compassion is empty talk of the dharma practitioner. We are suffering without you seeing it, we are crying without you hearing us.' Yet another leaflet showed decapitated yak bodies, sheep heads and animal entrails next to plates with stews, boiled ribs and other dishes. The text read: 'When will you stop killing? Kind and compassionate people, because of the appetite of meat eaters, look at how our fate has changed! Please join those who eat white food!'[13]

These leaflets were signed by Khenpo Tsultrim Lodrö, the leader of Larung Gar Buddhist Academy in Serthar, Kamdze TAP, Sichuan.[14] They are part of the so-called anti-slaughter movement started by Khenpo Jigmed Puntshok, founder of Larung Gar, who urged the pastoralists to stop or limit the sale of livestock for slaughter. This movement and its impacts were analysed by Gaerrang (2012), especially in Ngawa, where the pastoralists took oaths not

13 'White food' or *karzi* denotes dairy products, *tsampa*, rice and other vegetarian food. Its opposite is 'red food' or *marzi*.

14 Larung Gar, full name: Larung Ngarig Nangten Lobling or Larung Five Sciences Buddhist Academy (the name exists also in other versions), was founded in 1980 as a 'religious encampment' or *chögar* and attracted thousands of monastic and lay practitioners. After the death of Khenpo Jigmed Puntshok, Larung Gar has been led by Khenpo Tsultrim Lodrö and several other figures. Khenpo Tsultrim Lodrö's writings were available in Golok bookshops and some also discussed vegetarianism – see Khenpo Tsultrim Lodrö 2004 (a small section translated into English, see mKhan po Tshul khrims bLo gros 2014). On Larung Gar and *chögar*, see Terrone 2008 and Germano 1998.

to sell livestock. It did not go unnoticed in Golok either and its message was given further currency by Tibetan singers who sung about animals' suffering. Drölma Dhondup, whose song 'Nomad's Life' was popular during the time of my research, sang: 'Nomads make pilgrimages to holy mountains and recite mantras but kill animals. That's sad. We aren't so poor. Why should we kill animals?' Video clips of some singers had rather drastic content: Namkha's 'Animals' Karma' featured scenes of slaughterhouse workers killing yaks, with showers of blood spurting from partially decapitated necks.[15] Several people I met had eliminated meat from their diet. Tsering Drölma was one of them. She felt sorry for the animals and preferred to eat *rtsampa*, noodles, vegetables and bread. Wide availability of these products made this possible, though her decision complicated life in the household.

Domkhok pastoralists rarely slaughtered more than five yaks per year. Only one family declared in the survey that they slaughtered six yaks (it was a household of six, all living in the mountains). To calculate the average number of yaks killed per number of householders is a complicated task, since many children study in boarding schools and do not eat at home and others live with their relatives in a township or town. Thus, if counting all household members, the ratio would be less than one yak per person per year. In practice, given the large number of children schooled, this rate is higher. It is likely that the decision of how many yaks to slaughter depends on a household's affluence: those with large herds slaughter more, while young couples put a priority on enlarging the herd and so kill fewer animals. However, the relationship is not that simple. Small households of young couples certainly slaughtered few yaks, but those who had many did not necessarily slaughter more. In any case, the numbers of yaks that the pastoralists in Domkhok slaughtered during my research contrast with reports from the past. Ekvall, from his research in the 1930s and 1940s, noted that a pastoral family in Amdo killed more than ten yaks and fifty sheep per year (1968: 48). Goldstein and Beall reported that in the late 1980s pastoralists in the TAR slaughtered an average of twenty-two animals per family (mostly sheep and goats) and rich ones even as many as fifty (1990: 99).[16] These numbers used to be higher in Golok as well, and the

15 For more examples, see Gaerrang 2011. On Tibetan music and its political content, see Jabb 2011, Morcom 2008 and 2011, Sulek 2017.

16 The pastoralists in the TAR mostly have sheep and goats. About ten per cent of their herds are yaks. Only by converting the numbers of slaughtered animals into kilograms of meat can we make cross-regional comparisons. The observations about declining numbers of animals slaughtered by the pastoralists may not apply to areas without caterpillar fungus. Goldstein noted that many households in the TAR kill more animals than in 1986 when he began his research.

households who during my research slaughtered one yak in summer and four in autumn, would have slaughtered eight in the 1980s and 1990s. All this can be explained by the decline in household sizes and by changes in people's diet, which is now more diversified and includes noodles, rice, potatoes and other products. However, it can also be explained through the wide availability of cash, which enables the pastoralists to buy meat in town and makes them economically less dependent on their own herds.

Pastoral Products and the Market

The products of pastoral production are used not only for the pastoralists' own consumption, but also for sale. They are sold at the markets in Golok or to companies located in more distant towns. The possibility of selling pastoral products on the private market emerged during the 1980s, when economic liberalization reached Golok. Prior to that, the pastoralists produced for the people's communes, fulfilled the production plans sent from higher administrative levels, collected work points and were reimbursed according to their work results. They owned neither the livestock they tended (apart from the *khazhu* animals), nor the land they lived on and could not autonomously decide how to manage the land and its products. The communes 'exported' pastoral products and imported goods unavailable in Golok. The situation changed when the communes closed down and when the Household Responsibility System was introduced. This made pastoral households autonomous production units that made their own decisions as to how to manage their resources and the products of their labours. Households had to supply specified quotas of their products to state trade agencies but were free to manage the surplus as they wished. The emerging private market enabled them to sell it to private entrepreneurs who colonized the space left by the retreating command economy.

During the period of my research, state trade agencies no longer existed and pastoralists sold their products on the private market. However, they complained frequently about the prices and recalled the boom times for certain products as an example of how profitable pastoral production could

At that time, it was rare for a family to have meat left in May, while now 'it seems normal for the middle to better-off households to have some even in early June' (Melvyn Goldstein, email comm., 17 November 2012). The number of animals slaughtered also depends on whether there are stocking rates introduced which force households to sell or kill a certain number of animals each year.

be.[17] Talking about sheep's wool, my informants remembered the period 1986-1988, when one *jin* cost thirteen yuan, while during my research it was only 3.5 yuan per *jin*.[18] Talking about sheepskins, they recalled 1995-1996 when a sheepskin fetched 280 yuan; 20-60 yuan could not compete with it. *Khulu* has also seen better times: in 1998-1999, one *jin* was twelve or thirteen yuan, double the price from when I was carrying out my research. People speculated about the reasons behind this decline. Their explanations usually had some political dimension. Jigmed Dorji, for instance, believed that the prices for sheep's wool were good in the 1980s because Deng Xiaoping's son invested in the wool trade and had contracts for selling it abroad. In discussions about the current situation, this political aspect was even more prominent. My informants blamed the government and some other influential lobbies for pushing down the prices either because it was in their interest to support production elsewhere or – in the most extreme opinion – because they wanted to 'finish off' the pastoralists, Tibetan pastoralists in particular.

Although the prices of pastoral products were a topic of pastoralists' complaints, for some products, the prices were actually higher. This was the case of yak hair or *rtsipa*, for which the market in Golok started only in 2007. The prices from 2010 (fifteen yuan per *jin* of the longest hair from the animal's tail and five yuan per *jin* of usual-length hair) were the highest ever paid. The emergence of this market was important because the pastoralists, who now seldom live in yak-hair tents, do not have much use for *rtsipa* at home. Another product that they did not sell before is milk. Instead of processing it into dairy products at home, they could now sell it fresh. Small Tibetan-owned dairy companies in Dawu opened a new economic niche for the pastoralists. But the owner of one of them complained that

17 100-200 yuan was paid for the hide of a female yak (2-3 years old) and 250-300 yuan for that of a male animal (same age). For yaks sold to slaughterhouses, the biggest males (500-600 *jin* weight) fetched as much as 6000 yuan, smaller males (300 *jin*) 2000 yuan, *drirkam* (300 *jin*) 2500 yuan, younger yaks (3-4 years old) 1100-1500 yuan. Butter was 14-15 yuan per *jin*; *chura* 11-13 yuan per *jin*; sheep were 350 yuan for an average animal (30-40 *jin*), and 800-900 yuan for rams (up to seventy *jin*). Sheepskins were twenty yuan for a skin of 2-3 year old sheep, 30-50 yuan for skins of fully grown females, and sixty yuan for the skins of the biggest males. Lambskins fetched 10-11 yuan.

18 The marketing manager of a carpet factory in Xining assured me that they use only Tibetan wool. In fact, local wool has been replaced by wool from Australia and New Zealand. Mutton served in Golok restaurants was often imported too: Australian mutton was several yuan cheaper. It was also near to impossible to find a Golok sheepskin for a collar for my robe: only when I feigned despair did the tailor organize to get hold of one from his neighbour. What the tailors normally used were sheepskins from Xinjiang.

cooperation is not always easy, as in order to sell milk the pastoralists need to deliver it to town. For some, this is too great an effort, especially given the price they get: 3.5 yuan per *jin*. In fact, milk is one of few products that the pastoralists have to deliver to town. In other cases, it is the buyers or middlemen working for larger companies who visit their settlements. The situation has diametrically changed compared to the past when – as it was until the 1950s – pastoralists from Golok had to travel several weeks to Ngawa and Labrang in order to barter their products for barley, tea and other goods. Today, people can sell their products without leaving home. This is most obvious with livestock: rather than venturing to town, the pastoralists wait for the agents sent by slaughterhouses, some of them located in the cities of Xining, Lanzhou and Linxia, to knock on their door. It is similar with butter and cheese: some pastoralists deliver to shops, but others wait for the Han or Hui wholesalers to come to Golok in the autumn.[19]

It has been observed by scholars working on the Tibetan plateau during the 1990s and early 2000s that the pastoralists became increasingly integrated with the market. The level of this integration differed and areas closer to town, with industry and transport networks, produced more for the market than remoter ones (Manderscheid et al. 2004: 36). It could be expected that Domkhok would also benefit from the proximity of town and would be closely integrated with the market. However, it was not the case, at least not for the market in pastoral products. Many families relaxed their ties with the market and sold only small quantities of their products. They explained this in several ways. First, low prices made selling some products unprofitable. This applied to yak hair and sheep's wool, but also yak hides and sheepskins. One man asked, 'How can it be that a sheepskin costs fifty yuan? And a pair of shoes is 400 yuan and it falls apart immediately? I'd rather throw the sheepskins away and let them get rotten!' Such statements were common and some pastoralists admitted that they throw yak hides away and do not shear their yaks or pluck *khulu*. Shearing sheep and yaks is time consuming and many people saw this time and labour investment as disproportionate to the prices they receive. It is striking that the same logic applied to animal skins, which are a ready by-product of animal slaughter and can be sold untanned. The second explanation referred to the declining productivity of the herds and was mentioned in the context of dairy products. Although some households sold large quantities of butter and cheese (300 and 250 *jin*, say), as much as forty per cent of households did not sell any butter and thirty-seven per cent no cheese. These households were neither

19 This 'door-to-door buying' has also been observed by Manderscheid et al. 2004: 32.

particularly large nor owned fewer livestock than others. Yet, they declared that they consume all that they produce. If they had a surplus, they said, they would prefer to give it to their relatives or donate to a monastery rather than sell it at the market.

The last sentence is indicative of the pastoralists' attitude to the idea of *selling* their products. For some families it was not a matter of not being able to produce more, but not *having* to do so. As one man explained, 'We don't *have* to sell butter and cheese. If you have many women at home, you can do it. Otherwise it's too much work.' Another man said: 'Before, we had no money, so we milked even as often as three times a day. Now we have money and we can afford to do the milking just once. We don't *want* to sell butter.' A woman, explaining why she does not sell butter and cheese, remarked: 'We milk the *dris* once per day. Only families with very few yaks milk twice. People live from *yartsa* nowadays.' Her words reveal that it is the new financial situation that allows the pastoralists to reduce their sale of dairy products. When people do not have to earn money from selling butter and cheese, they can leave more milk to the calves, thus investing in the strength of their herd. As a teacher from Golok reflected: 'The less we milk, the more milk the calves get. It doesn't matter whether the quality of the grassland declines or not. Leaving more milk for the calves is always a good deed.'

Similar arguments appeared in discussions about selling livestock to slaughterhouses. Among the households surveyed, ten per cent sold twenty or more yaks per year, seven per cent sold ten to fifteen yaks and forty-eight per cent sold two to six yaks. The remaining group, as much as thirty-five per cent of households, did not sell any yaks, sometimes for several years. Several young couples, who had only really just started building up their herds, did not have enough fully-grown yaks to sell. Some other households refrained from selling yaks for one year after a close family member passed away; this is a common practice in Domkhok. The majority, however, explained their decision with arguments referring to religion and economy. A twenty-eight-year-old pastoralist said: 'Compassion is part of Tibetan culture. We shouldn't kill animals. It's our lifestyle that makes us kill yaks. But at least we don't have to sell them.' His family slaughtered only one yak per year and bought meat in the town. Other informants were even more explicit: 'We can afford to keep the yaks now', another man stated. Jigmed Dorji responded in a similar tone, 'We have money. We don't have to sell yaks.' His family slaughtered five yaks per year. But they did not sell any: 'We're trying to eat less meat and more vegetables and *rtsampa*. It's a *digpa*, after all.' Yet, he admitted that this would be difficult if his family depended on their yaks as the main supply of income.

The decline in the sale of certain products can indicate the improved economic standing of Domkhok families. Although the households who did not sell animals or dairy products were not always the most affluent ones, their decision showed that they had other sources of income. This income made them free to decide which products to sell and in what quantity. In the case of dairy products, the pastoralists recalled that they reduced their sales between 2003 and 2007. The situation with yaks was similar. Several families had not sold yaks since 2005; Jigmed Dorji stopped selling yaks in 2007, and some of his neighbours in 2008. All these dates are located in the period of the booming caterpillar fungus economy.

Growing Yak Herds

Reducing the sale of yaks creates a risk that the pastoralists' herds could grow too big for the land that supports them. However, it is not easy to tell how many yaks the pastoralists in Domkhok have. Talking to my informants about the numbers of livestock they own was as challenging as discussing their income from caterpillar fungus or land leases. The official statistics do not shed much light on this topic either: they are notoriously erroneous and bring up contradictory data.[20] The 2008 statistics inform, for example, that the number of livestock in Domkhok counted 41,003 head, including 9,462 yaks, 31,032 sheep and 509 horses (MCSY 2009: 44ff). However, an internal document read to me by a township official and reporting numbers from 2009 listed 7,742 yaks, 28,351 sheep and 499 horses. In Domkhok at that time there were 369 households. This implies that an average household owned slightly more than twenty yaks and eighty sheep. Even the township official who read out these numbers called them impossible: a family with so few animals would be too poor to survive, he said. According to his and his colleagues' estimations, given during a private conversation, only very poor households had fewer than twenty yaks, mid-affluent ones had fifty or more and rich ones even as many as 200 (Table 6). This is close to what my survey showed: the smallest herds my informants declared seldom had fewer than fifty yaks, the average herd counted seventy or eighty yaks and

20 One informant said of the local statistics, 'These are fake numbers. For example, schools: normally they're lazy and do nothing, and when they hear that there's a control coming, they go crazy and gather kids from all around the place [to show how many attend the school]. When the control is over, they send them back home. It's just like that. With the environment it's the same.'

some families admitted they had over a hundred. The same people, when speaking about general numbers of livestock in their township and not about their own herd, gave even higher numbers: they said that families with fewer than a hundred yaks were scarce, most households owned a hundred yaks or more, and wealthy families had more than 180.

Table 6 Estimated livestock ownership, Domkhok Township

Households	Officials				Pastoralists			
	% of the Households	Yaks	Sheep	Horses	% of the Households	Yaks	Sheep	Horses
Poor	5%	10-20	0	1	20-25%	0	0	0
Middle-affluent	85% and more	50 and more	50-100	2	50%	100 and more[21]	300-400	4-6
Rich	10% or less	200-280	300	3-6	25-30%	180	600-700	8-9

However, it is not the individual ownership that matters, but the proportion held between households of various levels of affluence. By the average estimation made by the pastoralists, half of local households belonged to the middle group (and each owned around 100 yaks), and a quarter (25-30 per cent) owned about 180 yaks. Thus, the total number of yaks in Domkhok would reach around 35,000, a few times more than what the official statistics say. It should be stressed that estimations made privately by local officials present numbers only slightly lower than the pastoralists' estimations and at least two times higher than the published statistics compiled by the officials themselves. All these figures are afflicted by errors of over- or underreporting. The county statistics report numbers lower than the actual ones on the ground. The township statistics are still low, but closer to the reality. In informal conversation, the same township officials who participate in producing the official data report higher numbers to show their township as being successful and affluent. The pastoralists, on the other hand, are interested in hiding both their own poverty and any excessive wealth: they use round numbers and vague expressions to describe the size of their herds. In this situation, other sources of information, those well informed but not directly involved, are helpful. Such a source was a local veterinarian who

21 My informants stressed that the yak numbers represent the *real* situation and the sheep numbers an *ideal* situation which is no longer observed, since so few households have sheep.

admitted that as much as forty per cent of yaks in Domkhok were those beyond the official numbers. The 'regular' animals received vaccination at public cost. All the others were vaccinated at the cost of their owners: the numbers of vaccinations he distributed and sold are informative about the number of yaks really living in the township.

The fear of disclosing the size of one's herds is understandable in the atmosphere created by current environmental policies and campaigns in China, which blame the pastoralists for fostering the process of land degradation. The pastoralists' livestock management system is described as irrational and destructive, and it is stated that it should be replaced with a modern one, based on scientific and rational methods; the rhetoric used in this context is a faithful copy of the one applied to caterpillar fungus. It is stated that numbers of livestock should be adjusted to the scientifically calculated carrying capacity of the land, which says that a given area can only support a certain number of livestock. In Domkhok, this is one yak per thirty-two *mu* and one sheep per eight *mu* of land.[22] The local authorities have to ensure that no one exceeds the limit. If this happens, the owner of the herd can expect fines, confiscation of livestock or withdrawal of their land lease contract. Such legal steps are seldom taken, but they create an atmosphere that discourages people from talking about the numbers of animals they have.

That this environmentalist rhetoric can be used by the Chinese state to justify direct interventions into the pastoralists' life is clear in case of the state sedentarization programmes implemented in Golok. The first of them, called Turning Pastureland into Grassland (Chin. *Tuimu huancao*) was announced in 2003 and stipulated that in order to restore the ecological balance on the grassland, restrictions should be placed on the scale of its use. Populations in the areas declared ecologically degraded or under threat of ecological degradation had to move to permanent settlements in or near the town. The second programme, Ecological Migration (Chin. *Shengtai yimin*), was launched in 2004 and had a further goal: to alleviate poverty and improve the standard of life in the pastoral regions.[23] This programme also implied that the pastoralists had to leave their land. Depending on the status of this land, which was covered either by a permanent, or temporary

22 The stocking rates are calculated with so-called sheep units (SU). A basis for the SU is one adult female sheep, which needs four kg of hay per day. Other animals are 'converted' as follows: one yak = five SU, one horse = six or seven SU (cf. Miller 2001; Yan 2005; Goldstein 1996). The SU has its equivalent in Mongolia (the sheep stocking unit or SSU; see Humphrey and Sneath 1999: 42).

23 This should not be confused with *Youmumin dingju*, the Nomadic Settlement programme, which had a similar goal, but did not imply that the pastoralists had to leave previous settlements.

grazing ban, they could or could not expect to return there in the future. In order to begin a new life, they received a house as well as some financial aid from the state. However, this aid was often insufficient, especially because before moving to town most pastoralists sold their herds and lost their subsistence basis.[24] It was a common concern among town dwellers and the pastoralists alike that the lack of financial support and other incentives that would help these migrants integrate with their new environment would push them into poverty, alcoholism and crime. People argued, similar to scholars who wrote on this topic, that poverty and psychological distress are a high price, which these people pay for accomplishing the goal of environmental protection.[25] The pastoralists themselves stressed that removing the livestock from the land does not automatically bring about its recovery and they observed that the state institutions that are responsible for implementing the sedentarization programmes do not care about implementing or controlling the grazing bans. A member of a local environmental NGO remarked:

> People have to sell their yaks and move to town. So, the problem is seemingly solved and the land saved. But don't imagine that any officials check whether the quality of the land then really improves. They just calculate that it will improve, for example, within ten years. But they don't know anything about it. They just calculate an abstract number. Not all nomads care about the land, but the Party uses the 'environment' as an excuse to move the nomads to town and destroy their culture.

The pastoralists in Domkhok are aware of the current policies not only because they observe the impact of these on neighbouring communities. They are also targeted by information campaigns propagating environmental protection. The most immediate visual side of these campaigns is billboards in Dawu and along the township roads. Whereas those in Dawu show a range of topics, such as family planning, social harmony and education, those in Domkhok speak only about ecology. They show mountain landscapes and grazing animals and say (in Tibetan and Chinese) things like, 'Conserving water and soil is the way to improve the environment' or 'Protect the natural environment and continuously increase development of social economy'.

24 These programmes were attractive to some pastoralists whose children or grandchildren worked or studied in town. Informants from various parts of Golok reported that some people volunteered to move, others were selected in a lottery. Some of them returned to the highlands later or moved to another house and rented out the new one as a shop or to other families. On the financial situation of the relocated pastoralists, see Sodnamkyid and Sulek 2017.

25 On this topic, see Yeh 2005; Zukosky 2007; Foggin 2008; Du 2009; Xun and Bao 2008.

Environmental protection was also a topic of meetings organized in the township where people were lectured about their role as guardians of nature. As Tendor said, they leave these meetings with phrases such as 'We nomads need to protect the environment' ringing in their ears.

To check whether the pastoralists do not own too much livestock, the Agriculture and Animal Husbandry Bureaus and other government units in the township, county and prefecture organize periodic controls on the grassland. However, as with the controls on the use of caterpillar fungus resources, the pastoralists receive information about these controls in advance. As one man revealed, his relative in the county calls him and says: 'Hide some yaks on this or that day'. The news spreads fast and the yaks disappear into the more remote valleys: it is impossible to count livestock dispersed over such wide territory, just as it is to check the numbers of the caterpillar fungus diggers working on it. The low-level administrative bodies responsible for implementing these controls are also interested in producing data that shows their successes in implementing the state policies. All township officials I interviewed were Tibetans from pastoral families, mostly from Machen. They identified themselves as pastoralists and most of their relatives continued the pastoral life in the highlands. Not surprisingly, these officials were interested in defending the local inhabitants' interests against the demands of the higher levels of government. Thus, the numbers featuring in official statistics are often a result of collaboration between the pastoralists and their representatives in the township or relatives working at higher administrative levels.[26]

Pastoralists' preference for large herds, a practice known from many parts of the world, has been explained by scholars in cultural and economic terms (Farooquee 1998; Levine 1999; Yan et al. 2002; Manderscheid 1998). Large herds are an insurance against livestock losses: the larger the herd, the bigger the chance that part of it survives outbreaks of disease or weather disasters. The size of the herd also compensates for the low productivity of the animals: if the animals reach reproductive maturity late, only a certain percentage of females have offspring in a given year. The herd is also a 'bank on the hoof' in which people keep their economic assets. Finally, it serves in identity and status building: the pastoralists are people who raise livestock, and their number indicates economic and social standing. In the

26 A man who was an accountant in Namtso in the TAR described this process: low-level officials produce inaccurate data and higher-level officials 'correct' it, adding or detracting something based on their understanding of what their colleagues have done before. David Holler, pers. comm., Berlin, 28 September 2012.

case of Golok, these reasons can be added to another one. The tendency to accumulate livestock has been observed, especially during the 1990s – i.e. in the post-collectivization period when the pastoralists received back livestock that had been confiscated from them before. The experience of forced collectivization is alive in individual and collective memory and the accumulation of livestock after the closing down of the communes could be a reaction to that traumatic experience. The 1990s were recalled by my informants as the golden age of the pastoral economy. During the period of my research, this golden age was part of history. The pastoralists had increased their dependence on non-pastoral income and livestock was not as crucial for their economic survival as it used to be. But the more the pastoralists depended on non-pastoral income, the more they needed their herds to preserve their identity. Since the pastoralists in Golok thought of themselves as primarily yak breeders, it was the yaks that received special attention. Sheep did not receive the same treatment.

Disappearing Sheep

In the pastoral economy of Golok there was one more branch that used to tie the pastoralists to the market with particularly strong bonds: sheep farming. Sheep are a reliable source of income and can be sold almost anytime between late spring and winter; they grow fast, reach reproductive maturity early and regularly have offspring. In comparison, yaks need a longer time to grow and are only sold in autumn. They bring a large income but only once a year; one cannot count on them in a financial emergency. The link between sheep and commercial production was strong in the people's communes, where sheep farming was encouraged (Manderscheid 2002: 279) and remained so when the Household Responsibility System was introduced. Yet, later developments showed that sheep are not the local pastoralists' animal of choice.

During my research, most households either did not have sheep or had only those with *tsethar* status. The sight of sheep flocks in Domkhok became so rare that when I travelled with a retired county official, he stopped the car to take a photograph of a woman taking a flock of sheep to graze. 'That's how it used to be before', he said. He estimated that within the first decade of the 2000s the sheep population in the area declined by 40-60 per cent, and that in the 1990s less than five per cent of households were *without* sheep. My survey revealed that by 2010 this situation had reversed and that *not having* sheep had become the norm: only fifteen per cent of households still had them. The remaining households had either never had sheep (nineteen per cent) or had sold them off (sixty-six per cent).[27] The former group consisted of young couples who on their marriage did not receive sheep from their parents or in-laws, while the latter made a conscious decision to sell their flocks off.

The reasons why people sold their flocks off varied. The pastoralists generally said that raising sheep is difficult for reasons ranging from those related to climate and vegetation to demographic and administrative factors. The biggest problem was the lack of labour force. Herding sheep is labour intensive and most households are not large enough to manage it. This work was usually done by children in the past, but today most of them are in school. Domkhok, a township located close to the town, is an easy target for controls from the state institutions supervising children's education. But this proximity to town also encourages parents to invest in their children's education. As a local doctor said, Domkhok is 'the first place after Dawu to make controls and enforce policies. But people also get information from the town and this increases their awareness of the importance of education. As a result, the number of children they send to school is higher than elsewhere.' Among the surveyed households, almost half had all their children at school, twenty per cent had more than half of their children at school (two out of three or three out of four), eight per cent had half of their children at school (one out of two or two out of four) and four per cent had less than a half (one out of three). The children attended the primary school in Domkhok or a primary, middle- or high-level boarding school in Dawu. One family had a son at university. Parents who did not send any children to school were looked down upon by their neighbours, who believed that some degree of education is indispensable in today's world. Many parents hoped that it would help their children find employment in the public sector or private business later. There is no chance that they would be able to support themselves in the pastoral economy and it is their parents' responsibility to facilitate them with a start into other avenues than pastoral life, as Tsering Drölma believed.

The pastoralists also blamed the climate and quality of grassland for making sheep production unprofitable. Compared to other parts of Qinghai, Golok is not a place for sheep, as my informants noted. Qinghai Television regularly broadcast award ceremonies where the title for best sheep breeder

27 This excludes *tsethar* sheep. My informants who only had such sheep would declare that they had none and listed only those sheep that contributed to their budget. On 'disappearing sheep', see Sulek 2010a.

would always go to someone in Malho or Mtsojang TAPs: the impressively sized sheep there aroused jealousy in Golok and people commented that their sheep would never reach that sort of size. Climate and altitude were part of the problem. Another issue was the declining productivity of the land. As the pastoralists said, even if theoretically they had a large area of pastureland, it did not produce enough fodder: 'One needs plenty of land to have both sheep and yaks. But our land seems to be shrinking. The grass is getting thinner and thinner every year. The sheep didn't have enough to eat.' The last problem that my informants mentioned was a ban on private ownership of firearms introduced in 1996: 'First in 1958 and now again they took our rifles away. How shall we fight the wolves?', one elderly man asked. All these problems have had a cumulative effect; they reinforced each other and led to the conclusion that sheep production is not viable. As one man said:

> We have no grass and no land. Sheep are difficult to raise. To be honest, they are very hard to raise. Wolves attack them a lot. Yaks are stronger and can fight back. The sheep die a lot because this is a high place. It is different in lower-lying areas, where breeding sheep is an excellent business.

Apart from investigating why some pastoralists sold off their flocks, it is worth asking why others did not. These persons presented arguments of an emotional or 'cultural' character. They declared their attachment to the pastoral tradition and said – like Jigmed Dorji – that they enjoy breeding sheep. The households that continued sheep farming were not substantially different from those that had disposed of their flocks. They were neither bigger, nor had more and better land. They faced similar problems to their neighbours but overcame them by either employing extra herders or leasing additional pasture. The case of two brothers who lived side by side and yet came to different conclusions on the matter illustrates it well. One of them sold his 170 sheep because he did not have enough land and the family was too small to manage the work. The other brother had 200 sheep (and around a hundred yaks), the same amount of land and only two working people at home. Though he admitted that breeding sheep is difficult, he did not hesitate to spend 10,000 yuan to hire a herder and another 5,000 yuan to lease pastureland from a neighbouring *richung*. His case shows that even in adverse conditions sheep production can be continued, but it requires commitment, creative solutions and sometimes significant financial input. However, it can also bring profits: from selling fifty sheep his family earned over 17,000 yuan per year. During the survey, they said that they planned to enlarge the flock and buy another forty animals.[28]

What helped the sheep breeding pastoralists to overcome the shortage of labour was the practice of employing herders (*zogdzy*). This is what Jigmed Dorji did. He only had daughters and preferred that they stay around the home rather than herd sheep and yaks in the mountains. The first herder he employed was Sampe, now his son-in-law. Sampe came from a farming family in Tunde and arrived in Golok in 2007. He was recommended to Jigmed Dorji by one of his villagers, who dug caterpillar fungus in Soglung. Sampe was seventeen back then and wanted to study, but his family's financial situation forced him to go and look for work. Working as a herder in Soglung he earned 10,000 yuan per year and had his winter clothes and shoes paid for. He also dug caterpillar fungus. This was the main reason he wanted to come to Domkhok. Many households suffer from a lack of working hands, but it also became a status symbol to have someone to help at home. Tsering Drölma and Tendor also had a herder from Tunde. They did not have sheep and the amount of work in their household was smaller, but they decided they needed an extra pair of hands. Tsering Drölma had children to look after (by the end of my research they had their second child) and Tendor could not manage all the work by himself. But larger families that had enough young and able-bodied men also hired herders. Several families from Domkhok moved to Dawu and left their livestock entirely in the hands of hired herders; they visited their land to check whether the work was being done properly. In spring, many of these herders dug caterpillar fungus together with their hosts. Some of them seemed to be ready to forgo their salary and took caterpillar fungus as their only payment. Jigmed Dorji paid Sampe for his work and agreed that he could also dig fungus, but some pastoralists were less generous. 'Most *zogdzis* come here because of *yartsa* and not money. If they get cash, they can't dig. It's either or', one man said. Just as with the immigrant diggers, the cooperation did not always go well. One woman recounted what had happened to her neighbour:

> Gonto's family took on a herder, and he stole all six of their horses. Since then I'm afraid to have a herder. They can be nasty. They blackmail you and demand that you give them five calves out of every ten born. My herder also came to me with such a demand. I said: 'No way'. If they stay with you, they want to dig *yartsa* for free. You have to pay them, as well as buy them clothes and shoes. They come here because of *yartsa*. Some

28 These two brothers and their economic strategies were discussed in Sulek 2010a: 16ff.

> don't even ask for money. They stay for five or six days and disappear later. You'll never see them working. It happened to me too. I've had maybe forty herders! Chinese, Tibetans, Muslims. Tibetans are the best.

The *zogdzi* employment market is connected to the development of the caterpillar fungus trade. The income that the pastoralists earn from it makes hiring herders affordable. This arrangement serves both sides: the pastoralists receive help with their work, and the herders get access to the grassland during the caterpillar fungus digging season. That the emergence of this employment market is linked to the caterpillar fungus economy is confirmed by 'The Executive Programme' (see Appendix) regulating the use of the caterpillar fungus resources. The document recognizes the practice of employing herders as part of the '*yartsa* problem'; it forbids this and states that 'all the herders' committees and farms in townships with caterpillar fungus resources have to register and turn away outsiders who have entered our county under the pretext of assisting in work and grazing' (Appendix: 11).

For many pastoralists, investing money in hiring herders was a conscious step that allowed them to preserve their tradition. However, it is much more than the power of sentiment that has made them continue to breed sheep. Jigmed Dorji believed that diversifying one's income is a safer strategy than relying on one source alone, especially such an unpredictable one as caterpillar fungus. It was clear to him that it was because people supported themselves with the caterpillar fungus economy that they had stopped breeding sheep. The income from sheep production cannot compete with the money from caterpillar fungus. Caterpillar fungus money is earned within one season, while animal husbandry is a year-round job: the herds require a constant input of labour and preparing the products for sale also takes time. The case of sheep wool demonstrates this. In order to shear 200 sheep, Tendor calculated, one needs the cooperation of several men. The work, if done by ten people, takes up to six hours. Adding to this the trip to town to sell the wool, makes a full day of work, but brings in only 1,400 yuan (in 2010), a sum that many people judged not worth the effort. Working for a small income in order to preserve the pastoral tradition is admirable, but not everyone was ready to do it. For instance, one woman remarked: 'Having both yaks and sheep? It's too much work. Why should we make extra work for ourselves if we have enough money?' The pastoral tradition was important for my informants, but they thought they practised it sufficiently by breeding yaks. Yaks supplied products important for the pastoralists' diet and lifestyle and were vital for their identity. Sheep, which acted as quick cash producers, were replaced in this role by caterpillar fungus and became redundant from an economic point of view.

The fate of sheep is connected to the rise of the caterpillar fungus economy: this is evident from the way people answered the question about whether they would sell their flocks if they did not have caterpillar fungus. Many of them said, ‘No. What would we live from then?’ The income from caterpillar fungus is thus an extra factor that has changed people’s optics of breeding sheep. The trend for selling off flocks developed in a certain demographic and environmental, but also economic context. It is a result of the interplay of many factors, but without the income from caterpillar fungus these factors would not have such importance. Only with this income could the pastoralists allow themselves to sell their flocks. That the phenomenon of ‘disappearing sheep’ is a consequence of the pastoralists’ improved financial standing is clear when one analyses its chronology. The first recorded cases of the selling of whole flocks come from 2000. The trend of selling flocks made a spectacular jump in 2006 and slowed down by the end of the decade. This was confirmed by the largest slaughterhouse in Qinghai, which in 2006 processed over three million sheep, mostly from Golok and Yushu.[29] After 2007, these numbers declined. My research confirmed it: none of the households I surveyed sold their flocks after 2008.

Returning to the environmental policies implemented in Golok, it is worth asking whether selling off sheep is connected to the politics of pressuring the pastoralists to reduce their livestock numbers. Surprisingly, none of my informants suggested that this was the case or signalled that they (or anyone else they knew) were forced to sell their animals. Confronted with a direct question such as whether their decision to sell their sheep followed the authorities’ wishes, people showed their surprise. Interestingly, the same people were very outspoken and even critical about the state’s interference in their lives, but when it came to selling off their flocks of sheep, they emphasized that they made the decision autonomously. A Golok singer, critical of the state’s interventions in pastoralists’ lives, was also convinced that it was not the state that stood behind the disappearance of sheep from Golok grasslands:

> People earn money from *yartsa*. They don’t want yaks and sheep. They want a comfortable life. Many people say that raising sheep is difficult and the money from it too small. *Tuimu huancao* [the Turning Pastureland into

29 The company processes about 2,400,000 sheep per year. In 2006, they processed around 3,500,000 sheep (interview, Xining, 15 June 2010).

> Grassland programme] could have had some influence but a very small one. Just look at *yartsa*. The government forbade digging, but people are still doing it. You can't stop them. If people really wanted to raise sheep, they would have found a way to do so.

The meetings and lectures on environmental protection in the township created an atmosphere that favoured the decision to sell off the sheep. But in a situation when nearly the whole population engaged in economic activities proscribed by the caterpillar fungus-related law openly ignored it and successfully evaded its implementation, such dutifulness in following the policies of reducing livestock numbers would be a turn up for the books. The decision to sell off sheep can only superficially be taken as a sign of the local pastoralists' adherence to the livestock reduction policies. This becomes clear when one realizes that selling sheep did not lessen the pressure on the land. As several informants admitted, by selling their sheep off they won extra space on the grassland. This space left by the sheep was filled by yaks: 'If you want to have sheep, you need plenty of land and good grass. My land was not enough, so I sold them. But I bought more yaks instead', one man said.

Lazy Nomads

The declining enthusiasm for the pastoral economy was widely debated by the pastoralists, who raised questions about their identity and pastoral life. People who sold more livestock and livestock products criticized those who sold less or none, claiming that they were not 'real nomads'. Tensions between neighbours were felt in many situations. In Soglung, before I visited the last house in the valley, which was inhabited by Sonor and his family, people warned that he would not let me in. He is a 'bad guy', they said and added that 'he prefers talking to yaks'. Sonor, a man his forties, was indeed strongly committed to pastoral work and stressed that his children will continue it. His family sold the largest quantity of dairy products in the valley. He also declared that in 2010 they sold twenty-seven yaks to a slaughterhouse. Whereas his neighbours criticized him for being a 'bad Buddhist' and a 'backward man' (since none of his children went to school), he answered them tit for tat and said that they were not real pastoralists: 'I think that being a nomad means not only that you keep livestock but live from it. If someone lives from *yartsa* and keeps livestock for himself, is he a nomad? I don't think so.' Similar opinions were voiced by people who continued breeding sheep about those who had stopped doing it. Jigmed

Dorji remarked: 'For me a nomad is someone who makes a living raising yaks, sheep and horses. Otherwise you aren't a nomad. I don't want to sell my sheep. On the contrary, I want to buy more. Why do people sell off sheep? I think it's because they have money from *yartsa* and have become lazy.'

The concept of laziness recurred in discussions about many aspects of pastoral work. Talking about yak hides or sheepskins, one man said: 'People don't sell them anymore. They throw them away. Otherwise they would have to tan them or bring them to town to sell. They don't do it because they're lazy.' Speaking about the *drekang* boxes in which new-born lambs were kept, a young woman observed: 'You don't find them anymore. People got lazy.' This loss of the ethos of pastoral work was one of the fundamentals of the critique of the social effects of the caterpillar fungus boom: it taught people that they can earn money without any effort. This opinion was voiced by many pastoralists, both young and old, but even more strongly by monks and other educated and influential folk. As a monk from Rarja monastery stated:

> Traditionally, nomads kept livestock and lived off the money from selling livestock and livestock products. But the quality of the people really declined. One needs to do just a little bit of work to dig *yartsa* but earns a lot of money. People got lazy. They don't want to have livestock. They don't want to work. This makes me angry. *Yartsa* really has a bad impact on our society. For example, there was some sense of protecting environment before. There's a saying, 'Tibetans depend on the yak. The yaks depend on the grass.'[30] This kind of awareness has been lost. People now are only interested in earning money.

Before the caterpillar fungus boom, pastoralists engaged with the market mainly through the sale of livestock and livestock products. Later, their engagement with the market of pastoral products declined. This can be explained in different ways. Gaerrang, in his analysis of the anti-slaughter movement, showed how pastoralists took oaths to reduce or stop the sale of livestock for three or more years. He argued that this decision challenged the state vision of development and that the pastoralists, by taking a step out of the market, fostered an alternative vision of development based on their own understanding of the world and a value system that contests capitalist development (2011: 32, 41). This diagnosis is only partly true. In

30 'Tibetans [lit. black haired] depend on the [black hair of the] yak / the yaks depend on the grass [or grassland]' or *mgonag rtensa funag re / funag rtensa penfu re* (*mgo nag rten sa spu nag red / spu nag rten sa spang spu red*).

Domkhok, the pastoralists relaxed their bonds with the market, but only with the market of pastoral products. Their decision not to sell livestock did not imply a complete renunciation of their commercial activities. On the contrary, it was their commercial success in another field that allowed them to stop selling yaks and reduce the sale of other products. The same applies to slaughtering yaks for home consumption: the pastoralists slaughtered fewer animals because they bought meat in town. They could manifest their religious sentiments because they were financially protected. Losing part of their income did not endanger their financial situation: they earned enough from caterpillar fungus to be able to afford this freedom. By selling their flocks of sheep and reducing the sale of yaks, the pastoralists challenged the state vision of the pastoral economy as focused on commercial production of meat for the market. But this did not make them *less* integrated with the market. They were so strongly connected to it through the caterpillar fungus economy that even while not selling anything else they did not become 'market outsiders' who defy the market, commercial production and consumer life.

However, the pastoralists' economic decisions are not only about money. Many informants perceived the digging of caterpillar fungus to be problematic. Not only was some degree of ethical ambiguity associated with the practice, but some also considered money earned from it as dangerous. Building one's fortune on such an activity can be risky and some people said that this wealth must be symbolically repaid. Seen in this context, the pastoralists' decision not to sell yaks for commercial slaughter can be interpreted as the manifestation of an 'economy of sinning'.[31] This denotes a mechanism of thinking that makes people measure their economic actions according to the positive or negative value or *digpa* that they create, and that makes them aware of a need to balance the accounts. In this particular case, the pastoralists reduce the sale of yaks to abattoirs, a practice that they perceive as ethically negative, while engaging in another activity that is also, at least potentially, negative: digging caterpillar fungus. Thus, they minimize their *digpa* accrued from the field of pastoral production to compensate for its growing in another field: the caterpillar fungus economy. My informants often said that they could *financially* afford to keep yaks instead of selling them. They could perhaps add that in terms of *digpa* they *cannot* afford to sell yaks when their engagement in the caterpillar fungus economy makes their position volatile and exposes them to sometimes difficult to predict negative effects.

31 For more on this topic, see Sulek 2016.

9 Spending the Money

Tibetan pastoralists are famous for their tents of black yak-hair cloth stretched over a complex structure of long ropes and external poles. Travellers of the past compared them to spiders, but Tsering Drölma said that they rather resemble turtles.[1] These tents are an important identity marker, not only for the Tibetan pastoralists as such, but for people inhabiting various different parts of pastoral Tibet. The tents that the people of Metsang have, as my informants stressed, are different from those the Goloks use. It is possible to find three types of tent in Domkhok, which differ in outline, shape of their covering and size of the living space inside. The *rinag* is square in its ground plan, large and spacious, and its cover is sewn from vertical stripes. The *chobra* is rectangular and its cover made of horizontal stripes. The third type, *ramari*, mixes the features of the other two: it has the plan of the *chobra* and a cover made as in a *rinag*. This is the sort of tent Tsering Drölma had. Her uncle's family lived in a *rinag*.

Jigmed Dorji's tent was one of the oldest in Domkhok: it was sewn by his grandfather in the 1940s. Although it had been repaired countless times and there was almost nothing of the original model left, in the memory of its users it was the very same tent that had been in the family for generations (Figure 21). Everyone in the township could tell whose tents were the oldest. However, in spite of their large sentimental and historical value, the black tents were not in common use. The pastoralists used them as an all-season dwelling until several decades ago. Even in the people's communes, as they noted, the tents remained in use: although confiscated (as was much other private property) at the beginning of the collectivization, they were returned to their owners later.[2] Things started changing during the 1980s when the pastoralists built their first houses on their winter quarters.[3] They then lived in houses for the greater part of the year and moved into tents for the warmest months. In the 2000s, the retreat of the black tents went one step further and they started disappearing from the summer land as well: people replaced them with mass-produced wall tents which, they said, were lighter, simpler and cheaper

1 On Tibetan tents, see Manderscheid 2001; Schuyler 1996; Sulek in press.

2 This does not apply to people labelled 'class enemies' in the class division introduced in the late 1950s. These persons, often from families of religious and political leaders, faced much more complicated life scenarios during this period.

3 There were families in Golok who lived in tents in winter, but not in Domkhok. These were usually young couples who lived in a tent for two or three years before moving into a house. My informants in Domkhok took this as a sign of poverty.

Figure 21 Black tent, an important family heritage

to maintain.[4] During my research many families used only white wall tents. Tsering Drölma and Tendor used both: their daily life was concentrated in the black tent, whereas the new tent was largely used for putting up guests.

Crazy about Houses

Houses are not an entirely new form of architecture in pastoral Tibet. In many regions, pastoralists also built houses on their winter pastures in the past (Goldstein 1990: 64; Iselin 2011: 432; Ekvall 1977: 59). This has not been the case in Domkhok and my informants recalled that the first permanent shelters in the township were built there not for people but livestock. This was part of the state project of modernizing the pastoral economy, which included introducing new animal breeds, winter fodder and animal shelters. These shelters, people recalled, gave them their first chance to experience the benefits of having something more than a tent roof above one's head. Some informants admitted that they moved into abandoned animal shelters in the

4 Three yaks are needed to carry a tent: one for each half of the tent cover and one for the wooden parts.

early phase of decollectivization. Others said that people did it even earlier. But the first houses *for people* were built only after the communes closed: it was decollectivization that set the sedentarization process in motion.[5]

Two decades later, houses dominated the architectural landscape in Domkhok. The pastoralists stayed in tents in summer, but apart from that they lived in houses. More than that: only young, recently married couples had just one house. More established families had more than one, as the examples of Tsering Drölma's and Jigmed Dorji's families show. Jigmed Dorji's parents built their first house in 1985. The second one he built with his wife in mid-1990s, and in 2007 they finished building another. Tsering Drölma and Tendor, although much younger, had two houses: the older one, from the 1990s, which Tsering Drölma had inherited from her mother, and a newer one that they built shortly before my research began.

Pastoralists in the vast Golok highlands are not troubled by a lack of space for construction and rather than dismantling their old houses they merely add new ones alongside (Figure 22). The fact that all houses sit together in the pastoralists' living areas gives an impression of people having a *collection* of houses, but it also allows us to observe how the houses have changed. The main difference is seen in construction materials. The walls changed from being built of earth to concrete during the 1990s and early 2000s and to bricks in the newest houses. The size of the houses has changed too: whilst older houses were about 50 m^2, newer ones have at least 80 m^2. In the oldest houses, there were one or two rooms and only one stove. In the newer ones, three rooms and two stoves are common. In older houses, windows were small and few, whilst in the newer ones they cover a large part of the wall, letting more light in. Gable roofs have taken the place of older shed or flat roofs, and ceramic tiles have replaced corrugated metal and pounded-earth roofs. Rammed earthen floors have been exchanged for cement floors covered with PVC. Finally, newer houses have been aesthetically transformed due to their façade decoration: they receive a covering of ceramic tiles with motifs of lotus flowers, dragons and even portraits of Buddhist personages, such as the 10th Panchen Lama who 'guarded' the entrance to Tsering Drölma and Tendor's house. Though house construction techniques have evolved, many principles remained unchanged. The houses are still rectangular in their ground plan. Organization of the inner space is determined by the fact

5 Until the mid-twentieth century, many monasteries in Golok functioned in tents. Guru Monastery, the closest one to Domkhok, was also a tent monastery until the first buildings were built in the 1980s ('Phrin las 2008: 24; Buffetrille 1997: 78). Gruschke says that there was a temple hall there in the 1950s (2001: 75). My informants did not confirm it.

Figure 22 A colony of houses; all buildings belong to only two households

that the windows face in one direction, whilst the back of the house sits against a hillside or a mountain slope. In this respect, the houses replicate the manner in which tents are pitched, with their entrance facing the bottom of the valley, often to the south, and their rear walls facing the slopes behind them. Thus, these houses are not expanded in terms of their depth, and all rooms are arranged in one row.

Tsering Drölma and Tendor's old house had only one room. The furnishings were simple: a stove in the middle, a bench and small table next to the window, and cupboards on the rear wall (Figure 23). Cooking pots, tableware, boxes with *rtsampa*, cheese and butter are kept in the cupboards, as well as bigger food reserves such as sacks of flour or rice (Figure 24).[6] Next to the door there was a washing bowl and a mirror, containers for water, fuel and waste, and a freezer containing meat and vegetables. A large platform in the other corner was a family bed. A small television waited for those who wanted to watch the evening news, and a bulb hanging from the ceiling gave some dim light in the evenings.

6 This resembles how the space in tents is organized, with sacks of grain and butter piled at the rear wall (Norbu 1997; Manderscheid 2001). This arrangement was partly transmitted into the house.

Figure 23 In the old house

In the new house, the space was divided into two rooms with a gallery-like corridor in front of them. The corridor, with window panels along its sides, was a storage space and extra living area, albeit lacking heating and being modestly equipped – a metal bed in the corner and a washing bowl is all one could find there. This is where I slept, under shiny posters of Mao Zedong as well as Hu Jintao and Wen Jiabao, a required sign of political loyalty that could be found in many homes. From the corridor, one entered the big room, and from there another, smaller one. The big room was furnished with piety and with no concern about costs. There was a shiny stove, hand-carved furniture and expensive electronic equipment (Figure 25). Along the front and side walls ran low wooden benches and sofas covered with Tibetan carpets. A row of tables stood in front of them. The tables and seats were made in 'Tibetan style', as with the main piece of furniture on the back wall: a sophisticated wooden cabinet, with shelves and drawers, and extra space in the middle for a television. The cabinet was carved and painted with snow lions, wind horses and other creatures of Tibetan imagery. Along its upper edge, a gallery of paintings and printouts showing Buddhist deities and religious personages was arranged with a large portrait of the Dalai Lama in the middle. A large metal safe stood in the corner: this is where the family stored important documents and

Figure 24 **In the tent**

Figure 25 **In the new house; as well as all the sweetmeats and new furnishings, note the portrait of the Dalai Lama, a television stand and a metal safe in the corner**

valuables, as well as caterpillar fungus. Every family had such a safe: it stood in the most prominent place in the room, often close to the family altar. In the smaller room the furniture was less expensive, but the space was organized in a similar way. There was a stove and a cabinet on the wall with a collection of religious pictures, electric butter lamps and bowls with offering water. A bed in the corner gave this room a more informal character. The room was used when Tsering Drölma's mother visited. It was similar in many families: the small room, warmer than the big one, was used by a senior family member.

These two houses appeared to offer sufficient living space for Tsering Drölma, Tendor and their small daughter, but before my research ended the couple had started building their third house. This decision was connected to the state programme called Nomadic Settlement (Chin. *You mumin dingju*). Launched in 2009, this programme aimed at improving living conditions in pastoral regions. It entitled the pastoralists to receive a state subsidy to build a house, but – other than the sedentarization programmes – it did not require that they leave their original living place.[7] Formally, the programme was aimed at families without a permanent house or those who lived in a house made of rammed earth or wood, especially if the house was old and in a bad condition. However, there was no one in Domkhok without a house and the quality of their houses was already higher than the programme stipulated. But many of these houses were built of concrete that, after the 2010 earthquake in Yushu, was labelled a dangerous material. This is why, as the county officials explained, the programme covered all residents.

There were certain rules that the pastoralists had to follow to receive this subsidy. The house had to be built of brick and offer at least 80 m^2 of living space. It also had to be enclosed with a two-metre-high wall: this was a novelty, since the older houses were seldom fenced. Building of a house fitting these requirements cost, based on the accounts of the households I surveyed, between 110,000 and 150,000 yuan. The pastoralists raised this money themselves. Part of it was reimbursed only later. In Domkhok, this reimbursement reached 40,000 yuan: 36,000 were paid from the provincial budget and the rest from the prefectural budget.[8] The decision of whether this money could be reimbursed was made during an inspection tour organized by officials from the Animal Husbandry and Financial Bureaus and other

7 For more on this programme, see Ptackova 2011.

8 This is more than in the TAR (*c.* 15,000 and 24,000 yuan; Yeh 2013a: 239; Goldstein et al. 2010: 67) and Sichuan (20,000 yuan; Ptackova 2011: 7).

bureaus in the prefecture and the township. In October, when construction works stopped, the inspectors would travel through Domkhok to inspect the new houses. A county official showed me photographs taken during such a tour: some buildings were empty, others were filled with the first furniture, with pastoralists posing next to their houses and officials measuring the walls. But if the pastoralists had to raise the money to construct their houses themselves, a question arises about how they managed to do it. Tendor was open about this: he and Tsering Drölma used savings made from caterpillar fungus. He admitted that they had saved at least 50,000 yuan during one digging season. His neighbours declared similar sums (20,000-50,000 yuan) or higher (150,000 yuan) but believed that some families saved as much as several hundred thousand yuan a year. Whereas the size of their savings certainly varied, there is little doubt that the money from caterpillar fungus fuelled this construction. It did so either directly, when people had ready cash to spend, or indirectly, when they took a bank loan and paid it back with the income from the next caterpillar fungus season.

The pastoralists' interactions with the banks, as with many state institutions, are little studied. To what degree do the pastoralists use banking facilities? Do they keep savings in banks or rather distrust them? Scholars working in other parts of the Tibetan plateau have been alarmed that the various state housing programmes force people to take out bank loans, which they are unlikely to be able to pay back in time and that will damage their fragile household economies (Yeh 2013a: 246). Some pastoralists seemed to confirm this: they complained that they could not take out a loan for formal reasons or were are afraid to do so because 'taking out a loan is easy but paying back is hard'. Some people said that they do not store savings in banks because they simply do not have any. However, most pastoralists admitted that they had savings and keep most of them in a bank, believing that this is safer than at home. An employee of the Rural Credit Cooperative (RCC), which had its office in Domkhok, confirmed this.[9] According to him, the pastoralists had opened saving accounts and long-term deposits on a massive scale. Although there were some people who, as he said, preferred to 'bury money in the ground', most families had several bank accounts and often opened a new account each year. Nearly half of the pastoralists had ten and more long-term deposits with around

9 Rural Credit Cooperatives (Chin. *Nongcun xinyongshe*) opened in the 1950s and served people's communes. In the 1970s, they started functioning as 'grassroots banking institutions' providing services to the rural population (Tsai 2002: 152). In Machen, they had units in Domkhok, Xueshan, Dawu Zhuma, Danlag and Jumgo, today part of Rarja.

50,000 yuan in each. Taking out loans was for them less important: the pastoralists took out loans, as the RCC employees declared, but not on a big scale. The purpose was usually to buy a car, to fence pastures or build a house. The loans taken for the last purpose oscillated around 50,000-60,000 yuan and covered only part of the house construction costs. This means that the pastoralists still had to pay the rest by themselves. The lack of interest in bank loans suggests that they had enough of their own funds. Some informants openly stated that they 'have enough money and don't need to borrow it from a bank'. If the pastoralists took out loans, they did so in spring, but it is more revealing when they paid them back. The vast majority (even as much as 80 per cent, as the RCC employees attested) paid back loans in July and August, immediately after the caterpillar fungus season. Other people did it in October and November, after selling their livestock. Many clients paid the whole amount at once, even before due time. Residents of townships other than Domkhok, Xueshan and Dawu Zhuma – the three townships particularly abundant in caterpillar fungus – were less punctual, but they also paid their loans back after the caterpillar fungus season. This shows that summer is the new cash season in Golok and that this cash comes into the pastoralists' pockets in such amounts as to allow them to make substantial savings and investments rather than spend it only on their current needs.

The Nomadic Settlement programme was a state initiative. But what would have happened if this programme had not been announced? Would the pastoralists continue living in their old houses or build new ones? Jigmed Dorji's case suggests a possible answer. Although his family built their third house in 2007, two years later they started constructing another one. Their neighbours did the same. Some said that before the programme had started, they had already considered building a house or were preparing for it. Other people did not have such plans, but welcomed the subsidies as bringing 'free money' for a purpose that they generally found desirable. My informants were convinced that sooner or later they would build a new house anyway, with or without the state subsidy. The pastoralists 'are crazy about houses', as many people in Golok claimed, including the pastoralists themselves.

This enthusiasm for houses in the pastoral society may be surprising, but my informants spoke about houses as a major improvement in their lives, which they would not be able or willing to renounce. The houses were an important material achievement that made their life easier and more comfortable. It can be debated whether this view was influenced by the state-promoted vision of modernity, with the 'modern' pastoralists living in houses and not in tents. However, my informants did not associate the building of their first houses with the state intervention.[10] They never regretted living in a house and some considered building one on their summer land as well.[11] For them, living in houses was a function of increased security: they could live in houses because Golok had become safer, whereas earlier tents and mobility were strategic options in case of conflicts and raids.

This popularity of houses relates to the manifold functions that houses play. Being exposed, on a lasting basis, to the general gaze, the house 'expresses or betrays, in a more decisive way than many other goods, the social being of its owners, [or] the extent of their "means"' (Bourdieu 2005: 19). A house is the owners' business card and a testimony of their pecuniary repute (Veblen 1987[1899]: 54). The house reveals its owners' 'taste, the classification system they deploy in their acts of appropriation and which, in assuming objective form in visible goods, provides a purchase for the symbolic appropriations of others, who are thereby enabled to situate the owners in social space by situating them within the space of tastes' (Bourdieu 2005: 19). It plays a role not only in the space of material or economic differences, but in the space of tastes and cultural loyalties. By having a house of a particular kind, one communicates to the neighbours: we are like this, we can afford that, we are connected in such-and-such a way and this far advanced on the path to 'being modern'.

The incongruity between the number of houses and the size of the families inhabiting them, and the fact that in many cases people use only one of several houses they own, shows that a house is more than simply a place to live. Tsering Drölma's family was not an exception. Her new house impressed with its elegance but had an air of sterile emptiness. The stoves were cold and the house had the chilly feel of a building that is seldom heated. The door was locked; it was opened when guests arrived, to monks or officials, or during New Year celebrations and family events. On such days the floor was swept, a fire made and tea and food served. Jigmed Dorji's family also used their newest house during New Year or similar events. Other neighbours used the new house for guests, and stored their belongings, cooked, slept and spent time in one of the older dwellings. A woman, whom I asked about her new house, exclaimed: 'We don't even go inside!' It is clear that some of these houses are redundant

10 It is often assumed that the state forced the pastoralists to build houses (Clarke 1988; Manderscheid 2002). Some informants admitted that 'the government encouraged people to build houses', but said that there was no coercion involved, which – as they said – accompanied sedentarization programmes or the birth control policy.

11 I asked a leader of a big Golok *dewa* whether living in houses would destroy the 'pastoral identity'. He said that people would not be able to build a house on each pasture, and thus will their identity be preserved. Though some people said they would like to have a house on their summer land, no one had built one so far.

from the point of view of the pastoralists' housing needs. This is also visible when one compares the number of houses with the number of people living there: in some cases, the number of houses per family grew while the number of inhabitants declined as children moved out or grandparents passed away.[12]

That a house speaks of its owners is a banal observation. However, in Domkhok, it is not merely a house that does so, but the number of houses people own. Houses are an investment commodity: my informants believed that houses and gold are the best ways of investing their savings. The pastoralists' houses, unlike those in the township or county town, lack certain traits that a house as a good on the housing market has. They are not objects of buying, selling or renting. If the 'commodity phase' is episodic in most of products' life history, then these houses would be commodities in their dormant phase (Kopytoff 2003: 68). One day, they may be commoditized and their value converted into cash, but so far, their economic biography is brief. They are a commodity only potentially. This potential of becoming a good on the housing market remains limited, but their owners believed that a house is an investment which will at some point bring them attractive profits.

Domkhok during my research resembled a huge construction site. Yet, all the houses were similar, especially those built within the Nomadic Settlement programme. If a family wanted to distinguish themselves from among neighbours, they had to invest in the furnishings. Thus, not only the fact of having a house or a number of houses mattered, but also the collection and quality of goods stored inside. The function of a house as a *container* for consumer goods was highlighted by the pastoralists, who occasionally joked that 'what's inside is more expensive than the house itself'. How much it costs to furnish such a house is evident from example of the most important elements in every house's inventory. The equipment comprising a Tibetan-style wooden cabinet, tables and benches, a metal stove and a television costs between 11,000 and 27,000 yuan, depending on affluence and ambition of the owner.

Houses offer storage space which tents did not have. On the one hand, having a house forces the pastoralists to buy goods necessary to furnish it. On the other, it allows them to accumulate goods which they previously could not buy because of lack of space to store them. One man, when asked why people are so fond of houses, said that they need houses 'to put things inside'. Living in a house, the pastoralists can buy more 'things' to 'put inside' and – if the house is electrified – can also buy electrical home appliances which were beyond their reach before. One of these is a television, which became so important that wooden cabinets, a piece of furniture considered

12 For a detailed discussion, see Sulek 2012.

'traditionally Tibetan', now have a television shelf: cabinets without this feature simply do not sell, as the carpenters in Dawu told me. Televisions are one of the largest expenses the pastoralists bear: some families bought flat-screen models worth as much as 9,000 yuan. Watching television is a popular pastime and when there is nothing interesting on the Tibetan-language Golok or Qinghai television channels, people watch music videos of Tibetan singers: a VCD player was found in every house. Freezers and washing machines are also standard equipment (Table 7). Exceptions from this rule are young couples who are still furnishing their first house, as well as the elderly, single people and divorcees who live separately, but do laundry or store food with their relatives.[13] Because the pastoralists did not dismantle their old houses, they could not reuse older furniture or appliances, and every new house entailed the purchase of new equipment. As a result, nearly half the families had more than one television, sometimes one in each house.[14]

Table 7 Ownership of electrical and other appliances, Domkhok 2010

Item	% of households owning at least one
TV and DVD	100%
Radio and tape recorder	71%
Washing machine	86%
Freezer	90%
Sewing machine	82%
Solar panel	100%
Milk separator (manual)	97%
Milk separator (electric)	43%
Gas stove	55%
Electric stove	30%
Electric heater	46%

13 A similar list was made by Goldstein et al. (2003). In 1997-2000, in villages in the TAR that they studied, 43 per cent of households had a sewing machine, 25 per cent had a radio, 2.8 per cent a truck and 0.4 per cent a motorbike (Goldstein et al. 2003: 770). These figures are much lower than in Domkhok in the same period. Many goods from Goldstein's list were so common in Domkhok that they could not be used as indicators of affluence or poverty. Metal stoves were in every house and the few families who had a clay stove in their tents had it for sentimental or aesthetic reasons because, as they said, 'such stoves fit the tent better'. Other goods from Goldstein's list had lost popularity, for example radios and tape recorders. There were also some appliances that were only now gaining popularity: electric stoves, electric heaters and milk separators.

14 A decade earlier, 'in 2000, there were a number of motorbikes, but they were still something unusual and many people (or maybe even most of them) actually still rode on horseback. In the tents there was usually a radio/cassette player and quite often a milk separator. Solar panels used to boil water were a rarity' (Bianca Horlemann, email comm., 14 March 2013).

A desire to experience the benefits of having electrical home appliances explains why the pastoralists were so interested in electrifying their houses and even participated in covering its costs. During my research, most households were already electrified; power generators were used in those valleys that still awaited electrification or during power cuts.[15] However, power generators could not support all electrical equipment that the pastoralists had or wanted to have, and people petitioned the township to speed up the electrification process. Electric appliances made work in the household easier, but they also introduced new dynamics into people's lives. Freezers made it easier to store food. They also freed the pastoralists from the need to wait to slaughter the animals until it got really cold outside. Indeed, the freezers they bought were the largest models available: a bestseller had a 240-litre net capacity. Washing machines facilitated doing the laundry but they also forced women to fetch water more often: on 'laundry days' women had to do so several times, going down to the river and climbing back with heavy water canisters. The wide use of washing machines made people think about having running water at home. In one valley, the pastoralists hoped to build a pipe between a spring on a higher mountain slope and the houses. They declared in the township that they were ready to pay for it. In another valley, people said that the officials had promised them running water. They were getting impatient: 'How long do we have to we wait?', they asked. This shows how the introduction of electrical home appliances leads to infrastructural changes or makes them perceived as necessary or desired.

The need to furnish houses creates opportunities for the expansion of the market into areas where few consumer goods were sold before. But apart from the fact that a house offers storage space and allows its owners to participate in consumer culture, its presence has other implications for the inhabitants' lives. One of them is the change in the character of summer camps. In today's Domkhok, tents are like satellites of the houses rather than independent 'homes' in their own right. They do not have an independent being, but function together with the house. The connection between the house and the tent is maintained throughout the summer. The pastoralists do not take everything with them that they need for the summer months: they can fetch things from the house at almost any time. Tendor regularly rode his motorbike between the tent and the house and each time he went

15 In summer 2009, there were power cuts that lasted weeks. Golok consumed too much electricity and had to save it for winter. Growing numbers of electrical appliances meant increased consumption of electricity and problems with power supply.

to the township village or Dawu he also went to the house to bring or take something or check whether everything was fine. Other families also visited their house because senior family members stayed there. Whereas younger generations moved to the tents in summer, the elderly often stayed in the house in more comfortable conditions.

Having a house with many expensive appliances creates a security problem, especially in summer when the pastoralists move to tents and leave their houses empty in the lower parts of the valley. How do they solve this? Some families living close to the road, in places easily accessible for intruders, decided to remain in the house and hired herders who took their herds to the summer pastures. In most cases, however, the pastoralists moved to tents, but organized night patrols that regularly checked the situation in the lower parts of the valley. In Soglung, groups of three to six men left the tents at night, patrolled the valley on motorbikes, checked up on the houses and returned to the summer camp. Similar patrols were described by Norbu, who observed that in Golok Serthar and Dzachuka 'every nomad community carries out patrols known as *risher* over its entire territory' (1997: 68). These patrols looked for cases of trespassing, raids or fire, inspected the state of the grassland and river crossings (Ekvall 1961: 1256). My informants said that the problem of trespassing or livestock thefts had declined thanks to improved security and fencing. Even if such incidents really became less frequent, they still took place. Moreover, new reasons for organizing patrols emerged, such as searching for people digging caterpillar fungus without the pastoralists' consent or checking out the situation in lower parts of the valley. Although the men in Soglung spoke about these night patrols as a boring routine, burglaries really did happen. Shortly before my research started, thieves broke into one of the houses, stole the family valuables and beat the family's grandfather who suffered a serious hearing loss after this attack.

Vehicles of Change

Traditionally, the pastoralists in Domkhok depended on yaks and horses for their mobility, using them both as riding and pack animals. This situation started changing in the 1990s when the first privately owned motor vehicles appeared in the township. A decade later, cars and motorbikes were already so common that people could hardly imagine not having one. When I trekked between the pastoralists' camps, sometimes over long distances, I repeatedly experienced the same situation. A car with a pastoral family inside stopped, the driver rolled down the window and, after exchanging greetings, asked, with surprise mixed with compassion, 'Don't you have a car? Not even a motorbike?'

Pedestrians are a rare sight in the highlands, but people riding horses are not common either. This does not mean that the pastoralists do not have horses though. According to the official statistics, in 2008 there were 505 horses in Domkhok (MCSY 2009: 44ff). This gives an average of three horses per two families. As with other livestock, the actual numbers were higher. The local officials estimated, in a private conversation, that an average family owned two horses and more affluent ones had four or five. My survey gave even higher results: three or four horses were the average and some families had five, six or seven.[16] Some of these horses were local champions that won races, diplomas and respect for their breeders, and could be worth 20,000 yuan or more. All horses, those with titles and those without, were nurtured with care, and both barley and wheat for fodder and veterinary care were a firm point in household expenses. Horsemanship is an important part of local know-how and my informants questioned me extensively about horse breeds in Europe.

Apart from featuring in people's discussions and becoming a focus of attention during seasonal races, horses seldom appeared in the pastoralists' daily life: they spend time grazing in the mountains and are only ridden on a few occasions. When excessive snow and rainfall made roads unpassable, horses were a convenient solution. They were ridden to look for stray animals or visit neighbours, especially by women, who – at least during the years of my research – did not ride motorbikes alone. When her husband was in town, Tsering Drölma kept a horse close to the tent: when she heard disturbing noises from the pastures or saw the herds behaving nervously, she rode bareback to check what was happening. Some people rode horses to community events, either out of respect for tradition, as they said, or for image-related reasons. Apart from that, horses were no longer 'working animals' and were kept for 'entertainment', as several men stated. In their practical role they had been replaced with motorbikes and cars. The pastoralists used them to travel to the township village and county town, to transport provisions and goods for sale. They also drove motorbikes around the pastures to scout the land for trespassers, inspect

16 People concealed the number of horses for the same reason they hid the number of *tsethar* animals: these animals were unproductive in a market sense and 'took up space' on the grassland without contributing to the local economy. They would be the first candidates to be 'removed' from the land to decrease the pressure on it.

fences and take the herds to graze. Finally, cars and motorbikes played an important role in the pastoralists' migrations – a vital aspect of their life and economy.

Elderly people claimed that in the 'old society' the Metsang pastoralists migrated five or six times a year. The land belonging to them was bigger; the township was formed from only part of their territories.[17] Today, the Metsang migrate between the summer land where they camp in tents and the rest-of-the-year settlement where they have houses. Few families have a house on their autumn land or stay in a tent during this period. Most bring their livestock to the autumn land, while they themselves return to the house. For Tsering Drölma and Tendor, the migration meant moving between a lower and higher elevated part of Soglung: their summer and winter land are six kilometres apart. The majority of households in Domkhok have their different seasons' pastures in one mountain valley. Only a few less lucky families have their summer land further away, in another part of the township.[18] Tsering Drölma and Tendor have about 1000 *mu* of summer (*yarsa*), 4000 *mu* of winter (*gunsa*) and 1000 *mu* of autumn (*tonsa*) pastures. (This is similar to what their neighbours have, but there are also households with only 2000 *mu* who have to lease land from others.) Their winter and autumn land they use together with Tsering Drölma's uncle, Jigmed Dorji. Only the winter and autumn pastures are fenced: they are in lower Soglung, close to the road. In the higher part of the valley the land is not fenced. The officials from town do not come to this altitude to check whether the fencing policies are properly implemented and people prefer to use their land all together.[19] Tsering Drölma and Tendor

17 They said that their land was three times bigger, but parts of it were lost to Xueshan and Dawu Zhuma. As a result, people in these two townships migrated four times a year and had spring pastures, which in Domkhok was impossible because of lack of space.

18 During the land allocation, the families drew lots as to who would get the land where: all the land belonging to one production team was in the 'drawing pot' and some households 'drew' pastures in different valleys. Marriages and other arrangements complicated the situation further. For example, a family who had their summer pastures 20 km away from their house gave their plot to the township in order for a school to be built. In exchange, they received pastures in another place. Families who travel ten or more kilometres between their house and the summer pastures are few, and most people stay in the same valley, moving up and down depending on the season.

19 Fencing is widely discussed in the literature (Miller 2000; Bauer 2005; Bauer and Yonten Nima 2011; Wu et al. 2012). Whether the fences exacerbate conflicts over the land or reduce their number is not clear: Yeh argues that violence on the grassland increased after the land was fenced (2003: 502), but Pirie says the opposite (2005: 22). During my research, I heard only one openly negative opinion about fencing. This was from a singer whom I interviewed about social and environmental problems in the region: 'Before 1959 we did not have desertification.

camp with five other families.[20] Such encampments, called *rukor*, act as mutual aid groups.[21] The neighbours – often but not always relatives – help each other with work, such as slaughtering yaks, shearing, searching for lost animals. Given the declining size of the pastoral families, this help is important. It also extends to other seasons and the families help each other with babysitting, in looking after each other's tents and houses when the neighbours are away, and during family events and religious rituals. Finally, they also assist during migrations, when pitching and striking tents and packing the cars.

The Metsang *dewa* perform their migrations at around the same time. The dates of the migrations are decided on the day of a community gathering devoted to the worship of Amnye Wayin. This gathering, called *chibsang*, takes place on the fifteenth day of the fifth Tibetan month. On that day, men (women cannot participate) climb the mountain range to the place where a 'fumigation altar' (*bsangchy*) is located, perform the fumigation ritual and offer Amnye Wayin a horse or a yak. As a monk from Guru Monastery explained:

> On the day of the ritual, you climb the mountain carrying *lungrta*, *bsangrtsi*, the *mirtag* rope and the arrows.[22] If you offer a yak you don't take it to the mountains, but if you offer a horse, you need to saddle and harness it nicely and take it with you. If monks are there, they chant and purify them. In the past, people also brought rifles, bows and arrows, to purify them. Later on, horse races take place. There's also a meeting, during which people settle conflicts, plan how to use the grassland and when to change pastures.

Even in the 1980s Martod was still fertile. Now it's degraded. Why? I think it's because of fences. The yaks know by themselves where to graze. But if enclosed behind a fence, they go crazy and trample the land in a fury. It seems to me that the state wants to gather people in one place, because it's easier to control them.'

20 Prior to collectivization, the pastoralists camped in circles during summer and dispersed during the rest of the year; see Duncan (1964: 217) and Rock (1956: XXVII). This was done for security reasons: in summer the animals were in the best shape and people kept them in the middle of the circle at night to guard them from thieves. My informants said that this is not practised anymore because Golok had become much safer.

21 On *rukor*, see Levine (s.a.) and Clarke 1992.

22 This includes a mixture of roasted barley and rice, *rtsampa*, juniper twigs, butter, milk tea and tea leaves. *Rlungta* are paper printouts with the image of the wind horse symbolizing welfare and luck which are thrown into the air during the ritual (see Karmay 1988d). The arrows symbolize auspiciousness and are used in Tibet and the Himalayas during marriage ceremonies and in rituals summoning good fortune and long life (Gerke 2012). *Mirtag* rope is tied to them after they are stuck into *labtse* (ritual cairns) dedicated to mountain gods.

After the gathering, the *richen* administration sends out a circular letter calling on the families to change their pastures by the indicated date. It also warns that staying on at the winter pastures longer or returning there too early will be met with legal consequences: the winter pastures must be able to recover during the short summer to have enough fodder in winter. In 2010, Tsering Drölma and Tendor moved to their summer land on the third day of the sixth Tibetan month. They drove their livestock to the autumn land on the tenth of the eighth month, and to the winter land on the twentieth of the tenth month. This corresponds with 14 July, 17 September and 26 November. These dates, as Tendor said, fall around the same time every year: in mid-July people are usually on their summer land, in September they move their herds to the autumn land and in November bring them to the winter land.

Tsering Drölma and Tendor change their living quarters twice a year, but their livestock follows a more complicated route. The pastoralists split their herds into groups which are sent to graze in different parts of the pastureland, depending on the particularities and state of vegetation there. Thanks to this approach, the grassland resources can be used in the optimal way, adapted to the needs of particular kinds of animal. The animals stay on different pastures for different lengths of time and change them according to their own rhythm. Only the move from the winter to the summer pastures is done by all livestock together. Then, their paths diverge. Sheep, for instance, graze together in summer and autumn. In winter, the males and unpregnant ewes stay on the winter land, but the pregnant ewes move to a fenced-off area called *rtsamdzod*, close to home and reserved for lambing. Pregnant *dris*, those *dris* that calved the year before, as well as their calves, move according to a similar pattern: they graze on the autumn pastures and return to the winter land together with the sheep. Male yaks and barren females remain on the higher pastures longer and move to the autumn pastures only when the other females and calves have left. They return to the winter land in February or March. Though it is much colder in the upper parts of the valley, yaks are resistant to low temperatures, and – apart from the calving *dris* and calves – they need less care from their owners. In the past, the households split, too, with the younger members setting up satellite camps in the mountains to look after the livestock there. Now, this job is either done by hired herders or grazing is managed in such a way that no satellite camps are needed.

Nowadays, moving camps is easy and most families transport their tents and other belongings by car and not with yaks. The majority of summer camp sites are accessible by road; in more difficult terrain the pastoralists transport their tents and other belongings by car up to a certain point, then load them onto motorbikes. The whole migration takes three to six days, during which people pitch their tents and organize their living space for the following months. While they drive back and forth, the camps are guarded by their dogs: huge Tibetan mastiffs roam the space freely, protecting the tents from intruders.[23]

At the time of my survey, nearly all households had a motorbike and more than half had a car.[24] Cars are family vehicles, but motorbikes are the property of particular family members and large families often had a motorbike for each son; at least one third of households had two motorbikes or more. Yet, the number of motorbikes does not always correspond with the number of active users. Jigmed Dorji's brother, whose family consisted of him, his wife and an unmarried daughter with a child, owned four motorbikes, although only he could ride. There were fewer cars around, yet there were also some exceptions here. One family owned five cars, but – as they said – they planned to sell some of them. In Golok, cars were often bought second hand and circulated in the local society, changing owners. This allowed people with a smaller budget to afford a car, even if it was not the newest model or not in perfect condition.

Golok in the early 2000s had a thriving market in cars and motorbikes (Figures 26 and 27). The scale of change that local society underwent may be difficult to conceptualize because little data about ownership of motor vehicles in the past is available. My informants recalled that they bought their first motorbikes in the late 1990s and their first cars in the 2000s. They often remarked that they changed them frequently and 'couldn't stop buying them', as one man joked. Some men reckoned that within one decade they had bought and sold ten motorbikes and four to six cars. These were usually second-hand vehicles, often unsuitable for driving in difficult terrain at 4000 m altitude. This situation altered during the course of my research. Whereas the most popular motorbike remained the Honda Wuyang, the taste

23 In 2009, when we climbed the mountain range to reach one informant's tent, we were rescued at the last minute by a man who came with a car to look for us. He had been alarmed by the news that the dogs near the tent to which we were heading attacked their owner, who came from the house bringing some kitchen equipment. The woman was hospitalized. Many persons, especially women and children, do not leave the tent or house alone at night, when the dogs are unchained.

24 This invites comparisons to regions without caterpillar fungus. Goldstein reports that in 2003-2005 the first privately owned motorbikes and in 2008 the first mobile phones appeared in Phala, Shigatse Prefecture, TAR (Goldstein 2009). By 2009, almost half of households owned a motorbike; many had a mobile phone, and some trucks or tractor-pulled carts (Goldstein 2012: 208). This shows a significant change compared to the past but is still modest compared to Domkhok (Goldstein 1991, Goldstein and Beall 1989, 1990).

Figure 26 Parking lot in front of the monastery in Dawu

Figure 27 Pilgrimage around Amnye Machen; these days often done by car

for cars underwent changes reflecting the rising affluence of the society. In 2007, when my research began, the Volkswagen Santana was a much sought-after model, although unsuitable for the local terrain. In later years, growing numbers of people drove SUVs, better adapted for driving off road and having bigger storage capacity, which is important for the pastoralists who use their cars to transport goods to sell and to bring provisions from the town. Chinese cars were everywhere, but Japanese and Korean brands were popular as well, and many men wanted to own a car from Europe.

Cars and motorbikes enable easy transportation, which is important in a region of such huge expanse. They afford the pastoralists easier access to healthcare, allow them to see their children in boarding schools and to visit monasteries and pilgrimage sites. But all this would not be possible without the expanding network of roads. My informants stressed that twenty or thirty years earlier they were isolated from the rest of Golok, let alone the world. During my research, most valleys were already accessible by road. However, at this altitude roads deteriorate quickly and their poor condition was one of the biggest challenges to pastoralists' daily life. Travelling on these roads at night was often risky and during rain or after snowfall entirely impossible. Road construction, at least in theory, is in the hands of the local government, but in practice only the roads connecting the town and the township seat with the pastoralists' houses were paid for from the public budget. As one official stated: 'We build roads to people's houses. If they want to have a road up to their tent or to repair the old one, they have to pay for it.' I observed how the pastoralists organized themselves to accomplish such tasks. In Soglung, even the best SUVs got stuck in the mud covering the roads after the slightest rain and so people had to undertake arduous motorbike rides across the mountains to reach other valleys. They would enquire as to whether the township could finance the repair of the road. The answer was always negative and so people pooled their money to do this work. They financed not only repairs to existing roads, but also the building of new ones: almost all summer camps in Soglung were accessible by road. All these roads had been built on the initiative of the pastoralists and financed by them, and all had been completed during the years of the caterpillar fungus boom.

The expanding road network, coupled with the proliferation of cars and motorbikes, has had implications that can be felt on many levels of the pastoralists' lives. Elderly pastoralists recalled that before 1958 people in Golok were more mobile, at least as measured by frequency of changing camps and pastures. Administrative reforms and allocation of the land to households limited this mobility and the establishment of the house as the main dwelling impacted people's lives – binding them to their land in new ways. Contact with the house was kept up during summer when people stayed in tents, and it was cars and motorbikes that made this possible. Even more significant effects were visible in the pastoralists' contacts with the town. Lilian Iselin observes that motor vehicles gave people 'access to a wider radius of action within the same time constraints given by the requirements of pastoral life' (2011: 440). Although the scope of the pastoralists' movement across their land decreased, their contacts with places beyond the world of their daily pastoral activities intensified. Whereas they now travelled less in terms of the physical distances they covered with their livestock, they moved more often between the highlands and the township seat, county or prefecture town and larger cities. Motor vehicles made new spaces accessible to the pastoralists in new ways, and previously remote towns became better integrated into their life. The degree of this integration differs and people living further away from the town had more sporadic contacts with it. The frequency of contact with the town varied between family members, age and gender groups: younger people visited the town more often than their grandparents, and women travelled less than men. Men are less bound by their house- and pastoral work and often used any excuse to go to town. Women did not drive and, similar to older people, they relied on their relatives as drivers. In a sense, the burgeoning numbers of motor vehicles, at least at this stage, had not enhanced the mobility of women.

In spite of these differences, pastoral households as a whole had more frequent contact with the town than before. Most households sent someone shopping at least once a week. Asked how often he went to town, Tendor replied: 'Every day or every second day maybe. Anytime we need something.' Shopping was not the only reason to go: he went there to charge a mobile phone, see a doctor, go to a barber or a bank. The pastoralists travelled not only to Dawu, but also to Xining, Rebkong, Barkham and Lhasa: for shopping, medical treatment, entertainment or pilgrimage. Cars and motorbikes allowed them to travel with increased speed and comfort, but also facilitated contact with the urban markets. Every trip to town is a potential shopping trip. 'When you're in the town, one thousand [yuan] goes away just like that', one man said, clicking his fingers; 'even if you don't need anything, you always buy something'. Cars and motorbikes not only bring people into contact with the urban consumption centres but are themselves consumer goods which 'consume' people's money. Whole streets in Dawu are covered by garages and other shops selling motorbike parts, and carwash tents serve not only town dwellers, but also drivers from the highlands. Purchasing petrol and paying for vehicle repairs were a large expense in family budgets.

My informants found it difficult to estimate how much they spent on these purposes, but the sums they declared started from 10,000 yuan. This shows that cars and motorbikes burn not only petrol but the pastoralists' money too.

Consumer Months

The caterpillar fungus economy supplied the pastoralists with income that they could spend on consumer goods. Their first shopping destination was Dawu, the main consumption and services centre in Golok. During the years of my research, Dawu underwent significant changes in terms of the goods and services it offered. The first supermarkets opened and introduced loyalty schemes and newsletters announcing sales, promotions and new products. They sold items that were certainly not in daily use in Golok: instant coffee, chocolate and whisky. Street sellers started to offer not only the usual stuff such as blood sausages and boiled yak heads, but also bananas, mangoes and prawns, which travelled a long way to reach the Tibetan plateau. New types of restaurants and bars appeared too. In the Tibetan-owned bar Sa Golok Nam Golok, guests could drink Johnnie Walker, Chivas Regal and Rémy Martin for 598 yuan per bottle, while a new teahouse nearby served coffee in Starbucks cups. In another restaurant, food was served by waitresses wearing Lhasa-style robes: something unheard of in Dawu before, where most eateries had the usual three or four dishes and did not even have a menu. The main street was dense with boutiques offering a wide variety of goods: wide-open doors invited in the customers and loudspeakers placed on the doorstep announced discounts. Red balloons swaying in the air and bits of coloured paper from fireworks scattered on the pavement showed whichever shop had just opened.

The connection between the boom in the caterpillar fungus economy and the outburst of consumerism seems obvious. Nevertheless, it is worth asking whether it is really the money from caterpillar fungus that enables the pastoralists to enjoy this consumer life. To check whether this is the case I conducted a questionnaire in a number of shops in Dawu that asked about the seasonal changes in the market, numbers of customers, turnover, but also products which sold particularly well and the customers' likes and dislikes.[25] The questionnaire showed that there was a distinc-

25 The questionnaire covered eighteen shops selling home appliances, furniture, clothes, jewellery, motorbikes and mobile phones. I also interviewed itinerant coral sellers as well as

tive rhythm to the shopping, whose intensity increased and declined at particular times of the year. June, July and August were the most intensive shopping period: customers bought more and spent more than during the rest of the year. With some slight exaggeration, one can almost say that there were two seasons in Golok: shopping and non-shopping ones. The contrast between them was so big that many store owners said that there was 'nothing going on' in their business apart from during the summer. One man remarked that he could practically close his shop for the other months, when he hardly made any profit. In fact, some shops did close for one or two months in winter. Furthermore, shopkeepers used to take holidays to celebrate New Year or escape the pervasive Golok colds, but now they could have a longer winter break without hurrying back to reopen their business.

The summer explosion of shopping in Dawu was observed in shops selling very different sorts of products, ranging from furniture to jewellery. Furniture was perhaps the most unusual of these products to enjoy popularity in Golok, a region where tent-dwelling pastoralists had barely had any furniture before. The fashion during my research required people to buy Tibetan-style furniture, imported to Golok or made by carpenters in Dawu. In the past, people bought so-called 'Chinese' furniture, the likes of which could be found in any other part of China.[26] In summer and early autumn, shopping for furniture went on at a massive scale: 'Some nomads buy everything at once for 9000 or 11,000 yuan. Sometimes we run out of stock within one day', one seller said. The sellers admitted that this type of shopping was related to house construction, but that was not all: 'The pastoralists have caterpillar fungus and can afford to build houses. And when they build houses, they need furniture.' A relation between the customers' place of origin and the amounts of money they spent supports this argument: pastoralists from Xueshan and Domkhok bought whole sets of furniture, while those from other townships, less fertile in caterpillar fungus, bought only individual pieces. The same was observed in shops selling home appliances. The owner of one of these estimated that in a single summer month he could sell between thirty and forty televisions, seventy or eighty VCD players, and forty or fifty freezers and washing machines.

a local dentist who worked on inserting gold dental crowns, which were popular in the early 2000s (many pastoralists wore such 'golden teeth', usually set symmetrically on both sides of their mouth) but during my research were losing popularity. Most of the shop owners and sellers were Han. I also conducted interviews at the car market in Xining.

26 'Chinese' is used in Golok to refer to mass-produced goods. Any piece of modern clothing or furniture, lacking Tibetan features, the pastoralists called 'Chinese'.

Another, smaller shop saw between twenty and thirty buyers a day. In summer, both shops extended their opening to twelve and fourteen hours and stayed open until late at night.

Electrical appliances are not things that one buys every day, but shopping for other products, for example clothes or shoes, exhibits the same seasonal rhythm. Tibetan tailors in Dawu have a period of intensive work before Tibetan New Year, which their customers want to greet wearing a new robe: around Losar almost everyone in Golok wears a brand-new robe and looks like something from a fashion catalogue. But winter cannot compete with summer, when horse races take place. Such community events are a good occasion to wear something new: fashion in Golok changes fast and the customers want to keep up with it. Summer, as the tailors emphasized, is when 'people have money and want to spend it'. These were almost the same words I heard in a shop offering 'Chinese' fashion: one seller estimated that in summer he sells three times more shoes than in winter. As a woman with a shop selling outdoor clothing said:

> On one day in summer we sell as much as during a whole month in winter. Practically speaking, we have only two months of good business. But in summer, Tibetans spend a lot on clothes. When we came here in 2005, they wore robes. Now they're becoming modern. Our shop grows together with the *yartsa* business. We're certainly not moving anywhere else.

The fact that the pastoralists have money and like to spend it, as the sellers said, renders Golok a place high on the list of locations where commercial success is almost guaranteed. 'There is no better place than Golok', as one Han woman said; 'one only needs to speak a bit of Tibetan'. The only 'problem' the sellers complained about was that the 'nomads [had] got smarter' and 'it's not so easy to sell them rubbish anymore'. Another new issue for the traders was the growing competition. Dawu, almost a commercial desert twenty or so years ago, had suddenly become crowded. Instead of just two furniture shops which were there in the 1990s, now there were twenty. The same applied to shops selling mobile phones, which had gained in importance thanks to the expansion of network coverage. Towards the end of my research, China Mobile and China Unicom already covered 60 per cent of Domkhok and were still expanding (MSGW 2010). In 2007, in order to make a phone call in Soglung one had to climb a neighbouring mountain to get any signal; three years later many people received signal

Figure 28 Mobile phones are ubiquitous; a wedding party in Dawu

at home.[27] In most households I surveyed, the number of mobile phones equalled the number of adult family members, perhaps with the exception of the oldest generation (Figure 28). Instead of a handful of shops selling second-hand mobile phones without warranty, as was the case before, Dawu now has a number of spacious telephone shops in the very centre of the town. These sell diverse models, including the first smartphones, some of them priced at as much as 2,000 yuan. The difference between summer and the rest of the year was also visible there: a Tibetan seller in one of these shops calculated that he sold up to twenty mobile phones per day in summer and perhaps only two or three a day in winter.[28]

27 At the beginning of my research I used my mobile phone mostly as a watch; the network coverage outside Dawu was almost non-existent. Imagine my surprise when one day my normally silent phone rang as I sat in a tent conducting the survey. This was my father calling to say that Michael Jackson had died. This anecdote captures the dimension of changes and new connectivity that have come in with the expansion of the telephone network.

28 He also sold video cameras, which the pastoralists used to record religious ceremonies, horse races and other events. Cameras do not end the list of desired hi-tech goods. One year, I met a teenage pastoralist herding sheep in Domkhok who said he wanted to buy a laptop. A year later, he already had one. He could neither read nor write but was strongly committed to learning how to do so.

Summer is also the season for buying cars and motorbikes. Customers ready to invest in a new motorbike could visit a new shop that sold shiny Hondas and Yamahas brought directly from the factory. Its owner estimated that during his first summer in Golok he sold around sixty motorbikes. While buying a motorbike in Golok is easy and the market for second-hand vehicles is large, in order to purchase a car, the pastoralists usually went to Xining. There, at the largest car market in Qinghai province, everything was available and for whatever budget. The sellers I interviewed at this market recognized Tibetan pastoralists, mainly from Golok and Yushu, as a distinctive group of customers. They said that the pastoralists prefer to buy their cars there not only because of the prices, but because transactions at the market were less formal than in a shop and so the customers did not feel intimidated by their lack of fluency in Chinese. The pastoralists start buying cars in mid-June: 'In summer the weather is good. Nobody comes here in other seasons', as one seller explained. His colleague added: 'The nomads come in summer because this is when they have money from caterpillar fungus.'

Many products find customers throughout the year, but some are bought only in summer. This is the case with jewellery. There are several jewellery shops in Dawu. In one of them, on Gesar Square, Buddhist statues jostle with Tibetan jewellery manufactured exclusively for this shop, as the owner stated. The shop was popular, especially for leather belts inlaid with metal and coral plates as well as for elaborate sash ornaments worn by Golok women.[29] The 'high season' in the shop starts in June; three-quarters of the turnover comes from summer. Similar observations were shared in a larger jewellery shop on Unity Street where, behind a massive metal door, was concealed a spacious room with an army of personnel and a row of glass cabinets full of heart-shaped pendants, chunky earrings and gold plates. These gold plates (a 50 g plate cost over 16,000 yuan) were popular among the pastoralists as a capital storing asset. The shop owner, a Tibetan from Kumbum, recalled: 'In 1997, when we opened the shop, people couldn't afford to buy gold. Only government officials bought it. Today, the economy is better and most people buy gold. There're still some women who prefer silver and we have something for them, but not much.'[30] Dawu is a perfect place for business, he claimed, because 'Golok men like spending money on

29 The belts cost up to 4,000 yuan. In terms of sash ornaments, *zhobzung* (an anchor-shaped ornament, being a festive version of a hook which women use to hold a bucket when milking) cost 1200-4200 yuan and *logzur* (resembling the sun and a crescent moon) were 1200-4800 yuan.

30 A monastery near Xining founded in a place where Tsongkhapa Lobzang Drakpa was born. It is the second-most famous monastery (after Labrang) in north-eastern Tibet. It is an important trade hub, and many traders who came to Golok had worked there before.

Figure 29 'Goloks like corals'; two women returning from shopping. The plastic bag protects the hat against dust.

Figure 30 The author with her informants

jewellery for their wives'. The shop earns half its income in summer; and all of this comes from the pastoralists, as other customers do their shopping in other seasons. Summer is also when Dawu is visited by itinerant coral sellers. These men, usually Hui, roam the market with a suitcase and several strings of coral beads hanging around their neck.[31] Many of them have worked in Kumbum, Yushu and Sichuan before, but moved to Dawu because, as they said, 'Goloks like corals'. In spite of high prices (a coral necklace can cost 40,000 yuan), coral is worn by women and also young men, not only on festive occasions but every day (Figure 29).

Summertime shopping is a new phenomenon in Golok. Prior to the caterpillar fungus boom, spring was the 'lean' season, when grassland resources were scarce, livestock exhausted after winter and provisions short. Summer was the time of abundance and recovery, when the herds rebuild their strength, when *dris* reach the peak of their milk production capacity and when the harvesting of livestock products starts. This was also a period of very intensive work. But the selling of dairy products and livestock, which brought pastoralists a cash income, took place in the second half of the year. This was when people had money and could spend it on food and other goods. This pattern is still partially visible: between September and November the pastoralists buy supplies of flour, rice, *rtsampa*, oil, tea, salt and sugar for the coming winter. These supplies are renewed when they run short, and people buy fresh products such as meat, vegetables, as well as different processed foods, during their visits to town. Intensive shopping also takes places before Tibetan New Year, when people buy alcohol, soft drinks, bread, cigarettes, fruit and other delicacies which will decorate their tables. Although the shops can be reached almost any time, autumn wholesale shopping remains part of the household calendar. But whereas the pastoralists buy the major part of their basic food provisions in autumn, shopping for other goods takes place in the earlier part of the year. The caterpillar fungus boom led not only to the increased participation of the pastoralists in the market for consumer goods, but also to the shifting of the shopping season. Shopping as such is no longer tied to the calendar of selling pastoral products; having another source of income, the pastoralists can respond to their needs and the offers of the market at other times of the year. Theoretically speaking, shopping for the household's annual food provisions could also be shifted. Whether this will happen remains to be seen.

31 It was an unusual experience to see these men vanish into thin air when they realized that I wanted to talk about their jobs. Selling coral is a high-risk enterprise and people told me that one seller had been murdered in his hotel room in Golok recently. People claimed that coral sellers are afraid of their customers, who may take revenge on them for selling fake coral.

Conclusions

The phenomenon of the caterpillar fungus boom has taken place in one of those regions of the world that is considered marginal from the point of view of the global economy. Somewhere out there, in the thin air and bleak landscape of the Tibetan plateau, dwell the pastoralists with their yaks and sheep, living in tents and changing camp with the seasons. This image is partly true: there are yaks, some sheep and horses, and even some tents, and people move widely over their land, but not in so spontaneous and uncontrolled a way as many would think. Now, suddenly, these pastoralists have been thrown into the whirlpool of a much bigger economy. It is not true that prior to this they lived in isolation, enclosed in their small economic pond. After Golok became part of the PRC, the region was indeed subjected to a series of political and economic reforms and began to function within the command economy. Yet, pastoralists were not isolated from their neighbours, markets and trade networks before this development. They had long exported and imported goods, even if the scale of these operations was so small that from today's perspective, which is accustomed to seeing things on a large, 'global' scale, it was almost invisible. Nor is it true that the pastoralists used to rely only on pastoral production. On the contrary, they lived from a combination of activities that generated a more or less regular income. Although the trade they practised before 1958 was largely based on barter and that done in the people's communes was carried out in conditions of cash shortage, it was not the case that the pastoralists were unfamiliar with money. They had it in their hands more seldom than they do today, but money was by no means an unknown concept.

Keeping all this in mind, one can properly consider the situation in Golok during the years of the caterpillar fungus boom. The pastoralists, who had *largely* relied on pastoral production, had limited but nonetheless some trade contacts, and produced certain goods sold to other regions both close by and farther away, suddenly turned out to be in possession of something very lucrative and sought after. This was a wonder drug that consumers in the cities were ready to pay large sums of money for. It was a luxury product that many people in China could not afford, but which grew in great abundance in Golok and on other parts of the Tibetan plateau. Digging it up was easy and did not entail complicated activities as with procuring other expensive medicinal products – for example deer antlers, let alone musk pods. In spite of some official restrictions, digging caterpillar fungus was allowed, at least for the pastoralists on whose land it grew. Thus, the pastoralists, who had previous experience with digging it in the people's communes, embraced the opportunity that the new surge of interest in this product created. They became important actors in the commodity chain that started on the Tibetan grasslands and ended in elegant stores in Shanghai, Beijing and Guangzhou.

This book shows that the caterpillar fungus boom in Golok led to the emergence of a whole economy encompassing a broad field of income-generating activities. Prices paid for caterpillar fungus in Golok are a fraction of those seen on the shopping boulevards of Chinese cities, but the money the pastoralists earn from it makes a substantial difference to their budgets. Furthermore, as well as digging and selling the fungus themselves, pastoralists started to lease out their land to diggers, which also brought them income. These two revenue streams combined changed their material fortunes, improving their economic standing and turning many of them – those who live in areas particularly fertile for the fungus – into affluent people. What this influx of money does to the pastoral society has been one of my main interests. But the caterpillar fungus boom as a topic of study has more to offer and is of broader relevance than in economic terms alone. Some aspects of this phenomenon have only been briefly mentioned in this volume but are worth bearing in mind because they show it in a wider context and address bigger questions related to this particular economy and the way in which people in Golok perceive it.

The economic boom that transformed this previously rather marginal medicinal resource into a major source of income for the local population commenced in the very recent past, which most adults remember. This gives these individuals the unique perspective of someone who has taken part in this phenomenon and is affected by it, but who can also reflect upon the way it began and developed. Thus, the pastoralists can see the situation both from the inside and the outside, and although they are deeply involved in this economy, they retain a position as observers – often detached and somewhat distrustful of the mechanisms governing it. This double perspective is a result not only of the short history of the boom, but also of the fact that the popularity of caterpillar fungus is not a Golok invention. The current market career of this product started outside the region and is down to processes happening far away. Thus, the pastoralists supply external markets with a product whose value they do not fully identify with, at least not on a wide scale. If a commodity is expensive, it must be worth its price because of its scarcity or superb quality, as many consumers believe. But caterpillar fungus is not scarce in Golok. On the contrary, it grows in huge quantities that can be measured in thousands of kilograms. The pastoralists are not consumers but producers of caterpillar fungus. Therefore, they can be sceptical about the fungus's powers and even cynical about its consumers' needs. Nor are they the architects of the boom, but suppliers who answer a market call from the outside. In this situation, they can either internalize the market narrative about this wonder drug or distance themselves from it. In whichever case, they must deal with the question of where the prices come from and what the sources are for the popularity of this product, which has become so crucial for their economic well-being. Many pastoralists explained this popularity as being due to the credulity of the Chinese consumers who are ready to believe any fantastic story about caterpillar fungus, or other reasons that make the latter an ideal target group. Popular jokes about why caterpillar fungus (in its Viagra-like role) is popular among Han men can be interpreted as an attempt to rationalize the situation. 'We Tibetans don't need it', many men in Golok would tell me with a knowing grin. 'But Chinese men do. And they must need it a lot, if they pay so much for it.' Thus, although financial dependence on caterpillar fungus among the pastoralists is high, the level of identification with the product, its qualities and market career is low. This makes studying the boom interesting, because it informs not only about the material side of this phenomenon, but also about the subjectivity of people involved, and shows how they position themselves against it.

There is no uniform way of thinking in Golok about the caterpillar fungus boom. It has been shown that in the local discourse even topics as remote from politics as weather may gain a political inflection. As can be expected, the economy is also interpreted through the prism of politics. Any decline in the prices of pastoral products was explained by the pastoralists as being connected to the machinations of the central government and related business circles. The caterpillar fungus economy also became a carrier of political meaning. It developed in a specific political context and is interpreted in terms used to describe historical and contemporary relations between Tibet and China. This economy is often envisioned as a field of competition and contest between differently empowered actors and not as a place of neutral economic activities and relations between equal partners or groups. Seen from this perspective, caterpillar fungus is not a 'neutral' commodity, but is invested with various meanings depending on a person's political (and other) views. According to some pastoralists, caterpillar fungus is yet another natural resource of the Tibetan plateau exploited for China's and not Tibetans' benefit: 'Tibet is like an empty bowl. Resources are gone and Tibetans are just trying to lick at the leftovers. Medicinal resources in China come mostly from Tibet, but we don't know how to use them and don't have the technology. The Chinese have it.' Some of my informants saw all this as a historical injustice; others thought that Tibetans themselves were to blame due to being incapable of making proper use of the resources on their land. However, for many pastoralists there was a clear difference between caterpillar fungus and other resources: the profits from it went, even if not entirely, into their wallets. So, not only did the Chinese state or large companies benefit from this trade; rural inhabitants of the Tibetan plateau did too. Many pastoralists wished that their profits were bigger, but they nonetheless enjoyed their five minutes of economic fame and wanted to exploit it to the maximum.

The word 'exploit' is important in this context and indicates the existence of a debate about who has the upper hand in the caterpillar fungus economy and who exploits whom. Is it the pastoralists who are dependent on the market demand for caterpillar fungus, and once it ends will they lose their financial security? Or is it the great masses of consumers who depend on the pastoralists, because the latter produce something that the former badly need? To use a metaphor inspired by caterpillar fungus biology, who economically parasitizes whom? This question appears simple, but my informants answered it in different ways. On the one hand, caterpillar fungus was for them a symbol of the situation of Tibet within the People's Republic of China, which is treated – so they said – as a pool of mineral resources and raw materials to exploit. On the other, it symbolized the pastoralists' struggle to gain control over the situation and to exploit it to their own advantage.

The caterpillar fungus boom offered a chance for rural Tibetans to capitalize on their land's resources on a scale hitherto unknown. Taking this opportunity meant reasserting their position vis-à-vis the Chinese state which, willingly or not, they are part of. Many pastoralists spoke about their disadvantaged position compared to inhabitants of other parts of China. They spoke about the state authorities, who did not represent their interests and who, as some people believed, purposely acted against them. Exploiting the opportunities created by the caterpillar fungus boom thus meant the pastoralists getting 'their due' and taking symbolic revenge on the state and those of its citizens who were in a more privileged position, economically and otherwise, and who were dependent on caterpillar fungus from the Tibetan plateau. The joke about why caterpillar fungus is popular among Han men has more than anecdotal value – it serves the pastoralists to demonstrate both their superiority over the Han and their agency in a situation that they do not entirely control.

Even if using the opportunities created by the caterpillar fungus boom necessitated engaging in activities declared illegal by the state, the pastoralists were ready to do it. They did not consider the Chinese state and its decisions to be fully legitimate, especially if these policies ignored their interests. It was a common view that one can or should contest or ignore them. Many practices that emerged with the caterpillar fungus economy are characterized by this disobedience factor, which makes them resemble James Scott's everyday forms of resistance from his analysis of the 'weapons of the weak'. This resistance takes the form of 'passive noncompliance, subtle sabotage, evasion and deception', it protects peasants' interests and nibbles away at unpopular government policies (Scott 1985: 31). However, it does not 'contest the formal definitions of hierarchy and power', is 'informal, often covert, and concerned largely with immediate, de facto gains', and is based on 'implicit disavowal of public and symbolic goals' (Scott 1985: 33). This description fits the situation in Golok, with one reservation: many examples of Scott's everyday resistance are perhaps passive compared to the sort of protests that confront the state in a direct or armed way, but the term 'passive' cannot be applied to caterpillar fungus-related practices flourishing in Golok in the shadow of the law. This book has shown how actively this economic field is constructed and maintained: non-compliance with the law regulating the exploitation of this resource may have a political dimension or a *post factum* added justification, but it implies much more than just foot dragging.

Many scenarios described in this book take place with the tacit consent of the local – prefectural, county and township – authorities, who turn a blind eye to certain activities or who interpret the law for the benefit of the people. Thus, my findings expose the weakness of thinking with polarized categories of 'people' and the 'state' or imagining the latter as a uniform bloc of single-minded actors who identify with state ideology and represent its interests. Observing the discrepancy between the formal regulations to which caterpillar fungus-related practices should be subjected and the general tolerance for not complying with them, one can talk about a sabotage of the law by the authorities themselves or a sabotage of the state by its organs at the lower level. Scott's analysis of everyday resistance can therefore be extended to other groups of actors than peasants or pastoralists and include people with a rural background who occupy official positions within the state apparatus and who, in a situation of negotiating interests, stand on the side of their local communities and not that of the state whose power they are supposed to represent. The tolerance for the caterpillar fungus economy or the negligence in executing the law reveals that generalizations fail and that big pictures consist of small pieces. They show that impersonal constructs such as 'state' or 'government' are a mosaic consisting of people who make their own choices and foster their own interests. These people are sometimes difficult to access for foreign scholars, but this should not be an excuse to assign them identities, views or agendas in absentia. The example of Golok illustrates that seemingly powerless pastoralists in fact possess significant power manifested in their ability to influence and cooperate with the local-level government units staffed by people from their communities.

There is much irony in the story of the caterpillar fungus boom and its impact on Golok society. The caterpillar fungus economy is based upon a wild natural resource that is *gathered*. Historically, the pastoralists in Golok lived mainly from pastoral production. They gathered wild grassland products to enrich their diet or be sold for profit. Under the impact of the new economy, these proportions changed and the pastoralists' livelihood strategies reversed the situation of the past. Instead of making a living from animal husbandry and supplementing it with gathering, the pastoralists started to earn most of their income from gathering, while livestock provided them with food products, raw materials and other means of subsistence, and cash only on a limited scale. The pastoralists became gatherers or – in case of leasing out their land to diggers – organizers or managers of gathering. It is a paradox that it was the economic reforms and freeing of the market from the constraints of the command economy that allowed this shift. In China's new economy, gathering has become so lucrative that the income from it covers households' financial needs and makes other occupations seem a waste of economic energy. It is ironic that in the conditions of a supposedly 'upgraded' economy the pastoralists have 'degraded' their production system, expanding that sector which had limited relevance before and was even supposed to be in retreat. Thus, the pastoralists have reversed the order of things as envisioned by some early theorists of culture who saw socio-economic progress as proceeding along a single evolutionary line in which one stage followed another. Long forgotten in mainstream anthropology, this evolutionary thinking has a strong place in the Chinese state narrative about social and economic progress and the hierarchy of who is more or less 'developed'. The gathering economy was ascribed a particular place in this unilinear narrative, one that belonged to the past rather than the present. Recent developments have evinced a renaissance of this economy and proved that not only can it live but even thrive in the conditions of contemporary society.

The main argument of this book is that the pastoralists are active agents of change; they are the creators and sponsors of the transformation of their own socio-economic lives, which they accomplish with the money they earn from the caterpillar fungus economy. They see this transformation as a series of improvements which bring their life and environment closer to what they perceive as optimal or ideal in a given moment. Some aspects of this transformation could also be considered by the state as 'improvements', and they agree with the officially promoted vision of 'development' and 'modernity'. However, one should not assume that the pastoralists unconsciously reproduce the models propagated by the Chinese authorities. The changes observed in Golok only partly agree with the official masterplan for how to 'modernize' the pastoralists. Increased sedentarization and consumerism are those aspects that can be positively valued by the central authorities, but in other areas the pastoralists' aspirations and choices disagree with what the authorities promote. Yak production is a good example: the pastoralists have limited their commercial sale of yaks. As a result, the yak population has grown. This is linked to the caterpillar fungus economy and the material comfort it has brought, but – from the perspective of the state aspiring to control the size of the herds and force the pastoralists to produce more for the market – it can be seen as regress or 'anti-development'. Such changes which diverge from the official plan and sometimes even conflict with the law are not always visible to agencies of state power and are not immediately evident to anthropologists either.

An important question concerns the place of the state in this picture. How is it possible that the caterpillar fungus economy remains largely beyond its control? It has been shown that digging caterpillar fungus should theoretically be regulated and controlled, and that many practices related to it are even penalized. However, this theory is seldom put into practice. There are two possible explanations. The first returns to the simple observation that to formulate a law is easier than to implement it. Digging of the fungus remains uncontrolled because of such prosaic reasons as the size and topography of Golok: it is financially and logistically impossible to mobilize the power necessary to effectively police a land so vast and rugged that even the pastoralists inhabiting it cannot always control things. It would be easier if people controlled themselves or each other – something the authorities promote by promising financial rewards to people who report their neighbours for violating the law. Although my informants expressed their anxiety that local animosities or antipathies might lead people to report each other, such a strategy by the state does not seem viable as long as all or nearly all households are involved in this economy: no one is 'clean' or 'safe' unless they have the protection of relatives working on a high prefecture government level.

While the first answer explains the lack of state control over the caterpillar fungus economy with the lack of means to implement it, the second questions the state's desire to control the situation. Leaving the caterpillar fungus economy uncontrolled and unregulated may actually serve the state and officials' interests. It is an established way of thinking in the PRC that a key to social and political satisfaction and stability lies in the economy: people content with their material situation are believed to be satisfied in non-material terms as well, and less prone to outbursts of political dissatisfaction, social unrest and inter-ethnic aggression. Not intervening too directly into their lives and livelihood practices can be considered beneficial to power holders, and for social harmony and political stability. Moreover, leaving the caterpillar fungus economy uncontrolled is good for the public budget. It reduces the expenses that the government would have to bear to maintain the mechanisms of control and enables savings on other levels. After all, the pastoralists earn and spend money, and they spend it partly on things that represent goals desired by the state. Thus, they remove from the state the responsibility for financing these goals. If it agrees that the pastoralists undertake a quasi-development project of their own, the state no longer has to 'develop' them or invest in such an objective. From the perspective of the state officials, the situation could not be better.

Returning to the question of who exploits whom in the context of the caterpillar fungus economy, it is perhaps possible to say that everyone is doing it to everyone else at the same time or, to put it differently, that everyone uses this economy for her/his own purposes – sometimes at the expense of other actors or in contradiction of their interests. It is tempting to ask who is smarter: the pastoralists who evade the law to maximize their gains and use them in the ways they choose, or the state officials who accept what is happening because it serves their own interests. The ambitions and interests of the pastoralists and of the state in its different levels overlap in some fields and conflict in others. There are also blurred zones where their respective interests are not transparent to each other. Again, it can be said that both sides are smart in their own way or that – paradoxically – both are smarter than each other in different ways and distinct fields. What seems to be a parasitic relationship driven by antagonism of interests can also be interpreted in terms of symbiosis, in which different actors offer each other a certain space for movement. This does not imply balance or harmony, but rather an ongoing process of active adjustment, in which both sides test each other's limits and can move only as far as the other allows. This does not imply conscious consent either; rather, the secret lies in the complex structure and diversity of various actors' interests and the existence of those blurred zones. Hidden agendas, which people are so keen to use to explain what is happening around them, are closely related to the existence of the blurred zones: they are the navigation tools through these areas of lack of transparency.

The caterpillar fungus economy is interpreted by the pastoralists within a political context and ascribed with political meanings. Yet, it is not only politicized but also demonized by some actors in Golok who stress the negative sides of this phenomenon. Some Buddhist monks and other influential people argue that this economy has led to an increase in the pastoralists' dependence on external sources of income, deprived them of their autonomy and self-sufficiency and made them 'lazy'. Digging caterpillar fungus has a deleterious impact on the grassland, as some environmental activists say, and in doing so the pastoralists destroy the economic basis of their pastoral existence. Finally, the caterpillar fungus boom is said to throw the pastoralists into the jaws of consumerism and 'modernity'. Clearly, the pastoralists are envisioned by some – more formally educated and not dependent on the pastoral economy – members of their society as rather innocent and short-sighted folk unaware that once one steps into this world in which everything is cash and consumption driven, it is not easy to leave. Phenomena such as consumerism and 'modernity' lack any intrinsic moral characteristics, but many Golok critics of the caterpillar fungus economy condemn them for being 'Chinese'.

Modernity as such is not a problem, especially in their lives, but that form of it constructed and desired by the Chinese state is. The fact that one may enjoy modern technologies and consumer goods does not mean that one automatically allows these pleasures to others. Common laments that I heard from Buddhist monks and other cultural activists about pastoralists living in houses and driving cars instead of preserving tradition show that in these informants' eyes not all social groups are equally entitled to enjoy such comforts or, ideally, that they should be disinterested in having them. By being 'modern' and active consumers, the pastoralists strengthen their bonds with the state, so the critics stress and claim that this is one of the negative effects of the caterpillar fungus boom. Thus, while the pastoralists see their involvement in the caterpillar fungus economy in terms of contestation of state programmes and ideas and think that they earn at the expense of their co-citizens, the critics see it differently. But the pastoralists with whom I worked viewed these concerns as exaggerated. Moreover, the pastoralists emphasized that such critical voices come from people who do not live from the land and do not face the dilemma of whether to engage in this economy. It is difficult to think that while the Chinese authorities can obviously be ascribed a 'political' agenda, those Tibetan circles which criticize the caterpillar fungus economy for bringing the pastoralists – the custodians of 'Tibetan tradition' – into closer contact with the Chinese state cannot be ascribed one as well.

A question that I was frequently asked during my lectures about the caterpillar fungus boom was what happens if the market collapses? The pastoralists, as people argued, have got used to money coming into their hands as easily as if it was falling from the sky (or in this case growing from underground) and, once this source of income dries up, they will lose financial security and be left fragile if not helpless. When the boom becomes a closed chapter of economic history, what will the pastoralists do? Only life can bring answers to these questions, but it is worth asking why we might expect that the pastoralists are not considering such a scenario. It is common to imagine that the pastoralists are not very compatible with the contemporary world, lacking in understanding of how political and economic institutions work, not shrewd enough to deal with them and thus doomed to lose out economically or in other ways. In consequence, pastoralists – and Tibetan pastoralists are a good example – end up depicted in the literature, as the introduction to this book shows, as victims of the machinations of more influential and more politically and economically empowered groups in society.

The caterpillar fungus boom appears superficially to confirm this picture. The widely observed fact that the pastoral regions of the Tibetan plateau have got more affluent, that men drive fast cars and women wear sumptuous jewellery seem to suggest that the money the pastoralists earn they – rather short-sightedly – 'waste' on consumer goods. However, one of the weaknesses of observation as a method of studying the world is the fact that it is based on what is immediately visible to our eyes, and this realm is limited. This explains why descriptions of the impacts of the caterpillar fungus boom focus on the rise of consumerism as the main side-effect of this phenomenon. Motorbikes, cars and jewellery are hard to miss, but investigating people's bank accounts and other assets comes with bigger difficulties. Inferring from a surge of consumerism in regions such as Golok that the pastoralists 'waste' their money instead of thinking about the future is logically flawed and shows a lack of understanding of how much people really earn. In some cases, much or even all of what people earn is spent on current needs and material goods. In other cases, as this book describes, what the pastoralists actually spend is only part of what they earn. Sometimes, while talking to my informants, I also expressed a concern for what will happen when market demand for the fungus ends. 'We're prepared for that', some people answered. The fact that they accumulated savings, as this book illustrates, indicates that they might indeed be preparing for such an eventuality. Actually, I thought, why are we inclined to assume that they do not think about the future? Perhaps the reason is a deeply rooted cultural belief that they are not able to do so?

This book tells the story of the pastoralists on the Tibetan plateau and their life in the first decade of the 2000s. But the information presented here has a relevance greater than just for that region. It illustrates a situation in which a new form of income becomes available in a region where few such sources existed before, and shows what mechanisms this sets in motion, what forces it stirs up and what effects it brings. Tibetan pastoralists as the main actors of this ethnography can be replaced by other pastoralists in other parts of the world or other rural and even non-rural populations. The role of caterpillar fungus can also be filled by another natural resource or alternative source of income. The scenography can be changed as well, and instead of a pastoral region in China the drama could be set in an entirely different political and economic system.

Returning to the opening paragraphs of this book and the description of Golok in the middle of caterpillar fungus season, we may add one more reason why this economic boom is interesting: it can be considered a contemporary incarnation of the well-known gold rush phenomenon. Anyone who has read Jack London's writings about the Klondike Gold Rush at the turn of the twentieth century knows of the magic spell of this phenomenon which drove people to the limits of the world as they knew it to risk privations and dangers in the hope of striking it lucky. Reading even brief descriptions of what was happening in Golok during the caterpillar fungus boom, one has to think of the similarities between these phenomena. However, there is something that makes the case of Golok different. Unlike in many gold rushes, where Western colonizers monopolized the economic gains while indigenous populations were reduced to the position of observers or suppliers of cheap labour and services, in Golok it is the *locals* who are masters of the situation. The land is in their hands and the caterpillar fungus too. Hence, this ethnography does not offer a narrative of migrant diggers who risk hardships and sacrifice their savings to travel to Golok to dig caterpillar fungus, although their story should also be written. Rather, it tells of the pastoralists whose land has suddenly gained value and attracted attention the likes of which it never had before. It focuses on *their* role in this economic boom and the consequences that it has for *them*.

Afterword: A Note on Methodology

This book is based on eleven months of field research conducted between 2007 and 2010. Carrying out this research was not easy. The study of practices located on the borders of the law is in itself challenging. Working in a politically troubled region adds further difficulty. The research was done on a very low budget. Finally, it was 'partisan research': I conducted it all without an official research permit.

During those years I was in Golok for several months at a time or twice a year. With each visit I felt increasingly at home. I was visiting the same settlements, mountains and valleys. The unmapped space from before the research became zoomed-in and revealed its details in high resolution, becoming *my* place too and turning me into a sort of local patriot. A similar process occurred in my contacts with people.

The repetitive character of my research resembles what Helena Wulff calls 'yo-yo fieldwork' (2007: 139); it comprised a series of comings and goings between my home country and Golok and between different locations within Golok. These repeated visits were valuable: with each one I gained credibility in the eyes of my informants, who grew familiar with me and my work. I observed changes in people's biographies and their reactions to new developments. And they could do the same: my biography was evolving too. Thanks to the changing circumstances under which we met every year, neither side of this contact was frozen in time, stopped at a certain biographical point. Repeated encounters added movie-like motion to a series of still images. We became for each other *lines* and *processes* instead of fixed points, even if endowed with a personal identity.

Being in Golok, I lived quite a normal life alongside the other people there. In Domkhok I did what my hosts did and helped – for better or worse – in their pastoral and domestic work. I milked yaks, tended livestock, collected dung for fuel, dug caterpillar fungus, cooked, babysat children. In Dawu I engaged in daily town life: I performed *kora* around the monastery, ate in the most crowded eateries, spent hours on the main square or at the market.[1] I kept encountering my informants in unexpected contexts: the head of the Religious Affairs Bureau at a smelly meat stall or a respected monk buying

1 *Kora* is a ritual of circumambulation performed around sacred sites or objects, such as mountains, shrines or monasteries, by Buddhists clockwise and by followers of Bon religion counterclockwise. In Dawu it is performed around Aku Chöyon's monastery on the edge of the town.

fancy trainers. These meetings helped us move beyond the surface level of the researcher-informant relationship, which was particularly important with such informants as government officials for whom the 'official' and the 'private' were distinct, if not mutually exclusive, spheres. Gradually, we stopped being research partners and became acquaintances.

I also seized every opportunity to take part in social and religious gatherings and family events and tried to make myself useful. I was there during horse race festivals, singing contests, school anniversaries, religious rituals, New Year celebrations, weddings and house warming parties. I was a community photographer during social and family events, and always carried a thick file of photographs to distribute among people. I taught English, both privately and at school, translated letters from abroad, filled in visa applications and passed English language exams on other people's behalf. Our meetings ceased to be just research oriented and with many people I built friendships or at least (to use anthropological terminology) a good rapport.

Good rapport is crucial for all anthropologists, but it is especially important for those scholars who study phenomena located on the borders of the law or which are ethically ambiguous and for those who work in regions where foreigners are not very welcome with the authorities or the local population. Anthropological research is not a neutral activity, and in many parts of the world the local governments are not keen on the idea of foreigners roaming the land uncontrolled and collecting data for unknown purposes. China is no exception. So, good rapport, in the case of my research, was necessary for my informants to be willing to contribute information to this book. They had to be willing to do it, because this information concerned activities that were not always legal. Moreover, cooperating with a foreign scholar who was working without a research permit could potentially bring problems their way.

For a scholar working in the Tibetan areas of China, obtaining an official research permit is not easy. But even having one still does not guarantee success: it solves many problems but creates new ones. Official affiliation with a Chinese academic institution helps in organizing fieldwork, gives the scholar a pleasant air of security and reduces potential risks that can arise for the local people through cooperating with someone who does not have an official stamp of state approval. However, carrying such affiliation often implies that one is accompanied by a 'research assistant' delegated by the host institution, is thus closely observed and has limited freedom of movement. More importantly, being affiliated with a state institution and working in a region whose population does not fully recognize the authority of this state endangers the scholar's credibility and discourages people from talking about sensitive or controversial topics.[2]

In light of this situation, I decided to work under the guise of being a tourist. This gave me independence and afforded me transparency in contacts with my informants. I studied practices that took place on the borders of the law and my goal was to see them from the perspective of the pastoralists. Their trust and interest in my work were therefore essential. My informants were aware that I did not represent any Chinese institution and was simply writing a book about Golok. They knew that I was doing it for myself, was not paid for doing so (I largely funded this research myself) and, finally, that I put myself at risk in conducting this work: I could have been expelled at any time for not having the right 'papers' and if that was the case this project, so dear to me, would never be accomplished. My informants knew that my research was risky and could decide to cooperate or not depending on their own reckonings. Although I certainly did not look like an emissary of the Chinese state (I did not speak Chinese, moved around on foot and had little apart from a toothbrush, a camera and my field notes), some people remained cautious. Some of them still asked who had 'sent' me there. This concern is understandable given that foreign scholars in China walk a path 'beaten by investigators or researchers whose goals were more clearly of a political nature' (Hansen 2006: 94). Before me, there were plenty of other investigators, ethnographers in the service of the state, officials counting livestock or gathering data for purposes other than academic knowledge.

Working without a research permit had its shadow side. I could conduct my research only in such time units as were allowed by a tourist visa. This explains why my field research was chopped up into three-or four-month stints. This 'unofficial' status also made my situation precarious. Until the end of my research I was not sure whether I would manage to complete it. This insecurity took its toll and had a concomitant high stress factor. I was always ready for someone from the police, the Public Security Bureau (PSB) or another institution to knock at my door, confiscate my field notes and other data and bring my work to a premature end. I routinely hid memory sticks and made sure not to leave any data on my computer. The price of this 'fieldwork conspiracy' was the loss of some valuable interviews which I accidentally deleted from my files. This feeling of constant insecurity and inner terror are well known to scholars working in Tibetan regions and can

2 This happened, for example, to Marius Næss, who was affiliated with the Forestry Bureau of the TAR: his informants assumed that he would hand over all information gathered to the government (2004: 12).

be illustrated by many examples. For instance, I panicked when a PSB officer rang at my door; I had forgotten that he had proposed going for lunch when I applied for an Alien Travel Permit to go to a neighbouring county and had not thought that *obviously* he would know my address. I was also extremely alarmed when police followed me to a meeting with an environmental activist who had previously had unpleasant interactions with the police: they said that I needed to visit a police station for some paperwork – purely a formality as it turned out (but I had the dubious pleasure of seeing a metal 'interrogation chair' complete with restraining belts for hands and feet). When, on the eve of an important state anniversary, a police car with its lights on parked for the whole night in front of my window, I did not go to sleep at all.

Such situations were many and, as with other anthropologists working in Tibet, I tried to watch my step in order not to cross a line behind which my activities could arouse suspicion. However, I did not know where exactly this line lay, especially as in authoritarian regimes anything can take on a political meaning. So as not to provoke fate, especially in the early stages of my research, I put myself under strong self-surveillance, which Emily Yeh describes from Lhasa (2003: 144ff). I avoided talking about politics and got anxious when people expressed their views too openly, asked me about the Dalai Lama or my opinion on China. But they wanted to talk about it and I had to get used to it. I also realized that I did not have to say much, as people – including those whom I barely knew – ascribed me with roles and opinions even before I opened my mouth. The events of 2008, when political protests swept through Tibet, followed by a military crackdown and arrests (which also affected people I worked with) and a later tragic wave of self-immolations, also done in political protest, helped me free myself from this self-censorship. Certain topics simply became impossible to ignore or not talk about. Regardless of how awful it may sound, these events had a 'liberating' impact on my relationships with people in Golok, which became much more outspoken if not, simply, sincere.

Working without a research permit did not mean that I conducted clandestine research. It was perhaps unknown to the embassy that issued my visa that I was not just a regular tourist, but the pastoralists whom I worked with were aware of the purpose of my stay in Golok. After a certain point, even the local officials knew it. I was obviously tolerated. I tried to stick to the official regulations as much as possible: I registered at the police station, as required in China, and applied for an Alien Travel Permit every time I visited areas 'closed' to foreigners. I did not hide. On the contrary, I was conspicuous. Being the only 'foreigner' for many hundreds of square kilometres, a woman in a Tibetan robe, riding a motorbike and speaking the local dialect, I could not fail to attract attention. People saw me, spoke about me and expected me at their homes. Since I was *visible*, I entered people's conversations long before I entered their houses. I was often greeted with words like, 'Oh, you're that woman who was here the last summer!' and realized that people on whose doors I knocked had been waiting for me. Paradoxically, my visibility might have helped me. I thought of myself as a transplanted organ: the new social body did not reject me.

Social anthropology as a scientific discipline has a well-known problem with objectivity and it long ago largely abandoned claims to produce objective knowledge and faithful representations of social worlds. Ethnographies do not tell 'ultimate truths' and my own ethnography as presented in this book has also been filtered through my perception. To reduce the risk that I would tell a story based on limited encounters with similar sorts of people, I consulted a broad group of different informants. I also used a variety of research methods, which included participant observation, interviews, questionnaires and a household survey. The first two methods are classical for social anthropologists who immerse themselves in the field and try to live, to some extent, as the 'natives' do and to conduct an 'engaged, contextually rich and nuanced type of qualitative social research, in which fine-grained daily interactions constitute the lifeblood of the data produced' (Falzon 2009: 1). What made my research different was the household survey, which collected data about the economic functioning of pastoral households. This became a core part of my analysis.

Looking from the outside, my research location appeared small and compact, but it split into distinct, interconnected spaces, between which I moved on a regular basis. I spent most of my time in Domkhok and Dawu but used every chance possible to conduct interviews in other counties as well. I also conducted interviews in Xining, the capital of Qinghai. My field research did not comprise uninterrupted single-site participant observation, rather it was mobile. Given the high level of mobility of people in Golok, stationary fieldwork would not have been the right choice. As I quickly learned, throwing an anchor into one mountain valley in Domkhok often meant sitting at home alone, while my informants were somewhere else: on a pilgrimage, doing shopping or visiting friends.

Not only did I move between places, I also shifted between different groups of people. I interviewed pastoralists, but also caterpillar fungus traders. Not only Tibetans, but also Han and Hui. I interviewed doctors of Tibetan medicine, leaders of the local *dewa*, local state officials, bank employees, shopkeepers, NGO activists, Buddhist monks and diviners, retired commune cadres and even local singers and people working in regional television. They all contributed something to this book. Retired commune cadres shared their memories of the caterpillar fungus trade during the people's commune era, bank employees spoke about the pastoralists' savings and monks complained that people, even though more affluent than they were before, had not become any more generous in their donations to the monasteries. This polyphony of voices – even if some of them are not directly quoted here – impacted my thinking about the caterpillar fungus boom and Golok society.

Moving between different groups of informants was important for me for several reasons. I wanted to see the community in the studied township as a possible focus, one which was particularly suitable for my analysis but which could, with any luck, be closely comparable with another one elsewhere. My ideal scenario was a study that was zoomed in on a certain township, but which allowed zooming the picture back out to see this location in a larger frame in which many other close-ups were possible. So, I entered the zoomed-in part of the map, only to take a step back a while later and look at it from afar: certain things are more clearly visible from a distance. This distance was measured in space, in kilometres, but also in terms of people's biographies, on scales subtler than metric units. Hence, members of the society who did not engage in the pastoral economy or the caterpillar fungus trade also provided valuable insights into the topic. Their role was to disrupt a potentially narrow focus on pastoralists and traders and help discern the ways and extent to which the caterpillar fungus boom had affected the area. My research was thus going deep and broad at the same time: Domkhok and its population formed the 'in-depth' part, while other locations and people gave it the 'breadth' dimension which helped embed my analysis in a wider context. Approaching the topic from several angles made it possible – staying with the photographic metaphor – to see the object in different lights. Each frame showed something new, but they all ultimately depicted the same object of my inquiries.

The major part of the information about the caterpillar fungus trade, people's income from it and other issues related to the economic functioning of pastoral households comes from the household survey. It is commonly believed that conducting this type of research is difficult if not impossible in Tibet, unless it is done under the auspices of a Chinese academic institution (Makley 2007: 16; Yeh 2005: 99). Carrying out such research implies that a scholar has to meet large numbers of informants and cannot keep the low profile so desired when one is working in a politically sensitive region. Scholars working in such places often rely on trustworthy or 'checked' informants recommended by their hosts and colleagues. This gives a sense of security, but also creates the risk that one may be confined to a single social circle or faction of society, and thus that one collects information which represents only this group and not a wider community. In order to avoid this risk, I worked in a systematic way and tried to make my research representative of the local society. The ideal of representativeness was for me something more than a matter of scientific reliability or producing data that could be used by others. I wanted to know whether what I observed was typical or isolated, marginal or shared by a larger group.

In order to conduct the survey, I had to select a sample of households. The sample I chose was homogenous and diverse at the same time. It comprised households in one township and one *dewa*, but the households differed in respect to other factors which could impact research results. They were big or small, belonged to young couples or well-established families, were situated in large or small valleys, fertile ones or those a bit less fecund, were well connected with town or more remote. This sample counted fifty families – about fourteen per cent of all households in Domkhok. Even among such a small group, experiences, situations and scenarios repeated themselves and the material got saturated. This indicated that including new cases would then be redundant and the data gathered could be used cautiously to infer broader conclusions about the local community.

The survey was composed of thirteen pages of questions covering issues related to economic functioning and family structure, the caterpillar fungus economy and pastoral production, living conditions, consumption and financial needs. The answers I received were not always accurate (and I did not expect them to be so), but I must acknowledge my informants' generosity in devoting their time to the survey, sometimes at the expense of other duties. Moreover, the time involved was quite significant: completing a full survey could take up to two hours. With one man, for example, I conducted it in the township village where he had come to supervise building of his house; I spread out my papers on the hood of his car, and he stood beside me answering the questions despite continuous calls from the construction site. I was seldom met with refusal and usually not permanent ones. In one case, a young man refused to take part in the survey, which made his neighbour scold him: 'She travelled so far and you don't want to talk?' When we next met, in different circumstances, this man turned out to be one of the best informants I ever had.

Gathering a representative sample for the survey (and conducting the survey as such) was possible thanks to my long-term acquaintance with the Domkhok community and it would not have been very successful (or perhaps altogether impossible) if I had not spent sufficient time there before. But even though I knew many people, I did not know everyone. The survey was an interesting exercise because it made me depart from my initial network of contacts and enabled me to meet people whom I normally would not have had a chance to talk to. This caused me to realize how much escapes our attention when we work in a less rigorous way. The tendency to work with 'safe' or recommended informants limits the scope of our experience and also creates a risk that the scholar becomes a captive of the social milieu or faction of her or his hosts. As Paul Rabinow noted, an anthropologist can imagine her- or himself as a neutral knowledge seeker but gets 'involved in politics and social divisions even before he [or she] enters the village' (1977: 92). This sort of experience is shared by many scholars, who recall being treated as 'a prize asset' by people they stayed with, 'paraded as a status symbol' or not allowed to freely choose their informants (Shore 1999: 31-32). I also realized at some point that I was surrounded by a glass wall. My hosts were sometimes overly protective and did not want me to socialize with everyone and discouraged me from visiting certain families, saying – often suspiciously – that they were not at home, were 'bad people' or implied that they were in the pay of the police. The survey helped me to change that situation.

What is even more dangerous, as it is not always perceived, is the fact that anthropologists often have a *Beuteschema* and are attracted by certain types of informants. Becoming aware of this and working with people who do not correspond with this model, who are not cooperative and often require more effort on the part of the scholar enriches the research. One of my most valuable informants was, for example, a retired Tibetan PLA soldier: after just one week of my stay in his house (where I was invited by his son) the ice broke and we started talking. Both of us needed time: he, a hardened Communist, and I, who came to Tibetan Studies largely through political engagement. After that we were best friends and he provided me with unique data.

The period when I was conducting the household survey was the most arduous part of my research as I carried it out on a small budget but had to interview people who were living dispersed over a large territory. There are no buses or shared taxis between Dawu and Domkhok and people rely on their own means of transport or ask relatives or neighbours for help. Private cars do take passengers, but the roads in Golok are not especially frequent with traffic and one can wait for hours to catch a lift. The only other solution was to rent a car. Not having the money to do this, I travelled in an improvised way. I borrowed motorbikes, hitchhiked and trekked from one valley to another. I climbed mountain slopes, trudged through mud, and forded streams and rivers on foot. This way of travelling influenced the hierarchy of relations in the field. It is common to imagine the anthropologist as coming from an economically privileged world and having the upper hand in relations with her or his informants. I did come from such a world but was not particularly privileged. My informants, when they inquired about my financial situation, usually pitied me. From their perspective, I was poor like a church mouse and people often offered me money. Things I did and the way I travelled surprised them. Rather than arriving in an air-conditioned SUV which sailed through the grass to reach the pastoralists' tents in the mountains, as some well-connected scholar affiliated with a Chinese academic institution might do, I painstakingly fought my way uphill (remember the altitude!) to reach people's camps and shouted from a distance to check whether the dogs (huge mastiffs, dangerous even to their owners) were chained up. I like to think that this way of travelling served my research well. I did not want people to talk to me because of any special status they could ascribe to me. If someone was not going to treat me seriously because I did not have the attire and entourage of a 'serious scholar', I was ready to pay such a price.

Travelling in this improvised way introduces (again) a large dose of unpredictability into field research, slowing it down, but also opening the space to new experiences and unexpected encounters. Trekking through the highlands let me experience moments of rare beauty. What can compare with sitting to rest on a frozen river, enjoying the silence of snow-clad mountains and the emptiness of a landscape with no sign of human habitation in sight? Waiting for lifts took up a considerable part of my time, and was sometimes frustrating, but the meetings on the road amply rewarded me for this hardship. Things in the field often happen by chance, but only if one gives them a chance to happen. And so, one of my favourite interviews, with a man who had been imprisoned for brigandry in the Republic of China, occurred after a ride with the only driver who gave me a lift that day: hearing his family story, I changed my plans and went to interview his dying father. I was the last person to whom he told his story. It is amazing how many 'byproducts' anthropologists produce alongside their main research. Such byproducts and stories of accidental encounters will be the material for another book.

When field research fails, it is often easy to tell (wrongly or rightly) what hindered it or made it impossible. When, on the contrary, one has a feeling that the research has gone well, it is an open question as to what contributed to its success. There are many things apart from anthropological skills and experience that can help. If there was something that helped me it was my enthusiasm – affection even – for Golok and its people. I expressed this enthusiasm for everything local, including dirt, the monotonous diet and the all-pervading cold, on endless occasions, and openly identified with the local manners, ethos and even those parts of local identity that were less spoken about – for example the banditry for which Golok is famous. My informants knew that I always wanted to work in this particular region and they appreciated it. Even police officers took pride in the fact that I travelled so far to visit Golok, which is otherwise not very popular, neither among visitors from within China nor those from abroad.

Another surprising factor was the language, or rather its lack: I did not speak Chinese. I explained to all Tibetans who tried to communicate with me in that language that 'I don't speak Chinese. I didn't come to China, but to Tibet.' This was quite obviously a political statement.[3] This linguistic handicap was paradoxically illuminating: I found myself in a situation similar to that of many rural Tibetans. Being in Golok, I survived by speaking Tibetan as most in-migrant shopkeepers or taxi drivers speak it, but I was helpless as soon as I left the region. Another political message that I produced, quite unconsciously, came through my clothing. Because I had had a bus accident on the way to my first period of fieldwork and lost my luggage, I had to shop for clothes in Golok. Since then I wore a Tibetan robe. I had never planned to do this and sometimes thought about it in terms of manipulating, if not faking one's identity for the sake of winning people's favour. It worked out differently though, and people took it as a sign of respect that I dressed in Tibetan style and not Chinese (as 'Western' clothing was called there). I was complimented for looking like a 'Tibetan woman' and did my best to learn the local etiquette and body language. The role of our appearance and body in contributing to our research is a significant topic. What played a role in my case was that I did not appear to be very different from my informants, in their own perspective, especially compared to other foreigners. People expressed their surprise that I was 'so normal', as they said. One day, when a couple of Swiss tourists passed through Domkhok (even the woman was taller than the tallest man in the township), one pastoralist asked: 'Why are you so small if they are so big?!' My size worked to my advantage.

3 Language also matters in other contexts. Gert Jan Veldwisch recalls from his work in rural Uzbekistan that even when he spoke Uzbek, people kept using Russian with him. Russian was the language of 'the centre', used by city people, old, male and educated: 'Learning Uzbek helped me shift my bias to the people in the margin; the younger people, the women, the uneducated, and the people in the rural areas' (2008: 172).

The last thing to mention is my personal experience. As Kirsten Hastrup says, fieldwork is a 'personal adventure and belongs between autobiography and anthropology' (1992: 119). This implies that an anthropologist is a person with a distinct biography, and that they must use '[themselves] and [their] own experience as primary research tools' (Watson 1999: 4). The fact that I come from a country that was once Communist and hence within the political orbit of the Soviet Union (associated by many Golok pastoralists with the source of all evil that befell them) opened many doors to me. I knew very well, from my own and my family's experience, that things are never black and white and was aware how complex the biographies of people living in such circumstances are. Poland's history, both the more recent and going further back, with over a century-long period of partitions, was a topic that people enjoyed as it had parallels with their own experiences. They could feel familiar with me. I was for them a source of information and a partner in discussions about politics, history and more. This 'Second World' background afforded me a position as someone who was from the West but not quite – an anomaly interesting for my informants who studied me as critically as I studied them.

Appendix

[Page 1:]
Documents of the Chancery of the People's Government of Maqin County
Announcement on distributing 'The executive programme of protective and administrative work of the Chancery of the People's Government of Maqin County in 2010 on Chinese caterpillar fungus resources'.[1]

To all (Township) People's Governments and all county-level subordinated departments:

Following research and general agreement, 'The executive programme of protective and administrative work of the Chancery of the People's Government of Maqin County in 2010 on Chinese caterpillar fungus resources' is hereby distributed so that it be zealously and directly implemented by you.

7 April 2010
(Chancery of the People's Government of Maqin County) Tags: administration of resources, caterpillar fungus, programme, distribution, announcement

CC: County Party Committee, County People's Assembly, County People's Government, County Chancery Disciplinary Commission, Chancery of Maqin County People's Government

7 April 2010 issued in 35 copies

[Page 2:]
'The executive programme of protective and administrative work of the Chancery of the People's Government of Maqin County in 2010 on Chinese caterpillar fungus resources'

In order to continue to enhance protection and management of Chinese caterpillar fungus resources, to thoroughly advance standardized management of caterpillar fungus collection activity, effectively keep a check on predatory picking of the fungus, which inflicts damage to ecological environment of the grassland, to further protect the area of the Three Rivers' sources [Yellow River, Mekong and Yangtze] and the grasslands' fragile ecological

1 In the following document, all place names are given in pinyin, not in their Tibetan version.

environment, ensure rational development and sustainable use of the caterpillar fungus resources, gradually create a harmonious society which economizes resources and is friendly towards the environment, stimulate overall and lasting development of the socialist economy based on thorough research and detailed assessment of the protection and management of the caterpillar fungus resources hitherto, according to the stipulations of the Grassland Law of the People's Republic of China (*Zhonghua Renmin Gongheguo caoyuan fa*), the Law of Qinghai Province on Contracts Pertaining to Grass Farms (*Qinghaisheng caochang chengbao banfa*), Regulations of the People's Republic of China on Protection of Wild Plants (*Zhonghua Renmin Gongheguo yesheng zhiwu baohu tiaoli*), Administrative Regulations on Protection of Medical Wild Herbs Resources (*Yesheng yaocai ziyuan baohu guanli tiaoli*), Administrative Methods of Collecting Dry Grass and Chinese Ephedra (*Mahuang*) (*Gancao he mahuang caiji guanli banfa*) and Administrative Regulations on Grasslands in Golok Tibetan Autonomous Prefecture (*Guoluo Zangzu Zizhizhou caoyuan guanli tiaoli*), Golok Prefecture 2008 Executive Programme of Caterpillar Fungus Resources Management (*Guoluozhou 2008 nian chongcao ziyuan baohu guanli shishi fang'an*) as well as the Executive Opinions on the Work on Protection and Management of the Caterpillar Resources in Maqin County between 2007 and 2010 (*Maqin Xian 2007 he 2010 nian chongcao ziyuan baohu guanli gongzuo shishi yijian*) and other relevant laws and regulations, in conjunction with the reality in our county, the following Executive Programme is drawn up:

1. Unify thinking, progress in deepening knowledge of the importance and urgency of administrative work on caterpillar fungus resources management.

The [Chinese Communist] Party and the state attach great importance to the work on development and protection of the ecological environment of the area of the Three Rivers' grasslands, investing significant capital [to this end], carrying out development and protection through restoration of pastureland back to grassland, protection of natural forests, sowing herbage grass, extermination of rats and worms and through other projects; pastoralists have obtained material benefits and profits from these measures. The 17th [Chinese Communist] Party Congress more explicitly came out with a request to create 'an ecological civilization', and at the provincial level the strategy of the 'ecological development of the province'
[Page 3:]
has been formally established. Therefore, it appears to be of particular importance that we further protect this area and its fragile grassland

ecology, rationally develop local caterpillar fungus resources and prohibit outsiders from coming into our county to pick the fungus. Our county is situated in the central zone of the Three Rivers Nature Reserve; its grassland ecological system is extremely fragile and it is easy to cause damage to this environment. If the ecological system of grasslands at high [altitude] and in such a cold [climate] is damaged then it is very difficult to restore. And this – significantly – not only directly affects the continuous development of the local ecological socio-economic environment but also limits local pastoralists' living standards and the area's social stability. Implementing protection and management work for caterpillar fungus resources must be based on scientific development and stimulating social harmony, aiming at protection of the grasslands' ecological environment and rational use of caterpillar fungus resources, to stimulate the use of these resources while protecting them at the same time, between stimulation of sustainable socio-economic development as well as harmonious coexistence between the humans and nature.

2. The key points on and the basis of political regulations for administrative and protective work on the caterpillar fungus resources.

The key points on and the basis of political regulations for administrative and protective work on the caterpillar fungus resources are: thorough and direct implementation of a scientific development approach, according to the request of the 17th [Chinese Communist] Party Congress to create '[an] ecological civilization' and the strategy of 'ecological development of the province' established by [this] province, to scientifically plan and manage the caterpillar fungus resources in conformity with the law, enhance the protection of the ecological environment of the grasslands, stimulate harmonious development between humans and nature, create a harmonious socialist society so that the concept of 'the ecological civilization' penetrates [everybody's] conscience and makes this region ceaselessly advance along the road of civilized development [under the motto of] 'production develops, life is prosperous, ecology is good, society is stable'. The political regulations basis is: the Grassland Law of the People's Republic of China, Regulations of the People's Republic of China on Nature Protection Areas, Administrative Regulations on Protection of Medical Wild Herbs Resources, The Law of the People's Republic of China on Forests, Regulations of the People's Republic of China on Protection of Wild Plants, Administrative Methods of Collecting Dry Grass and Chinese Ephedra (*Mahuang*), the Law of Qinghai Province on Contracts Pertaining to Grass Farms, Detailed Executive Regulations of

Qinghai Province on the Grassland Law of the People's Republic of China, Administrative Regulations on Grasslands in Golok Tibetan Autonomous Prefecture and
[Page 4:]
Golok Prefecture Methods of Caterpillar Fungus Resources Management and the stipulations of any other relevant laws and regulations.

3. Principles of the work on protection and management of the caterpillar fungus resources in 2010 and general requirements:

Enhancement of protection of the ecological environment [should be] beneficial to stimulation of structural economic adjustment and transformation into the growth model, bringing about faster and better development; [should be] beneficial to raising the level of knowledge of ecology within the whole society, stimulating creation of new, socialist pasturelands; [should be] beneficial to safeguarding long-term profits of broad masses of pastoralists, providing a good livelihood and room for development for future generations. Therefore, in the work on protection and management of caterpillar fungus resources it is necessary to stick to the following principles and requirements:

1. Stick to the principle of stimulating 'the ecological civilization' and 'establishing the ecological province'
2. Stick to the principle of overall planning of the harmonious development of humans and nature as well as creating a harmonious society.
3. Stick to the principle of linking ecological protection and increase the income of pastoralists.
4. Stick to the principle of strict control of resources and management of subordinated areas.
5. Stick to the principle of protection in conformity with law, scientific planning, rational collecting and sustainable use.
6. Stick to the principle of collecting [of the fungus] by the local people and strict prohibition of collecting by outsiders coming into our county.
7. Stick to the principle of forbidding the sale [or] contracting of caterpillar fungus collection rights on mountain [slopes], in ravines and on [river] banks.
8. Stick to the guiding principle of stability overriding everything [else], properly deal with the relationships between protecting social stability and increasing pastoralists' income.

9. Stick to the principle of sovereignty of grasslands and let the majority of pastoralists manage the caterpillar fungus resources by themselves, strictly prevent random picking and uprooting, and protect the ecological environment of the grasslands.

For the coordinated, stable and continuous development of the ecology, economy and society of our county, the purpose is scientific management, orderly collection, protection of ecology, increasing the income of the masses and the stability of the region. General requirements are: in 2010 continue implementing measures on forbidding collection of caterpillar fungus, continuing to take measures to provide for stability to be of the first importance,
[Page 5:]
ban [on fungus collection to be applied] to outsiders and restrictions to the locals, the masses to exercise self-control and the [local] government to strictly observe the ban on fungus collection; as for grasslands managed by the pastoralists themselves, at farms [operating] on a contract basis caterpillar fungus may be collected in a planned way; for those pastoralists who come from poor families among ecological migrants from Maduo County and this county upon consultation with the government in charge of a caterpillar fungus production area and under premises of ecological protection, rational development, scientific use and maintaining stability, collection of the fungus is allowed in a planned and organized way with villages (or brigades) being basic units; it is strictly forbidden that whoever from outside this county and any city dwellers from the prefecture and the county, cadres and staff workers (including relatives as well as children), pensioners, unemployed persons (laid off, awaiting employment, unemployed), people with minimum-livelihood husbandry roles, self-employed industrial goods producers and traders, students from caterpillar fungus non-producing areas and other people be allowed by any means to collect the fungus.

4. Detailed measures and tasks for the protection and management work on caterpillar fungus resources.

1. Measures for the work contained in 28 [Chinese] characters: strictly exercise control over sources, mobilize cadres and masses, let everyone take his/her responsibility, he who recruits advises, [collect] irrefutable evidence, deal with [trespassers] severely, do not slacken the grip.

2. Editing regulations on fungus resources protection schemes and annual collection plans. All townships with fungus resources should scientifically review resources protection schemes and annual collection plans related to the area, according to regulations [formulated in] The '2005 Golok Prefecture opinions on strengthening protection of caterpillar fungus resources' and in conjunction with situation on the local grasslands and with caterpillar fungus resources as well as the labour force. Those regulations [should then be] reported to Agriculture and Animal Husbandry Bureaus of the county [government] for examination and upon approval [should be] put into practice. Upon editing caterpillar fungus protection schemes and annual [collection] plans for the whole county, the county bureaus [should] report to the Agriculture and Animal Husbandry Bureaus at the prefecture [level]. The schemes must specify details such as area of distribution of caterpillar fungus resources, the area under the collection ban, collection periods and numbers of collectors allowed. Upon approval of those schemes by higher authorities they will be disseminated among the public in order to raise the [level of] transparency of [our] work.
3. Improve standardizing the caterpillar fungus market. According to paragraph 19 of the Regulations of the People's Republic of China on Protection of Wild Animals and Plants, which reads 'people selling and buying second-grade state-protected wild animals and plants must obtain permissions from departments of the People's Governments of a province, an autonomous prefecture or directly subordinated unit supervising administration of wild animals and plants

[Page 6:]

or from their plenipotentiaries'. According to paragraph no. 24 [of the same Regulations] 'those who violate these regulations and sell and purchase wild animals and plants [classified as] key state-protected ones, will be fined by administrative departments in charge of industry and trade or wild animals and plants, up to ten times the amount of their illegal incomes, based on degree of responsibility, [and their] wild animals and plants as well as illegal income will be confiscated.' Any activity amounting to purchasing of caterpillar fungus requires obtaining approval from relevant government departments and permission [in writing]. At the same time, business permission must be obtained from Industry and Trade Bureaus to carry out commercial activity at a specified market. As for our county, the commercial market for caterpillar fungus is located on Qixiang [Meteorological] Lane of

Dawu Town. It is strictly forbidden that buyers of caterpillar fungus enter production areas to carry out trade; those who purchase the fungus in the production areas, upon investigation and prosecution, will be fined in accordance with [regulations] pertaining to outsiders. Persons who perform state duties and staff members of enterprises without [written] permission to purchase caterpillar fungus are not allowed to deal commercially with the fungus.

4. Improve accuracy of administrative law on caterpillar fungus resources. As for the scope of the administrative law on collecting caterpillar fungus, it is necessary to fully respect the will of the masses from the [fungus-] producing areas and base [all thinking] on units such as a herders' committee or herders' teams to manage the collecting of caterpillar fungus and formulate concrete management methods in accordance with opinions [expressed] by the majority of people from the [fungus-] producing areas. In principle, the majority of pastoralists should manage contracted grasslands by themselves in order to reach the goal of controlling the sources. It is forbidden for anyone who enjoys the right to exploit the grasslands to dig caterpillar fungus in disputed areas between counties, townships, villages and pastoralists' pastures.
5. Act according to local conditions, combine [theory] with practice, establish management plans for the collection of caterpillar fungus. In grasslands management, stick to the principle of '[proper] managing [of] subordinate areas' and respecting the will of the majority of the population of the [fungus-] producing areas; implement the working system of controlling the sources and self-management of the masses; take up concrete management measures for macro-control on the county level and practical operation in townships and concrete management in villages; focus on working for coordination between service [for the nation] as well as supervision and examination; hand over the responsibilities for caterpillar fungus management to villages and brigades; establish management units [in the form] of herders' committees and production cooperatives, organize local populations to expand their activity in caterpillar fungus collecting;

[Page 7:]

let basic organizations in villages and brigades enjoy full sovereignty. All townships and related [administrative] units have to work out management methods suitable for their practical conditions and report them to the Leading Group for the Management Work on Chinese-Tibetan Medicinal Resources.

6. Establishment and management of the caterpillar fungus resources compensation fund. Standards for reception (that is, eligibility) for the caterpillar fungus compensation fund are discussed and established by a Township Congress of People's Deputies; 90 per cent of grasslands fund beneficiaries are masses in production areas, the remaining 7 per cent of the fund is paid to cooperatives and 3 per cent to a herders' committee. The fund is mostly utilized for management of caterpillar fungus collection and for daily management of grasslands. Methods of management of village funds at the township level and use at the village level strengthen collective management of financial resources. As for utilization of the grasslands compensation fund, it is up to a county's audit department to audit and the state of affairs is to be reported to the chancery of the county's Leading Group for Management of Chinese-Tibetan Medicinal Resources and the county government. When [cases] of violation of discipline and regulations are discovered, such as withholding or embezzlement, then the county government will investigate them to determine who is responsible.

5. Working demands and guarantee measures:

(i) strengthen leadership, pay attention [to the situation] to a great extent, earnestly provide solid organizational guarantees for work on banning the collection of caterpillar fungus.

Efforts to ban the collection of caterpillar fungus is complex and protracted work of a strongly political nature. All townships of a county [and] all departments should pay great attention [to it] and regulate and replenish on time the leading groups for work on protection and management of Chinese-Tibetan medicinal resources at all levels, strictly implement the policy of 'one hand grip' by the [local] government, and, according to the principle of 'the county is responsible for a township, the township for a village, the village for husbandry', strictly specify the scope of work and responsibility. At the time of caterpillar fungus collection, emergency leading committees in counties and townships should be formed for work on banning [unauthorized] collection with the principle of superiority and subordination between two immediately interdependent levels only.

(ii) Unite the way of thinking and knowledge, enlarge the dynamics of propaganda in favour of grassland environmental protection.

All governments at the county and township level and all departments must consider the work for protection of the grassland environment
[Page 8:]
and ban [illegal] collection of caterpillar fungus as their top priority and lead the preparatory work in this. Actively organize cadres and staff members of the area and/or department(s) to learn relevant laws and regulations, grasp the ideological essence [of this situation], unite the actions of cadres and the people. Mainly the following measures are to be taken:

1. Basic cadres of government organization at the township level [should] go deep into villages, brigades and pastoralists' lands, broadly launch multifaceted propaganda activity, cause the propaganda contents to penetrate into people's hearts, increase the responsibility and knowledge of the environment among the populace. Actively guide the pastoralists' initiative to attend to management work on caterpillar fungus collection protection, to shape a good situation in which all the people [are able to] participate.

2. According to the principle of rational arrangement and usefulness of simplification, this year the work on establishment of propaganda stations for environmental protection in our county will be integrated with establishment of propaganda stations for environment protection on the prefecture level. [This year] in conjunction with practical conditions in our county, one propaganda station will be founded in Dawu Town and two mobile propaganda stations along the Xining-Jiuzhi [Tib. Jigdril] Highway. As well as that, two propaganda stations for management and protection of Chinese and Tibetan medical resources in the Golok Prefecture will be founded along the Xining-Jiuzhi Highway and the Huaji Highway – that is, the Number 2 Propaganda Station for Management and Protection of Chinese and Tibetan Medical Resources in the Golok Prefecture next to the government of former Chamahe Township and the Number 3 Propaganda Station for Management and Protection of Chinese and Tibetan Medical Resources in the Golok Prefecture from Hebei Township to Lajia [Tib. Rarja] Town Deyanggou Pass. For every propaganda station there has to be one cadre in charge of liaison, who has to prepare three official notifications: an announcement to inform [the public] about names and posts of personnel [manning the station], information about the scope of activity [of those personnel] and a governmental public notice. The person who will man that post has to attend a pre-assignment training course arranged by the County's Leading Group for Management of Chinese-Tibetan Medical Resources and

assumes [his/her] duties only upon receiving a certificate. In other townships with caterpillar fungus resources, propaganda stations [for dissemination of knowledge about] environment protection are to be established at their boundaries and major intersections. The County's Leading Groups for Work on Management of Chinese-Tibetan Medical Resources should establish propaganda stations along state and provincial highways departments as well as representatives of the masses working to transfer personnel from security, traffic, land resources, environmental protection, water resources, agriculture, animal husbandry and so forth; these should be established in order to guarantee that all departments divide the labour, mutually and closely coordinate, cooperate, supervise one another, earnestly fulfil their working obligations of 'guarding the gate, taking care of the people, managing the affairs [properly], safeguarding stability, stimulating an increase of income'.

[Page 9:]
During their term in office, the County's Leading Groups for Work on Management of Chinese-Tibetan Medical Resources will provide a certain amount of economic allowance and the propaganda stations upon discovering any form of violation of regulations and discipline and after investigating them refer the case(s) to the County's Leading Group for Dealing with Violation of Discipline and Regulations in the Management of Caterpillar Fungus Resources. Unauthorized persons who enter our county [with the purpose] to dig caterpillar fungus will bear the expenses for their being diverted to another location without exception; all the propaganda stations for environmental protection established by townships must be built along township roads and not along state and provincial highways.

3. In April, when 'The All-Prefecture Environmental Protection Propaganda Month' is due, relevant units in charge of agriculture, animal husbandry, forestry, land resources, water management and environmental protection, security, industry and trade, grasslands legal affairs and so on all over the county will launch a movement for propaganda and education on a large scale to produce propaganda materials in both Tibetan and Chinese, fully utilizing propaganda tools such as radio and TV, propaganda vehicles, banners, scrolls and so forth, to bring home to everyone the importance of caterpillar fungus resources protection.

4. In the second week of March, the county leadership and officials from relevant departments [are required to] go to Xining, Haidong, as well as prefectures, counties and other areas where working people from the county are concentrated and will continue to propagate the policy of our county on banning collection of caterpillar fungus using publications, posters, distributed circulars, SMS and other forms of information and disseminating the policy through radio, TV and other media. At the same time, [it is necessary to] strengthen coordination and contacts with governments of neighbouring areas such as Xinghai, Gansu, Aba [Tib. Ngawa] Prefecture and so forth and request them to support out county's administrative work on banning collection of caterpillar fungus and assume overall and effective control [of those resources].

5. At the end of March, under the leadership of the County Bureau of Agriculture and Animal Husbandry, and with participation of other departments, training courses for key cadres working on banning collection of caterpillar fungus in counties, townships and villages will be specially arranged by party schools of county party committees in order to strengthen the important and urgent knowledge of work on banning collection of caterpillar fungus. The county political and legal department will continue to provide legal education to the pastoralist masses and create an appropriate environment for the ban on caterpillar fungus collection.

(iii) Clarify responsibility, create a goal responsibility system.

[Page 10:]
The work on banning collection of caterpillar fungus is to be continuously carried out by dividing it up and assigning responsibility, namely to: cadres at the county level take responsibility for propaganda stations in villages and townships, cadres at the village and township level take responsibility for key areas, ordinary cadres at the county, village and brigade level take responsibility for brigades and farms. They are fully responsible for the work [of their] liaison points (*lianxidian*) on banning collection of caterpillar fungus and departments directly subordinated to the county [government] – with an exception of personnel transferred to man all the propaganda stations and some personnel staying on duty [in departments] – have to take responsibility for work on banning collection of caterpillar fungus in villages and brigades and people in charge of this must go to farms, establish the [propaganda] stations and start their work of being on duty 24 hours. If not then all the problems and consequences resulting thereof will be blamed on the leadership of liaison points. The

County's Leading Groups for Work on Management of Chinese-Tibetan Medical Resources and leading personnel of liaison points at the county level sign a target memo on responsibility [for the work at the propaganda stations] with townships and [relevant] stations and if any problems arise in a given area then the personnel of that liaison point is responsible for those areas. Investigation [in cases of] responsibility [violation] will be strictly conducted in accordance with relevant regulations [as stipulated in] 'Maqin County viewpoints on consistent implementation of "the (temporary) methods of investigating responsibility for violation of the ways of management of the caterpillar fungus resources in Golok Prefecture"'. Memos on responsibility signed at specific levels [of administrations] between the county, townships, villages and farms introduce annual evaluation methods based on critical examination of outstanding performance being rewarded and negligence being fined. All the units directly subordinated to institutions at township and county level must hand over commitments in a written form related to their work on banning the collection [of caterpillar fungus] to the county's leading group for management of the caterpillar fungus resources and all the basic organizations have to fulfil their commitments to township governments; all the farms have to fulfil their commitments to the local basic organizations in turn meeting their commitments layer by layer. Prior to 4 October, upon surrendering of 'grasslands utilization permits' issued to pastoralists' farms on contracted grasslands all over the county, those permits will be registered by the issuing office and upon issuing a legal notarization proper arrangements will be made to issue those permits anew upon completion of the work on banning the collection of caterpillar fungus. As for pastoralists' farms that violate regulations for the work on management of caterpillar fungus resources protection, Grasslands utilization permits will not be issued and their rights to use the grasslands will be bestowed upon other beneficiaries against compensation.

(iv) Strengthen coordination, delegate responsibility, properly carry out the work on banning the collection of caterpillar fungus.

All the herders' committees and townships with caterpillar fungus resources can establish additional environmental protection propaganda stations along the township and village roads which are in their area of competence
[Page 11:]
and persuade outsiders to return to their original residence; at the same time, by 4 October of each year, all the herders' committees and farms in

townships with caterpillar fungus resources have to register and turn back outsiders who have entered our county under the pretext of assisting in work and grazing. Detailed registration has to be conducted in relation to work teams and traders who enter caterpillar fungus-producing areas. As long as major projects are not in question all work must be stopped with no exceptions [granted] and it may be resumed only after the period of caterpillar fungus collection is over. As for major projects, it is necessary to produce relevant certificates when lodging applications at the office of a leading group for management of caterpillar fungus resources. After approval, as a contract of responsibility is signed by the government of a township where a caterpillar fungus-producing area is located and leaders of a work team and upon imbursement of a guarantee only then may the latter party enter [the producing area]. It is necessary to strictly strengthen the management of working personnel, for instance if it turns out that the working personnel has dug caterpillar fungus in the mountains then fines have to meted out based on an average rate of 10,000 yuan per work team. It is also necessary to strengthen supervision over traders, as any purchase of caterpillar fungus must be conducted at designated markets only. Traders are not allowed to enter the caterpillar fungus-producing areas, otherwise they will be punished for illegal collection [of caterpillar fungus]. All the townships and herders' committees have to organize inspection teams [consisting] of basic cadres, party members and people's militia members to patrol grasslands in their areas and make sure that no outsiders are left in any house and any farm.

The People's Government of Dawu Town and directly subordinated police substations as well as the county's Industry and Trade Bureau must strengthen management of house rental in the Dawu area and forbid any rental to outsiders coming to dig caterpillar fungus. If a house is rented to an outsider coming to dig caterpillar fungus then the lessor will be severely punished according to management methods of house rental and all the costs arising thereof will be borne by the house owner. If state cadres and employees violate management methods of house rental and rent a house to an outsider coming to dig caterpillar fungus then they will be fined both economically and administratively. An informants' network [must be] built up and all informants providing verified reports will be awarded 30 to 50 per cent of collected fines.

(v) Strengthen management of vehicles and safeguard the safety of traffic and transportation.

First of all, between 1 April and 25 June each year the county's Leading Group for Management of the Caterpillar Fungus Resources will
[Page 12:]
task departments in charge of traffic management to implement the eighty-day long traffic regime on the roads in a caterpillar fungus producing area.

Second, departments in charge of vehicle management have to strictly address illegal activity such as driving without a permit, using private vehicles to carry passengers and overloading. It is also necessary to strictly check all kinds of motor vehicles (including motorcycles) and buses carrying outsiders into Dawu – if such things are discovered then departments in charge of traffic management, [acting] in accordance with the Law on Road Traffic safety and relevant laws and regulations, will detain motor vehicles – with the exception of those confiscated as illegally acquired – and deal with them only after the work on caterpillar fungus collection is finished. It is necessary to prevent accidents happening but once a traffic accident occurs then the person concerned will be responsible for all the losses.

Third, departments in charge of traffic management have strictly to follow the spirit of 'The circular of Qinghai Province on cancelling extra buses in the Three Rivers Area', administer work on passenger transport in the directly subordinated areas and deal severely with illegal transport [violations] in accordance with relevant regulations.

Fourth, within the period of the ban on collection of caterpillar fungus all vehicles [belonging to] institutions and departments directly under township and county supervision will be deployed and used by the Leading Group for Management of the Chinese-Tibetan Medical Resources.

(vi) Closely cooperate, do proper work on social stability at the period of ban on caterpillar fungus collection.

First, strengthen the work on elimination of antagonisms and disputes, set dates for elimination and arbitration in [cases of] antagonisms.

Second, establish and perfect schemes to deal with all kinds of mass accidents and guarantee that sudden accidents will be resolutely dealt with.

Third, the County People's Deputies Congress, Consultative Conference as well as investigation and audit departments have to fully play their supervisory role and together properly do their work on rational utilization of resources.

Fourth, the county's Leading Group for Management of Chinese-Tibetan Medical Resources Protection must sign responsibility contracts on management and security of caterpillar fungus resources with governments of townships [abundant in] the caterpillar fungus resources, leaders of liaison points at the county level as well as with personnel in charge of environment protection propaganda and have to safeguard political stability according to the principle of administration of a subordinated area and [specific] responsibility at [each] level.

(vii) Measures for award and fine in management of the caterpillar fungus resources.

In management of the caterpillar fungus resources establish report boxes and a system of control reports.

[Page 13:]
If a report on a [specific] situation proves to be true then according to the Grassland Law of the People's Republic of China, Regulations of the People's Republic of China on Protection of Wild Plants, the Law of the People's Republic of China on Forests, Regulations of the People's Republic of China on Nature Protection Areas, Administrative Regulations on Protection of Medical Wild Herb Resources, Administrative Methods of Collecting Dry Grass and Chinese Ephedra (*Mahuang*), Detailed Executive Regulations of Qinghai Province on the Grassland Law of the People's Republic of China, the Law of Qinghai Province on Contracts Pertaining to Grasslands, Golok Prefecture Methods of Caterpillar Fungus Resources Management and relevant provisions of other laws and regulations mete out punishment to all persons [found] responsible and according to relevant directives grant rewards to informants.

1. As for grasslands contactors individually soliciting outside labourers: in each case of soliciting, a township government or village committee (the contract-issuing party) at the time of confiscation of illegal gains will impose a fine amounting up to a tenfold value of the illegal income. In serious circumstances, stipulations of the 2nd subparagraph of the 24th paragraph of the Law of the Qinghai Province on Contracts Pertaining to Grasslands will be invoked: 'if the contracting party illegally bring the grasslands

under cultivation or carry out production and economic activity harmful to protection of grasslands then the contract-issuing party will revoke the right of grasslands' user'.

2. As for those who ignore the government's policy on banning the collection of caterpillar fungus, agreements on contracts reached by township, village or brigade cadres with individuals for use of mountains, gullies, banks or rentals to individual proprietors, will be treated as a swindling activity; cases of contractors using mountains, gullies, banks or rentals to individual proprietors and soliciting labourers to collect caterpillar fungus will also be treated as a swindling activity and if reports by informants prove true then those cases will be fined politically and economically; those who violate the abovementioned regulations will also get their right of user revoked. Those committing crimes will be handed over to judicial institutions to be punished. At the time of caterpillar fungus collection, all labour conflicts arising from collection will not be accepted by labour departments.

3. State cadres, staff workers and their relatives and children are not allowed to collect caterpillar fungus and moreover are not allowed to solicit outsiders to collect it. If a case of this nature is discovered then not only will the leadership of a unit be subject to investigation
[Page 14:]
but also, a political and economic punishment will be meted out to perpetrators [in form] of dismissal or demotion. Those who are supposed to retire soon, if they solicit outsiders to collect the caterpillar fungus, apart from getting their annual salary withheld, will also be punished in accordance with relevant regulations and rules. All those who bring in personnel unqualified for caterpillar fungus collection and arbitrarily enter the natural protection zone of Three Rivers' grasslands, apart from confiscation of income, will be fined up to 10,000 yuan contingent on the specific circumstances of the case in accordance with stipulations of subparagraph 2 of paragraph 34 and paragraph 35 of the Regulations of the People's Republic of China on Nature Protection Areas.

4. If any outsiders who should have cleared up and left are discovered in the caterpillar fungus resources area during the period of collection ban, then leadership of a liaison point at the county level for that area will be in charge and specific responsibility lies with the leadership at the township level. Outsiders qualifying for clear-up and return should be handled independently and no teams will be dispatched by the county [authorities]

to carry out clear-up and return procedures. As for some uncontrolled zones and areas difficult to access, expenses for dispatch of security personnel, armed police, cadres and vehicles will be borne by townships, herders' committees and pastoralists' farms. If outsiders are discovered on the grasslands, it does not matter whether they are vagabonds or persons taken in and cared for, all of them will be treated as solicited outsiders and pastoralists' farms contracting grasslands [work] will be fully responsible for them and all expenses resulting will be borne by those pastoralists' farms. If any problems arise then county liaison points and township leaderships will be responsible.

5. If personnel in charge of administering the ban on caterpillar fungus collection abuse their authority, neglect their duties and engage in favouritism or commit irregularities, each case will be punished administratively in accordance with the law and those guilty of committing crimes will be handed over to judiciary institutions for criminal investigation.

6. During the period of the ban on caterpillar fungus collection, the expenses of propaganda stations will be borne by the county's budget and all fines and confiscated amounts [of money] will be transferred to the county budget; misappropriation or retention of the funds and revenues is not allowed and income and expenses must be proceeded with properly.

7. If turf is extracted, ditches dug out, grass burnt, bush and sand-fixing plants cleared then severe punishment will be administered according to the Grasslands Law and the Forest Law; it is strictly forbidden to carry out any actions
[Page 15:]
harmful to the basic structure of the grasslands.

8. Poachers hunting wild animals will be punished according to regulations of the Law on Protection of Wild Animals.

9. It is strictly prohibited to carry any knives and if they are discovered then they will be confiscated on the spot and [the perpetrator] will be handed over to the judiciary institutions to be punished according to the law.

10. It is strictly forbidden to light a fire in the open, it is necessary to treat the problem of garbage properly, and particularly plastic and glass products; users are responsible for burning them or burying them deep underground.

11. All governments of townships abundant in caterpillar fungus resources will call mass meetings in all villages successively by every 5 April and through concrete consultation with the local population consistently implement measures and methods on management of caterpillar fungus resources protection.

12. All the propaganda stations must man their posts in a civilized manner and with ID documents and if there arise any problems related to improper working methods then county liaison points, team leaders and persons guilty of misdeeds will be held responsible.

13. In order to build a sound reporting system, counties and townships have to install report boxes, report telephones so that they are able to intercept illegal actions and violation of discipline in matters of caterpillar fungus collection on time. Once a report is received it should be investigated and checked in a timely manner and if it proves true then it should be treated according to relevant regulations; the informant should be given a specific amount of and secrecy for the informant's actions should be safeguarded. The county's report telephone number is: at the Office of the County Disciplinary Commission [of the Party Committee] – 8382935; at the Office of the County Government – 8386530; and at the County Agricultural and Animal Husbandry Office – 8383364.

(viii) When the work on collection of caterpillar fungus is finished then the county is to call a summing-up meeting to review the work over the past year. Awards and commendations will be granted to areas and individuals with outstanding achievements in their work on the management of caterpillar fungus resources. As for areas and individuals who have failed in their organizational and administrative work, criticism will practised in the form of a circular and some material punishment and at the time of annual evaluation at the end of the year a certain amount of points will be deducted. As for areas where serious accidents, mass incidents, loss of life and loss of property have happened during the period of caterpillar fungus collection, it is necessary to apply a veto and at the same time to investigate administrative and other responsibilities of party and government cadres.

[Page 16:]
It is necessary to deal severely with cadres and people who do not observe the policies and regulations, who at the time of the ban on the caterpillar fungus collection overtly agree [with the policy] and covertly oppose [it],

operate secretly or consciously do not interfere, do not care and do not ask [questions], do not punish or pretend to follow the policy and in fact oppose [that policy] and only pay lip service [to it]. If it is necessary to dismiss those people then dismiss them, if to punish them then punish, but absolutely do not be indulgent [of such activity].

In sum, strengthening management of caterpillar fungus resources protection and protecting the ecological environment of the grasslands of the Three Rivers' Area is a common obligation for the whole of society. It is also the concrete embodiment of the earnest and consistent implementation of the scientific development and the spirit of the 17th [Chinese Communist] Party Congress. We have to unify our thinking, to unite as one and guarantee orderly development of the work on management of caterpillar fungus resources protection in 2010 in accordance with the law.

Tibetan Word List

abra	a bra
Amdo	A mdo
Amnye Dzargen	A myes rdza rgan
Amnye Machen	A myes rma chen
arak	a rag
bar	bar
Bathang	'Ba' thang
Bayan Har	Pa yan khra la
beha	be sha
behing hamo	be shing sha mo
borig	Bod rigs
bsang	bsang
bsang chöd	bsang mchod
bsangchy	bsangs khri
bsangrtsi	bsangs rdzas
bu thuwa	'bu thos ba
chaka	ja kha
Chamahe	Khra ma hi
chagyi	khrag g.yos
Chebchu	Khyeb chu
Chemdry	Khyi 'bri
Chentsa	gCan tsha
chibsang	spyi bsangs
chitsog nyotson	phyi phyogs nyo tshong
chobra	skyo sbra
chögar	chos sgar
chom	khrom
Chongmar	Khrong dmar
chongrdel	grong rdal
chöyon	mchod yon
chura (*kompa*)	phyur ra (skom pa)
Danzhung	Tang gzhung/Dwangs gzhung
dara	ta ra
Darlag	Dar lag
Dartsemdo	Dar rtse mdo
Dawu	rTa bo
Dawu Gongma	rTa bo gong ma

Dawu Zhuma	rTa wo zhol ma
Dechen	bDe chen
Derni	Dhi gnas
dewa	sde ba
digpa	sdig pa
dom	dom
Domchu	sDom chu
Domkhok	sDom khog
drang	'bring
drekhang	dred khang
dri	'bri
drib	sgrib
Drichu	'Bri chu
drirkam	'bri skam
drok	'brog
drok u khang	'brog u khang
drokki sacha	'brog gi sa cha
drokpa	'brog pa
Drölma Dhondup	sGrol ma Don 'grub
droma	gro ma
Drogön Chögyel Phagpa	'Gro mgon chos rgyal phags pa
düchen	dus chen
dukpaekur	sdig pa 'khyer
Dunchong Dzamar	Dung skyong rdza dmar
dungpo	sdong po
dungpung	sdong phung
düpa	mdud pa
dyglag	sdug lag
dygwu	sdug bo'u
Dzachu	rDza chu
Dzachuka	rDza chu kha
Dzambala	Dzam bha la
Dzartod	rDza stod
dzayung	rdza yung
dzong	rdzong
Gabde	dGa' bde
gang tabshid	gangs da pyid
Gangri	Gangs ri
ganlha metok	gangs lha me tog
ganwel	gangs sbal

gartah	dga' rtags
Gelug	dGe lugs
go	mgo
Golok	mGo log
Guchu	dGu chu
guguhamo	khu khu sha mo
gunsa	dgun sa
Guru Tsodruk	sGur ru tsho drug
Gyalthang	rGyal thang
habzang	sha bzan
Hcharang	sKya ring
hchum	lcum
Hehe	He he
Huerig	Hu'e rigs
Huonkor	dPon skor
huonpo	dpon po
Jadrel Chöyon Rangdröl	Bya bral chos dbyings rang grol
jama	rgya ma
Jigdril	gCig sgril
jigten nyangwa	'jig rten rnying pa
jigtenpi lha	'jig rten pa'i lha
joby	gyod bug
julug	rgyu lug
Jumgo	rGyud mgo
Kache	Kha che
kakle	kag le
Kamdze	dKar mdzes
Kanlho	Kan lho
karma	skar ma
karmong	khur mong
karzi	dkar zas
Kham	Khams
khandroma	mkha' 'gro ma
Khangsar	Khang gsar
khazhu	kha zhos
Khenpo Tsultrim Lodrö	mKhan po tshul khrims blo gros
khoryug	'khor yug
khulu	khu lu
kowa	go ba
Kumbum	sKu 'bum

kuwa	rko ba
Labrang	bLa brang
Labrang Tashi Khyil	bLa brang bkra shis 'khyil
Larung Gar	bLa rung sgar
Larung Ngarig Nangten Lobling	Bla rung lnga rig nang bstan slob gling
le	las
lhadrima	lha 'bri ma
Lhari Tashi Thondröl Dokha	Lha ri bkra shis mthong grol rdo kha
lharta	lha rta
lhayak	lha g.yag
Lithang	Li thang
lozur	glo gzur
Losar	lo gsar
lü	lus
lugra	lug ra
lugtul	lug thul
Machen	rMa chen
Machen Gangri	rMa chen gangs ri
Machu	rMa chu
magpa	mag pa
Malho	rMa lho
mamo	ma mo
Martod	rMa stod
marzi	dmar zas
mda	mda'
me	dme
Mekogi	Me go gi
menrtsa	sman rtswa
Menthang	sMin thang
mentsi fanyik	sman rtswa'i dpang yig
metok serchen	me tog ser chen
mirig thondzi	mi rigs thon rdzas
mirtag	dmu thag
mo	mo
mthosa	mtho sa
Mtsojang	mTsho byang
Mtsonag	mTsho nag
Nagchu	Nag chu
nalo	rna lo

Namkha	Nam mkha
Nangchen	Nang chen
nangde	nang sde
nangme	nang dme
nartag	rna rtags
nga	snga
ngagpa	sngags pa
ngangan	ngang ngang
Ngari	mNga' ris
Ngawa	rNga ba
Ngorang	sNgo ring mtsho
norma	nor ma
Nyamnyi Dorje	mNyam nyid rdo rje
Nyenpo Yurtse	gNyan po g.yu rtse
nyinsa	nyin sa
Opame	'Od dpag med
Phagpa lama	Phags pa bla ma
pimo	pas mo
pshilde	phyi sde
pshirjid	phyi rgyal
ra	rwa
rab	rab
ramari	bra ma ras
Rarja Gon Ganden Tashi Jungne Shesub Darje Ling	Rwa rgya dgon dga' ldan kra shis 'byung gnas bshad srub dar rgyas gling
Rebkong	Reb kong
richen	ru chen
richung	ru chung
rinag	ras nag
Rjami	rGya mi
rlungta	rlung rta
Rongwo Lonchu	Rong bo blon chos
rtsa	rtswa
rtsa tabshid	rtswa da pyid
rtsabshid	rtswa pyid
rtsacha	rtsa khral
rtsala	rtsa la
rtsamdzod	rtsa mdzod
rtsampa	rtsam pa

rtsipa	rtsid pa
rukor	ru skor
Sa Golok nam Golok	sa mgo log gnam mgo log
Sakya Pandita Kunga Gyeltsen	Sa skya pandita kun dga' rgyal mtshan
sabchud	sa bcud
sabdag	sa bdag
saga dawa	sa ga zla ba
Sakya	Sa skya
Sangchu	bSang chu
sayig	sa yig
serkar	ser dkar
Serthar	gSer thar
shang	srang
shang	zhang
Shigatse	gZhis ka rtse
shunkor	srung 'khor
shybsa	srib sa
Soglung	Sog lung
Sumdo	gsum mdo
ta	tha
tabshid	da byid
Tarthang Tulku Kunga Gelek Yeshe Dorje	Dar thang sprul sku kun dga' dge legs ye shes rdo rje
terbdag	gter bdag
thima	drus ma
tonsa	gton sa
Tri Songdetsen	Khri Srong lde brtsan
Tri Songtsen	Khri Srong brtsan
tsar	tshar
Tsepame	Tshe dpag med
tsethar	tshe thar
Tsigorthang	rTsi gor thang
tsogpa	btsog pa
Tsongkhapa Lobzang Drakpa	Tsong kha pa blo bzang grags pa
tsowa	tsho ba
Ü-Tsang	dBus gtsang
wa	bal
wabryk	wa phrug
wanlag	dbang lag

Weri Dukmtso	Ber ri'i duk mtsho
Wranakh	sBra nag
wulug	sbo lug
Yamdrokmtso	Yar 'brog mtsho
yarsa	dbyar sa
yartsa gumbu	dbyar rtsa dgun 'bu
yaryma	yar ma
yulbdag	yul bdag
yulha	yul lha
Yunzhi	g.Yon bzhi
Yuthog Yonten Gonpo Sarma	g.Yu thog yon tan mgon po gsar ma
zhibdag	gzhis bdag
zhobzung	bzho bzung
zogra	zog ra

Bibliography

Abraham, Itty, and Willem van Schendel. 2005. 'Introduction. The Making of Illicitness.' In *Illicit Flows and Criminal Things. States, Borders, and the Other Side of Globalization*. Itty Abraham and Willem van Schendel (eds.), Bloomington: Indiana University Press, pp. 1-37.

Alai. 1997. 'Pilze.' In *An den Lederriemen geknotete Seele. Erzähler aus Tibet*. bKra shis Zla ba and Alice Grünfelder (eds.), Zürich: Unionsverlag, pp. 127-152.

AM (Aloha Medicinals). 2012. http://www.alohamedicinals.com/cordyceps.html, accessed 7 May 2012

Appadurai, Arjun. 1986. 'Introduction. Commodities and the Politics of Value.' In *The Social Life of Things. Commodities in Cultural Perspective*. Arjun Appadurai (ed.), Cambridge: Cambridge University Press, pp. 3-63.

Arora, David. 2008. 'The Houses That Matsutake Built.' *Economic Botany*. 62:3, pp. 278-290.

Bassini, Patrizia. 2012. 'Dress for Success. A Journey from Past to Present among Tibetans.' *Costume. The Journal of the Costume Society*. 46:1, pp. 92-110. Bauer, Kenneth M. 2005. 'Development and the Enclosure Movement in Pastoral Tibet since the 1980s.' *Nomadic Peoples*. 9:1-2, pp. 53-81.

Bauer, Kenneth M. 2014. 'High Frontiers: Dolpo Revisited: An Essay in Honour of Wim van Spengen (1949-2013).' In *Tibet Journal (Trade, Travel and the Tibetan Border Worlds. Essays in Honour of Wim van Spengen (1949-2013)*. Emilia Roza Sulek (Emilia Róża Sułek), John Bray and Alex McKay (eds.), 39:1, pp. 157-181.

Bauer, Kenneth M., and Yonten Nima. 2011. 'Laws and Regulations Impacting the Enclosure Movement on the Tibetan Plateau of China.' *Himalaya*. 30:1-2, pp. 23-37.

Berland, Joseph C., and Aparna Rao. 2004. 'Unveiling the Stranger. A New Look at Peripatetic Peoples.' In *Customary Strangers. New Perspectives on Peripatetic Peoples in the Middle East, Africa, and Asia*. Joseph C. Berland and Aparna Rao (eds.), Westport: Praeger, pp. 1-29.

Bessho, Yusuke. 2015. 'Migration for Ecological Preservation? Tibetan Herders' Decision-Making Process in the Eco-migration Policy of Golok Tibetan Autonomous Prefecture (Qinghai Province, PRC).' *Nomadic Peoples*, 19:2, pp. 189-208.

Bloch, Maurice, and Jonathan Parry. 1989. *Money and the Morality of Exchange*. Cambridge: Cambridge University Press.

Boesi, Alessandro. 2003. 'The *dByar rtswa dgun 'bu* (*Cordyceps sinensis Berk.*). An Important Trade Item for the Tibetan Population of the Li thang District, Sichuan Province, China.' *The Tibet Journal*. 28:3, pp. 29-42.

Boesi, Alessandro. 2005. 'Plant Knowledge among Tibetan Populations.' In *Wildlife and Plants in Traditional and Modern Tibet. Conceptions, Exploitation, and Conservation*. Alessandro Boesi and Francesca Cardi (eds.). *Memorie della Società Italiana di Scienze Naturali e del Museo Civico di Storia Naturale di Milano*. 33:1, pp. 33-48.

Boesi, Alessandro, and Francesca Cardi. 2009. '*Cordyceps sinensis* Medicinal Fungus. Traditional Use among Tibetan People, Harvesting Techniques, and Modern Uses.' *HerbalGram*. 83, pp. 52-61.

Bonacich, Edna. 1973. 'A Theory of Middleman Minorities.' *American Sociological Review*. 38:5, pp. 583-594.

Bourdieu, Pierre. 2005. *The Social Structures of the Economy*. Cambridge: Polity.

Bovingdon, Gardner. 2010. *The Uyghurs. Strangers in Their Own Land*. New York: Columbia University Press.

Brox, Trine, and Ildikó Bellér-Hann. 2014. 'Introduction.' In *On the Fringes of the Harmonious Society: Tibetans and Uyghurs in Socialist China.* Trine Brox and Ildikó Bellér-Hann (eds.), Copenhagen: Nordic Institute of Asian Studies Press, p. 1-28.

Buffetrille, Katia. 1997. 'The Great Pilgrimage of A myes rMa chen. Written Traditions, Living Realities.' In *Mandala and Landscape.* A.W. Macdonald (ed.), Delhi: D.K. Printworld, pp. 75-132.

Byams pa 'Phrin las. 2006. *Bod lugs gso rig tshig mdzod chen mo.* (*Great Dictionary of the Tibetan Medical Tradition.*) Beijing: Mi rigs dpe skrun khang.

Byg, Anja, Jan Salick, and Wayne Law. 2010. 'Medicinal Plant Knowledge among Lay People in Five Eastern Tibet Villages.' *Human Ecology.* 38, pp. 177-191.

Cai, Peiyun. 2010. *Countermeasures of Ophiocordyceps sinensis Resource Conservation in Qinghai Province.* Paper presented at the 2010 International Conference on *Ophiocordyceps sinensis.* Xining, China, 08-10.06.2010.

Cannon, Paul F., Nigel L. Hywel-Jones, Norbert Maczey, Lungten Norbu, Tshitila, Tashi Samdup, and Phurba Lhendup. 2009. 'Steps towards Sustainable Harvest of *Ophiocordyceps sinensis* in Bhutan.' *Biodiversity and Conservation.* 18:9, pp. 2263-2281.

Carrier, James G. 1995. *Gifts and Commodities. Exchange and Western Capitalism since 1700.* London: Routledge.

CD (*China Daily*). 2012. 'Six Arrested, 89 Surrender to Police in Township Attack.' http://www.chinadaily.com.cn/china/2009-03/22/content_7603753.htm, accessed 30 December 2012.

Chan, Anita, and Jonathan Unger. 1982. 'Grey and Black. The Hidden Economy of Rural China.' *Pacific Affairs.* 55:3, pp. 452-471.

Chan, Kam Wing, and Li Zhang. 1999. 'The *Hukou* System and Rural-Urban Migration in China. Processes and Changes.' *The China Quarterly.* 160:4, pp. 818-855.

Chan, Kan-min. 2009. 'Harmonious Society.' In *International Encyclopedia of Civil Society.* Helmut K. Anleier and Stefan Toepler (eds.), New York: Springer, pp. 821-825.

Childs, Geoff. 2004 *Tibetan Diary. From Birth to Death and Beyond in a Himalayan Valley of Nepal.* Berkeley: University of California Press.

Childs, Geoff. 2008. *Tibetan Transitions. Historical and Contemporary Perspectives on Fertility, Family Planning, and Demographic Change.* Leiden: Brill.

Childs, Geoff, and Namgyal Choedup. 2014. 'Indigenous Management Strategies and Socioeconomic Impacts of Yartsa Gunbu (*Ophiocordyceps sinensis*) Harvesting in Nubri and Tsum, Nepal.' *Himalaya.* 34:1, pp. 8-22.

Childs, Geoff, Melvyn C. Goldstein, and Puchung Wangdui. 2010. 'An Entrepreneurial Transition? Development and Economic Mobility in Rural Tibet.' *Himalaya.* 30:1-2, pp. 51-62.

Clark, Barry. 1995. *The Quintessence Tantras of Tibetan Medicine.* Ithaca NY: Snow Lion Publications.

Clark, Gracia. 1988. 'Introduction.' In *Traders versus the State. Anthropological Approaches to Unofficial Economies.* Gracia Clark (ed.), Boulder: Westview Press, pp. 1-16.

Clarke, Graham E. 1988. 'China's Reforms of Tibet, and their Effects on Pastoralism.' *Kailash.* 14:1-2, pp. 63-131.

Clarke, Graham E. 1992. 'Aspects of the Social Organization of Tibetan Pastoral Communities.' In *Language, History and Culture. Tibetan Studies. Proceedings of the Fifth Seminar of the International Association for Tibetan Studies, Narita 1989.* Shōren Ihara (ed.), Vol. 2. Narita: Naritasan Shinshoji, pp. 393-411.

Clarke, Graham E. 1995a. 'Development, Society and Environment in Tibet.' In *Development, Society, and Environment in Tibet. Papers Presented at a Panel of the 7th Seminar of the International Association for Tibetan Studies, Graz, 1995.* Graham E. Clarke (ed.), Vienna: Verlag der Österreichischen Akademie der Wissenschaften, pp. 1-45.

Clarke, Graham E. 1995b. 'Socio-Economic Change and the Environment in a Pastoral Area of Lhasa Municipality.' In *Development, Society, and Environment in Tibet. Papers Presented at a Panel of the 7th Seminar of the International Association for Tibetan Studies, Graz, 1995*. Graham E. Clarke (ed.), Vienna: Verlag der Österreichischen Akademie der Wissenschaften, pp. 97-119.

Clifford, James. 1988. *The Predicament of Culture. Twentieth-Century Ethnography, Literature, and Art*. Cambridge: Harvard University Press.

CNS (Census). 2010. 'Guoluo zhou 2010 nian di liu ci quanguo renkou pucha zhuyao shuju gongbao.' ('Golok Prefecture 2010 Sixth National Census Data Bulletin.') http://www.tjcn.org/rkpcgb/rkpcgb/201112/22724.html, accessed 15 January 2013.

Coales, Oliver. 1919. 'Eastern Tibet.' *The Geographical Journal*. 53:4, pp. 228-249.

Coméliau, Christian. 2002. *The Impasse of Modernity. Debating the Future of the Global Market Economy*. London: Zed Books.

Cook, Scott. 2004. *Understanding Commodity Cultures. Explorations in Economic Anthropology with Case Studies from Mexico*. Lanham: Rowman and Littlefield.

Costello, Susan E. 2002. 'The Economics of Cultural Production in Contemporary Amdo.' In *Amdo Tibetans in Transition. Society and Culture in the Post-Mao Era. Proceedings of the Ninth Seminar of the International Association for Tibetan Studies, Leiden 2000*. Toni Huber (ed.), Leiden: Brill, pp. 221-240.

Costello, Susan E. 2008. 'The Flow of Wealth in Golok Pastoralist Society. Towards an Assessment of Local Financial Resources for Economic Development.' In *Tibetan Modernities. Notes from the Field on Cultural and Social Change. Proceedings of the Tenth Seminar of the International Association for Tibetan Studies, Oxford 2003*. Robert Barnett and Ronald Schwartz (eds.), Leiden: Brill, pp. 73-112.

CPCMH (Committee of Pharmacopeia Chinese Ministry of Health). 1964. *Chinese Pharmacopeia* (1963 Edition), Part 1. Beijing: The People's Medical Publishing House.

Craig, Sienna, and Vincanne Adams. 2009. 'Global Pharma in the Land of Snows. Tibetan Medicines, SARS, and Identity Politics Across Nations.' *Asian Medicine*. 4, pp. 1-28.

Crapanzano, Vincent. 1986. 'Hermes' Dilemma. The Masking of Subversion in Ethnographic Description.' In *Writing Culture. The Poetics and Politics of Ethnography*. James Clifford and George E. Marcus (eds.), Berkeley: University of California Press, pp. 51-76.

Crouch, Lucas. 2010. 'Caterpillar Fungus of the Tibetan Plateau.' Independent Study Project Collection. Paper No. 916. http://digitalcollections.sit.edu/isp_collection/916, accessed 8 March 2012.

da Col, Giovanni. 2012. 'The Elementary Economies of Dechenwa Life. Fortune, Vitality, and the Mountain in Sino-Tibetan Borderlands.' *Social Analysis*. 56:1, pp. 74-98.

Devkota, Shiva. 2006. '*Yarsagumba* [*Cordyceps sinensis* (Berk.) Sacc.]; Traditional Utilization in Dolpa District, Western Nepal.' *Our Nature*. 4, pp. 48-52.

Devkota, Shiva. 2010. '*Cordyceps sinensis* (*Yarsagumba*) from Nepal Himalaya: Status, Threats and Management Strategies.' Paper presented at the 2010 International Conference on *Ophiocordyceps sinensis*. Xining, China, 08-10.06.2010.

dGa' ba'i rDo rje. 1998. *'Khrungs dpe dri med shel gyi me long.* (*The Immaculate Crystal Mirror of Source and Identification.*) Beijing: Mi rigs dpe skrun khang.

Don grub dBang rgyal. 1991. 'mGo log lo rgyus deb ther.' ('A Chronicle of Golok History.') In *mGo log rig gnas lo rgyus* (*A Cultural History of Golok*). Vol. 1. Xining: Srid gros mgo log khul u rig gnas lo rgyus rgyu cha zhib 'jug u yon ltan khang, pp. 1-178.

Don grub dBang rgyal, and Nor sde. 1992. *mGo log rig gnas lo rgyus*. (*A Cultural History of Golok*). Vol. 2. Xining: Srid gros mgo log khul u rig gnas lo rgyus rgyu cha zhib 'jug u yon ltan khang.

Dong, Caihong, and Yijian Yao. 2010. 'On the Reliability of Fungal Materials used in Studies on *Ophiocordyceps sinensis.*' *Journal of Industrial Microbiology and Biotechnology*. 38:8, pp. 1027-1035.

dPal bzang. 1991. 'rTa bo dme tshang gnyis kyi lo rgyus mdor bsdus.' ('A Brief Summary of the History of Dawu and Metsang.') In *mGo log rig gnas lo rgyus*. (*A Cultural History of Golok*). Vol. 1. Xining: Srid gros mgo log khul u rig gnas lo rgyus rgyu cha zhib 'jug u yon ltan khang, pp. 239-244.

Du, Fachun. 2009. 'Tibetan Herders without Animals and Pastures. Ecological Resettlement in Maduo County, Golok Prefecture Qinghai.' Paper presented at the conference *Sustainable Livelihoods in Tibetan Pastoral Areas: Global Change and Local Challenges*, Martin Luther University, Halle-Wittenberg, Germany, 02.12.2009.

Du, Fachun, 2012. 'Ecological Resettlement of Tibetan Herders in the Sanjiangyuan. A Case Study in Madoi County of Qinghai.' *Nomadic Peoples*. 16:1, pp. 116-133.

Duncan, Marion. 1964. *Customs and Superstitions of Tibetans*. London: Mitre Press.

Eisenstadt, Shmuel N. 2000. 'Multiple Modernities.' *Daedalus*. 129:1, pp. 1-29.

Ekvall, Robert B. 1961. 'The Nomadic Pattern of Living among the Tibetans as Preparation for War.' *American Anthropologist*. 63, pp. 1250-1263.

Ekvall, Robert B. 1963. 'A Note on 'Live Blood' as Food among the Tibetans.' *Man*. 175, pp. 145-146.

Ekvall, Robert B. 1964a. 'Peace and War among the Tibetan Nomads.' *American Anthropologist*. 66:5, pp. 1119-1148.

Ekvall, Robert B. 1964b. *Religious Observances in Tibet. Patterns and Function*. Chicago: University of Chicago Press.

Ekvall, Robert B. 1968. *Fields on the Hoof. Nexus of Tibetan Nomadic Pastoralism*. New York: Holt, Rinehart and Winston.

Ekvall, Robert B. 1977. *Cultural Relations on the Kansu-Tibetan Border*. Chicago: University of Chicago Press.

ENS (Environment News Service). 2005. 'Tibetans, Chinese Battle over Access to Medicinal Fungus.' http://www.ens-newswire.com/ens/jun2005/2005-06-02-0l.asp, accessed 29 August 2007.

EP (Everest Pharmaceuticals). [s.a.]. *Everest Pharmaceuticals*. [Advertising Brochure]. Xining: Qinghai Everest Aveto Pharmaceuticals Co., Ltd.

Escobar, Arturo. 1995. *Encountering Development. The Making and Unmaking of the Third World*. Princeton: Princeton University Press.

Esteva, Gustavo. 1992. 'Development.' In *Development Dictionary. A Guide to Knowledge and Power*. Wolfgang Sachs (ed.), London: Zed Books, pp. 7-25.

Estioko-Griffin, Agnes, and P. Bion Griffin. 1981. 'Woman the Hunter. The Agta.' In *Woman the Gatherer*. Frances Dahlberg (ed.). New Haven: Yale University Press, pp. 121-151.

Evers, Hans-Dieter, and Ozay Mehmet. 1994. 'The Management of Risk. Informal Trade in Indonesia.' *World Development*. 22:1, pp. 1-9.

Eyferth, Jacob. 2009. *Eating Rice from Bamboo Roots. The Social History of a Community of Handicraft Papermakers in Rural Sichuan, 1920-2000*. Harvard: Harvard University Press.

Fairbank, John King, and Merle Goldman. 2006. *China. A New History*. Cambridge: Harvard University Press.

Fan, Jiafeng and Jian Lin. 2007. 'Successful Cluster Cooperation in China Private Enterprises. Case Study on Artificial Caterpillar Fungus Industry in Xinhui City.' Paper presented at the 2007 International Conference on Management Science and Engineering. http://www.seiof-bluemountain.com/upload/product/201001/12644748107m76ok1q.pdfm, accessed 14 Juli 2013.

Farooquee, Nehal A. 1998. 'Development and the Eradication of Traditional Resource Use Practice in the Central Himalayan Transhumant Pastoral Society.' *International Journal of Sustainable Development and World Ecology*. 5, pp. 43-50.

Ferguson, James. 1990. *The Anti-Politics Machine. Development, Depoliticization and Bureaucratic Power in Lesotho.* Cambridge: Cambridge University Press.

Ferguson, James. 2000. *Expectations of Modernity. Myths and Meanings of Urban Life on the Zambian Copperbelt.* Berkeley: University of California Press.

Finan, Timothy J. 1988. 'Market Relationships and Market Performance in Northeast Brazil.' *American Ethnologist*. 15:4, pp. 694-708.

Fischer, Andrew M. 2005a. *State Growth and Social Exclusion in Tibet. Challenges of Recent Economic Growth.* Copenhagen: Nordic Institute of Asian Studies Press.

Fischer, Andrew M. 2005b. 'Close Encounters of an Inner Asian Kind. Tibetan-Muslim Co-existence and Conflict Past and Present.' V *Crisis States Working Paper* (No. 68), London: London School of Economics, http://www.crisisstates.com/publications/wp/WP1/wp68.htm, accessed 17 September 2011.

Fischer, Andrew M. 2008. 'The Muslim Cook, the Tibetan Client, his Lama and their Boycott: Modern Religious Discourses of anti-Muslim Economic Activism in Amdo.' In *Conflict and Social Order in Tibet and Inner Asia*. Toni Huber and Fernanda Pirie (eds.), Leiden: Brill, pp. 159-192.

Fischer, Andrew M. 2013. *The Disempowered Development of Tibet in China. A Study in the Economics of Marginalization.* Lanham, MD: Rowman and Littlefield/Lexington.

Foggin, J. Marc. 2005. *Promoting Biodiversity Conservation and Community Development in the Sanjiangyuan Region Proceedings of a Conservation Planning Meeting, with Agreed Priority Action Points.* [s.l.]: Plateau Perspectives. http://www.marcfoggin.com/wp-content/uploads/2012/04/Promoting-biodiversity-conservation-Foggin-2005-English.pdf, accessed 15 March 2014.

Foggin, J. Marc. 2008. 'Depopulating the Tibetan Grasslands. National Policies and Perspectives for the Future of Tibetan Herders in Qinghai Province, China.' *Mountain Research and Development*. 28:1, pp. 26-31.

Foggin, J. Marc, and Andrew T. Smith. 1999. 'The Plateau Pika (*Ochotona curzoniae*) is a Keystone Species for Biodiversity on the Tibetan Plateau. *Animal Conservation*, 2, pp. 235-240.

Forbes, Andrew D. W. 1986. *Warlords and Muslims in Chinese Central Asia. A Political History of Republican Sinkiang 1911-1949*. Cambridge: CUP Archive.

Foucault, Michel. 1972. *The Archaeology of Knowledge and the Discourse on Language.* New York: Pantheon Books.

Gaerrang (Kabzung). 2011. 'The Alternative to Development on the Tibetan Plateau. Preliminary Research on the Anti-Slaughter Movement.' *Revue d'Etudes Tibétaines*. 21, pp. 31-43.

Gaerrang (Kabzung). 2012. *Alternative Development on the Tibetan Plateau: The Case of the Slaughter Renunciation Movement.* PhD dissertation, University of Colorado at Boulder.

Gaonkar, Dilip Parameshwar. 2001. 'On Alternative Modernities.' In *Alternative Modernities*. Dilip Parameshwar Gaonkar (ed.), Durham: Duke University Press, pp. 1-23.

Garbyal, SS, K.K. Aggarwal, and C.R. Babu. 2004. 'Impact of *Cordyceps sinensis* in the Rural Economy of Interior Villages of Dharchula Sub-division of Kumaom Himalayas and its Implications in the Society.' *Indian Journal of Traditional Knowledge*. 3:2, pp. 182-186.

Geertz, Clifford. 1963. *Peddlers and Princes. Social Development and Economic Change in Two Indonesian Towns*. Chicago: University of Chicago Press.

Geertz, Clifford. 1973. *The Interpretation of Cultures*. New York: Basic Books.

Gelek, 1998. 'The Washu Serthar. A Nomadic Community of Eastern Tibet.' In *Development, Society, and Environment in Tibet. Papers Presented at a Panel of the 7th Seminar of the International Association for Tibetan Studies, Graz, 1995*. Graham E. Clarke (ed.), Vienna: Verlag der Österreichischen Akademie der Wissenschaften, pp. 47-58.

Gereffi, Gary, Miguel Korzeniewicz, and Roberto P. Korzeniewicz. 1994. 'Introduction. Global Commodity Chains.' In *Commodity Chains and Global Capitalism*. Gary Gereffi and Miguel Korzeniewicz (eds.), Westport: Greenwood Press, pp. 1-14.

Gerke, Barbara. 2011. *Long Lives and Untimely Deaths. Life-Span Concepts and Longevity Practices among Tibetans in the Darjeeling Hills, India*. Leiden: Brill.

Gerke, Barbara. 2012 (2013). "Treating the Aged' and 'Maintaining Health'. Locating *bcud len* Practices in the Four Tibetan Medical Tantras.' *Journal of the International Association of Buddhist Studies*. 35:1-2, pp. 225-257.

Germano, David. 1998. 'Re-membering the Dismembered Body of Tibet. Contemporary Tibetan Visionary Movements in the People's Republic of China.' In *Buddhism in Contemporary Tibet. Religious Revival and Cultural Identity*. Melvyn C. Goldstein and Matthew T. Kapstein (eds.), Berkeley: University of California Press, pp. 53-94.

Giddens, Anthony. 1990. *The Consequences of Modernity*. Stanford: Stanford University Press.

Giddens, Anthony. 1998. *Conversations with Anthony Giddens. Making Sense of Modernity*. Stanford, California: Stanford University Press.

Gladney, Dru. 1996. *Muslim Chinese. Ethnic Nationalism in the People's Republic*. Cambridge: Harvard University Press.

Gladney, Dru. 1998. *Making Majorities. Constituting the Nation in Japan, China, Korea, Malaysia, Fiji, Turkey, and the U.S.* Stanford: Stanford University Press.

Gladney, Dru. 2004. *Dislocating China. Reflections on Muslims, Minorities, and Other Subaltern Subjects*. London: Hurst.

GLPRC (The Grassland Law of the People's Republic of China). 2002. The Grassland Law of the People's Republic of China. http://english.agri.gov.cn/governmentaffairs/lr/ap/201305/t20130509_19608.htm, accessed 12 March 2013.

Gold, Thomas, Doug Guthrie, and David Wank (eds.) 2002. *Social Connections in China. Institutions, Culture, and the Changing Nature of* Guanxi. Cambridge: Cambridge University Press.

Goldstein, Melvyn C. 1994. 'Change, Conflict and Continuity among a Community of Nomadic Pastoralists. A Case Study from Western Tibet, 1950-1990.' In *Resistance and Reform in Tibet*. Robert Barnett and Shirin Akiner (eds.), London: Hurst, pp. 76-112.

Goldstein, Melvyn. C. 1996. *Nomads of Golok, a Report*. http://www.case.edu/affil/tibet/documents/NomadsonGolokaReport.pdf, accessed 15 March 2014.

Goldstein, Melvyn C. 2009. 'Successfully Sustaining the Traditional Nomadic Way of Life in the Face of Privatization, Fencing and Sedentarization. The Adaptive Strategies of a Community of Nomads in Western Tibet.' Paper presented at the conference *Sustainable Livelihoods in Tibetan Pastoral Areas: Global Change and Local Challenges*, Martin Luther University, Halle-Wittenberg, Germany, 02-04.12.2009.

Goldstein, Melvyn C. 2012. 'Change and Continuity in a Nomadic Pastoralism Community in the Tibet Autonomous Region, 1959-2009.' In *Pastoral Practices in High Asia. Agency of 'Development' Effected by Modernisation, Resettlement and Transformation*. Hermann Kreutzmann (ed.), Dordrecht: Springer Verlag, pp. 257-272.

Goldstein, Melvyn C. and Cynthia M. Beall. 1989. 'The Impact of China's Cultural and Economic Reform Policy on Nomadic Pastoralists in Western Tibet.' *Asian Survey*. 29:6, pp. 619-641.

Goldstein, Melvyn C. and Cynthia M. Beall. 1990. *Nomads of Western Tibet*. London: Serindia.

Goldstein, Melvyn C. and Cynthia M. Beall. 1991. 'Change and Continuity in Nomadic Pastoralism on the Western Tibetan Plateau.' *Nomadic Peoples.* 28, pp. 105-123.

Goldstein, Melvyn C., Geoff Childs, and Puchung Wangdui. 2008. 'Going for Income in Village Tibet. A Longitudinal Analysis of Change and Adaptation, 1997 to 2007.' *Asian Survey.* 48:3, pp. 513-534.

Goldstein, Melvyn C., Geoff Childs, and Puchung Wangdui. 2010. 'Beijing's "People First" Development Initiative for the Tibet Autonomous Region's Rural Sector. A Case Study from the Shigatse Area.' *The China Journal.* 63, pp. 58-75.

Goldstein, Melvyn C., Ben Jiao, Cynthia M. Beall, and Phuntsog Tsering. 2003. 'Development and Change in Rural Tibet. Problems and Adaptations.' *Asian Survey.* 43:5, pp. 758-779.

Goodman, David. 2004. *China's Campaign to 'Open Up the West'. National, Provincial and Local Perspectives.* Cambridge: Cambridge University Press.

Gregory, Christopher A. 1980. 'Gifts to Men and Gifts to God. Gift Exchange and Capital Accumulation in Contemporary Papua.' *Man.* 15, pp. 626-652.

Gregory, Christopher A. 1997. *Savage Money. The Anthropology and Politics of Commodity Exchange.* Amsterdam: Harwood Academic.

Grenard, Fernand. 1974[1904]. *Tibet. The Country and its Inhabitants.* Delhi: Cosmo Publications.

Grumm-Grzhimailo, Grigorii. 1899. 'Spisok slov narechiia Tangutov Panaka.' ('List of words of Tangut Panak Dialect.') In *Opisanie puteshestviia v zapadnyi Kitai.* (*A Description of a Journey to Western China*). Vol. 2. St. Petersburg: V. Kirschbaum, pp. 417-420.

Grumm-Grzhimailo, Grigorii. 1907. *Opisanie puteshestviia v zapadnyi Kitai.* (*A Description of a Journey to Western China*). Vol. 3. St. Petersburg: V. Kirschbaum.

Gruschke, Andreas. 2001 *The Cultural Monuments of Tibet's Outer Provinces. Amdo.* Vol. 1. Bangkok: White Lotus Press.

Gruschke, Andreas. 2008. 'Nomads without Pastures? Globalization, Regionalization, and Livelihood Security of Nomads and Former Nomads in Northern Khams.' *Journal of the International Association of Tibetan Studies.* 4. http://www.thlib.org/collections/texts/jiats/#jiats=/04/gruschke/, accessed 27 September 2010.

Gruschke, Andreas. 2011a. 'Nomads and their Market Relations in Eastern Tibet's Yushu Region: The Impact of Caterpillar Fungus.' In *Economic Spaces of Pastoral Production and Commodity Systems. Markets and Livelihoods.* Jörg Gertel and Richard Le Heron (eds.), Farnham: Ashgate, pp. 211-229.

Gruschke, Andreas. 2011b. 'Nebenerwerbsnomaden und Raupenpilzökonomie. Pastorale Existenzsicherung in Osttibet.' In *Nomaden in unserer Welt. Die Vorreiter der Globalisierung: Von Mobilität und Handel, Herrschaft und Widerstand.* Jörg Gertel and Sandra Calkins (eds.), Bielefeld: Transcript Verlag, pp. 126-137.

Gruschke, Andreas. 2011c. 'From Yak Herders to *Yartsa* Traders. Tibetan Nomads and New Market Options in Qinghai's Yushu Region.' *China Tibetology.* 1, pp. 95-118.

Gruschke, Andreas. 2011d. 'Konflikte im Weideland – Konflikte um Weideland?' *Geographische Rundschau.* 7-8, pp. 20-26.

Gruschke, Andreas. 2011e. 'Yushu Nomads on the Move. How can the Use of Pastoralist Resources be Sustainable?' In *Selection of Presentations from the International Symposium on the Human Dimensions of Ecological Conservation in the Tibetan Plateau Region held in Xining, Qinghai, China, on 21-26 August 2011.* http://plateauperspectives.org/wp-content/uploads/2011/12/QASS2011.pdf, accessed 19 March 2014.

Gruschke, Andreas. 2012. *Nomadische Ressourcennutzung und Existenzsicherung im Umbruch. Die osttibetische Region Yushu (Qinghai, VR China).* Wiesbaden: Reichert Verlag.

Guyer, Jane. 2004. *Marginal Gains. Monetary Transactions in Atlantic Africa.* 2004. Chicago: University of Chicago Press.

Hansen, Mette Halskow. 2005. *Frontier People: Han Settlers in Minority Areas of China*. London: Hurst.

Harrell, Stevan. 1995. *Cultural Encounters on China's Ethnic Frontiers*. Seattle: University of Washington Press.

Harris, Richard. 2010. 'Rangeland Degradation on the Qinghai-Tibetan Plateau. A review of the Evidence of its Magnitude and Causes.' *Journal of Arid Environments.* 74, pp. 1-12.

Hart, Keith. 1973. 'Informal Income Opportunities and Urban Employment in Ghana.' *Journal of Modern African Studies.* 11, pp. 61-89.

Hart, Keith. 2009. 'On the Informal Economy. The Political History of an Ethnographic Concept.' *CEB Working Paper No. 09/042*. Brussels: Centre Emile Bernheim. Research Institute in Management Sciences. https://dipot.ulb.ac.be/dspace/bitstream/2013/54329/1/RePEc_sol_wpa-per_09-042.pdf, accessed 16 January 2014.

Hastrup, Kirsten. 1990. 'The Ethnographic Present: A Reinvention.' *Cultural Anthropology.* 5:1, pp. 45-61.

He, Jun. 2010. 'Globalised Forest-Products. Commodification of the *Matsutake* Mushroom in Tibetan Villages, Yunnan, Southwest China.' *International Forestry Review.* 12:1, pp. 27-37.

Hermanns, Mathias. 1949. *Die Nomaden von Tibet. Die sozial-wirtschaftlichen Grundlagen der Hirtenkulturen in A Mdo und von Innerasien. Ursprung and Entwicklung der Viehzucht.* , Vienna: Verlag Herold.

High, Mette. M. 2013. 'Polluted Money, Polluted Wealth: Emerging Regimes of Value in the Mongolian Gold Rush.' *American Ethnologist,* 40:4, pp. 676-688.

High, Mette M. 2017. *Fear and Fortune. Spirit worlds and Emerging Economies in the Mongolian Gold Rush.* Ithaca: Cornell University Press.

Hillman, Ben and Lee-Anne Henfry. 2006. 'Macho Minority. Masculinity and Ethnicity on the Edge of Tibet.' *Modern China.* 32:2, pp. 251-272.

Ho, Peter. 2000. 'The Myth of Desertification at China's Northwestern Frontier. The Case of Ningxia Province, 1929-1958.' *Modern China.* 26:3, pp. 348-395.

Holler, David. 2002. 'The Ritual of Freeing Lives.' In *Tibet, Past and Present. Religion and Secular Culture in Tibet. Proceedings of the Ninth Seminar of the International Association for Tibetan Studies, Leiden 2000.* Henk Blezer (ed.), Leiden: Brill, pp. 207-226.

Holliday, John, and Matt Cleaver. 2008. 'Medicinal Value of the Caterpillar Fungi Species of the Genus *Cordyceps* (Fr.) Link (Ascomycetes). A Review.' *International Journal of Medicinal Mushrooms.* 10:3, pp. 219-234. http://www.alohamedicinals.com/Cordyceps_Ascomycetes.pdf, accessed 10 March 2012.

Holliday, John, Matt Cleaver, and Solomon P. Wasser. 2005. 'Cordyceps.' In *Encyclopedia of Dietary Supplements.* New York: Marcel Dekker and Taylor and Francis Group, pp. 1-13.

Holy, Ladislav, and Milan Stuchlik. 1983. *Actions, Norms and Representations. Foundations of Anthropological Inquiry.* Cambridge: Cambridge University Press.

Horlemann, Bianca. 2002. 'Modernisation Efforts in Mgo log. A Chronicle, 1970-2000.' In *Amdo Tibetans in Transition. Society and Culture during the Post-Mao Era. Proceedings of the Ninth Seminar of the International Association for Tibetan Studies, Leiden 2000.* Toni Huber (ed.), Leiden: Brill, pp. 241-269.

Horlemann, Bianca. 2007. 'The Golok of Tibet through Western Eyes. Fascination and Horror.' In *Tibet in 1938-1939. Photographs from Ernst Schäfer's Expedition to Tibet.* Isrun Engelhardt (ed.), Chicago: Serindia, pp. 91-102.

Horlemann, Bianca. 2012a. 'Tibetan Nomadic Trade, the Chinese 'Xiejia' System and the Sino-Tibetan Border Market in Stong 'khor/Dan'gaer in 19th/20th Century A mdo.' *In Studies on the History and Literature of Tibet and the Himalaya.* Roberto Vitali (ed.), Kathmandu: Vajra, pp. 109-142.

Horlemann 2012b. 'Tibetans and Muslims in Northwest China. Economic and Political Aspects of a Complex Historical Relationship.' *Asian Highlands Perspectives*. 21, pp. 141-186.

Horlemann, Bianca. 2015. 'Victims of Modernization? Golok Struggles against the Ma Warlords in Qinghai, 1917-1942.' In *Amdo and Islam. Past and Present*. Marie-Paule Hille, Bianca Horlemann and Paul K. Nietupski (eds.). Lanham: Lexington, pp. 153-177.

Hou, Liufei 2011. 'Ecological Service Value of Natural Grassland in Guoluo Prefecture of Qinghai Province.' *Animal Husbandry and Feed Science*. 32:6, pp. 23-24.

Hu, Jintao. 2007. 'Hold High the Great Banner of Socialism with Chinese Characteristics and Strive for New Victories in Building a Moderately Prosperous Society. Report to the Seventeenth National Congress of the Communist Party of China on October 15, 2007.' http://www.china.org.cn/english/congress/229611.htm, accessed 24 March 2014.

Huber, Toni. 1991. 'Traditional Environmental Protectionism in Tibet Reconsidered.' *The Tibet Journal*. 16:3, pp. 63-77.

Huber, Toni. 1999. *The Cult of Pure Crystal Mountain. Popular Pilgrimage and Visionary Landscape in Southeast Tibet*. New York: Oxford University Press.

Huber, Toni. 2002. 'Introduction. *A mdo* and its Modern Transition.' In *Amdo Tibetans in Transition: Society and Culture during the Post-Mao Era. Proceedings of the Ninth Seminar of the International Association for Tibetan Studies, Leiden 2000*. Toni Huber (ed.), Leiden: Brill, pp. xi-xxiii.

Huber, Toni. 2004. 'Territorial Control by 'Sealing' (*rgya sdom-pa*). A Religio-Political Practice in Tibet.' *Zentralasiatische Studien*. 33, pp. 127-152.

Huber, Toni, and Poul Pedersen. 1997. 'Meteorological Knowledge and Environmental Ideas in Traditional and Modern Societies. The Case of Tibet.' *Journal of the Royal Anthropological Institute*. 3:3, pp. 577-598.

Humphrey, Caroline, and David Sneath. 1999. *The End of Nomadism? Society, State, and the Environment in Inner Asia*. Durham: Duke University Press.

Hutchinson, Sharon E. 1996. *Nuer Dilemmas. Coping with Money, War, and the State*. Berkeley: University of California Press.

Hywel-Jones, Nigel L., and RNR-RC Yusipang. 2002. '*Cordyceps sinensis* in Bhutan. A Concept Note on the Sustainable Collection and Utilization.' Thimphu: RNR-RC Yusipang.

Hyytiäinen, Tiina. 2011. 'Repkong tantric practitioners and their environment: observing the vow of not taking life.' *Studia Orientalia* (*Himalayan Nature: Representations and Reality*). Erika Sandman and Riika J. Virtanen (eds.), 109, pp. 17-38.

ICT (International Campaign for Tibet). 2008. *Tibet at a Turning Point. The Spring Uprising and China's New Crackdown*. Accessed online https://savetibet.de/fileadmin/user_upload/content/berichte/Tibet_at_a_Turning_Point_for_online.pdf, accessed 27 September 2017.

ICT (International Campaign for Tibet). 2009. *A Great Mountain Burned by Fire. China's crackdown in Tibet*. Accessed online https://savetibet.de/fileadmin/user_upload/content/berichte/ICT_A_Great_Mountain_Burned_by_Fire.pdf, accessed 27 September 2017.

Ingold, Tim. 1987. *The Appropriation of Nature. Essays on Human Ecology and Social Relations*. Iowa City: University of Iowa Press.

Iselin, Lilian. 2011. 'Of Horses and Motorbikes: Negotiating Modernities in Pastoral A mdo, Sichuan Province.' *Journal of the International Association of Tibetan Studies*. 6. http://www.thlib.org/collections/texts/jiats/#!jiats=/06/iselin/, accessed 3 November 2012.

Iselin, Lilian. 2012. 'Neue Mobilitäten und neue Räume für Pastoralisten in Amdo? Strassen- und Telekommunikationsnetze auf dem tibetischen Hochplateau.' *Tsantsa. Zeitschrift der Schweizerischen Ethnologischen Gesellschaft*. 17, pp. 163-167.

Iselin, Lilian. 2014. 'Translocal Practices on the Tibetan Plateau. Motorized Mobility of Pastoralists and Spatial Transformations.' *Nomadic Peoples*. 18:1, pp. 15-37.

Iselin, Lilian. 2017. *Mobile Technologien und nomadischer Raum: Motorisierung, Mobiltelefonie und Urbanisierung in Südamdo, Osttibet.* Berlin: De Gruyter.

Jabb, Lama. 2011. 'Singing the Nation. Modern Tibetan Music and National Identity.' *Revue d'Etudes Tibétaines*. 21, pp. 1-29.

Jarvenpa, Robert, and Hetty Jo Brumbach (eds.). 2006. *Circumpolar Lives and Livelihood. A Comparative Ethnoarcheology of Gender and Subsistence.* Lincoln: University of Nebraska Press.

Kapstein, Matthew. 2004. 'A Thorn in the Dragon's Side. Tibetan Buddhist Culture in China.' In *Governing China's Ethnic Frontiers*. Morris Rosabi (ed.), Seattle: University of Washington Press, pp. 230-269.

Karmay, Samten G. 1998a. 'The Cult of Mountain Deities and its Political Significance.' In *The Arrow and the Spindle. Studies in History, Myths, Rituals and Beliefs in Tibet.* Kathmandu: Mandala, pp. 432-450.

Karmay, Samten G. 1998b. 'The Four Tibetan Medical Treatises and their Critics.' In *The Arrow and the Spindle. Studies in History, Myths, Rituals and Beliefs in Tibet.* Kathmandu: Mandala, pp. 228-237.

Karmay, Samten G. 1998c. 'Mountain Cult and National Identity.' In *The Arrow and the Spindle. Studies in History, Myths, Rituals and Beliefs in Tibet.* Kathmandu: Mandala, pp. 423-431.

Karmay, Samten G. 1998d. 'The Wind-Horse and the Well-Being of Man.' In *The Arrow and the Spindle. Studies in History, Myths, Rituals and Beliefs in Tibet.* Kathmandu: Mandala, pp. 413-422.

Khenpo Tsültrim Lodrö (mKhan po Tshul khrims bLo gros). 2014. 'Words to Increase Virtue.' Geoffrey Barstow (transl.), http://geoffbarstow.files.wordpress.com/2013/01/khenpo-tsc3bcltrim-lodrc3b6-words-to-increase-virtue2.pdf, accessed 28 March 2014.

Kopytoff, Igor. 1986. 'The Cultural Biography of Things. Commoditization as Process.' In *The Social Life of Things. Commodities in Cultural Perspective.* Arjun Appadurai (ed.), Cambridge: Cambridge University Press, pp. 64-91.

Kornman, Robin. 1997. 'A Tribal History.' In *Religions of Tibet in Practice.* Donald S. Lopez (ed.), Princeton: Princeton University Press, pp. 77-97.

Lama, Kunga. 2007. *Crowded Mountains, Empty Towns. Commodification and Contestations in Cordyceps Harvesting in Eastern Tibet.* MA thesis, University of Colorado at Boulder.

Lama, Yeshi Choden, Suresh K. Ghimire and Yildiz Aumeeruddy-Thomas. 2001. *Medicinal Plants of Dolpo. Amchis' Knowledge and Conservation.* Kathmandu: WWF Nepal Program.

Lee, Richard B., and Irven DeVore (eds.). 1968. *Men the Hunter.* Chicago: Aldin.

Levine, Nancy E. [s.a.] *Reconstructing Tradition. Persistence and Change in Golog Social Structure.* http://www.case.edu/affil/tibet/booksAndPapers/Segpaper.htm?nw_view=1307345682, accessed 6 June 2011.

Levine, Nancy E. 1998. 'From Nomads to Ranchers: Managing Pasture among Ethnic Tibetans in Sichuan.' In *Development, Society, and Environment in Tibet. Papers Presented at a Panel of the 7th Seminar of the International Association for Tibetan Studies, Graz, 1995.* Graham E. Clarke (ed.), Vienna: Verlag der Österreichischen Akademie der Wissenschaften, pp. 69-76.

Levine, Nancy E. 1999. 'Cattle and the Cash Economy. Responses to Change among Tibetan Nomadic Pastoralists in Sichuan, China.' *Human Organization*. 58, pp. 161-172.

Levine, Nancy E. 2015. 'Transforming Inequality: Eastern Tibetan Pastoralists from 1955 to the Present.' *Nomadic Peoples*. 19:2, pp. 164-188.

Li, Diqiang, and Jianwen Li (eds.). 2002. *Biodiversity of Three River Source Region.* Beijing: China Science and Technology Publishing House.

Li, Jianglin. 2016. *Tibet in Agony. Lhasa 1959*. Cambridge: Harvard University Press.

Li, Wenlian, Yuling Li, Xin Liu, Haifeng Xu, Zonghao Zhang, and Shaoli Ma. 2010. 'Guoluo zhou donchong xiacao ziyuan lilong yuguan liqian xi.' ('Utilization and Management of *Cordyceps sinensis* Resources in Golok Prefecture.') *Chinese Journal of Grassland*. 32, pp. 32-35.

Li, Yi, Xiaoliang Wang, Lei Jiao, Yi Jiang, Hui Li, Siping Jiang, Ngarong Lhosumtseiring, Shenzhan Fu, Caihong Dong, Yu Zhan, and Yijian Yao. 2011. 'A Survey of the Geographic Distribution of *Ophiocordyceps sinensis*.' *The Journal of Microbiology*. 49:6, pp. 913-919.

Lin, Hsiaoting. 2007. 'Nationalists, Muslim Warlords, and the 'Great North-Western Development' in Pre-Communist China.' *China and Eurasia Forum Quarterly*. 5:1, pp. 115-135.

Lipman, Jonathan. 1994. 'Ethnicity and Politics in Republican China: The Ma Family Warlords of Gansu.' *Modern China*. 10:3, pp. 285-316.

Liu, Jianbang (ed.). 1994. *Ganzi zangzu zizhizhou lingyezhi* (*Forestry History of Ganzi Tibetan Autonomous Prefecture*). Chengdu: Sichun kexue jisu chubanshe.

Ma, Youxiang. 2010. '*Ophiocordyceps sinensis*. Resource and its Management in China.' Print-out of presentation given at the 2010 International Conference on *Ophiocordyceps sinensis*, Xining, China, 8-10.06.2010, pp. 1-12.

Makley, Charlene E. 2007. *The Violence of Liberation. Gender and Tibetan Buddhist Revival in Post-Mao China*. California: University of California Press.

Malinowski, Bronislaw (Malinowski, Bronisław), and Julio de la Fuente. 2004. 'Ekonomia meksykańskiego systemu targowego.' ('The Economics of a Mexican Market System'). In *Dzieła* (*Collected Works*). Bronislaw Malinowski. Vol. 13. Warsaw: Państwowe Wydawnictwo Naukowe PWN, pp. 15-201.

Manderscheid, Angela. 1998. 'Life and Economic Patterns of Nomads on the Eastern Tibet Plateau. *'Brog pa* and *Sa Ma 'Brog* in 'Dzam Thang.' In *Development, Society, and Environment in Tibet. Papers Presented at a Panel of the 7th Seminar of the International Association for Tibetan Studies, Graz, 1995*. Graham E. Clarke (ed.), Vienna: Verlag der Österreichischen Akademie der Wissenschaften, pp. 59-67.

Manderscheid, Angela. 1999. *Lebens- und Wirtschaftsformen von Nomaden im Osten des tibetischen Hochlandes*. Berlin: Reimer Verlag.

Manderscheid, Angela. 2001. 'The Black Tent in Its Easternmost Distribution. The Case of the Tibetan Plateau.' *Mountain Research and Development*. 21:2, pp. 154-160.

Manderscheid, Angela. 2002. 'Revival of a Nomadic Lifestyle. A Survival Strategy for Dzam thang's Pastoralists.' In *Amdo Tibetans in Transition. Society and Culture during the Post-Mao Era. Proceedings of the Ninth Seminar of the International Association for Tibetan Studies, Leiden 2000*. Toni Huber (ed.), Leiden: Brill, pp. 271-289.

Manderscheid, Angela, Arvo Naukkarinen, Ning Wu, and Alfred Colpaert. 2004. 'From Subsistence to Market Economy. Responses of Tibetan Pastoralists to New Economic Realities.' *Rangifer*. 15, pp. 29-37.

Marshall, Steven D., and Susette Ternent-Cooke. 1997. *Tibet. Outside the TAR*. CD-ROM.

Martin, Dan. 2011. 'Greek and Islamic Medicines' Historical Contact with Tibet: A Reassessment in View of Recently Available but Relatively Early Sources on Tibetan Medical Eclecticism.' In *Islam and Tibet. Interactions along the Musk Routes*. Anna Akasoy, Charles Burnett, Ronit Yoeli-Tlalim (eds.), London: Routledge, pp. 117-143.

Marx, Karl. 1995. *Capital. An Abridged Edition*. David McLellan (ed.), Oxford: Oxford University Press.

Maurer, Bill. 2006. 'The Anthropology of Money.' *Annual Review of Anthropology*. 35, pp. 15-36.

Mauss, Marcel. 1966. *The Gift. Forms and Functions of Exchange in Archaic Societies*. London: Cohen and West.

MCSY (Machen County Statistical Yearbook). 2009. *Guomin jingji he shehui fazhan tongji ziliao huibian 2008.* (*Statistical Yearbook on People's Economic and Social Development 2008*). Dawu: Maqin xian tongjiju.

Meng, Yijian. 2012. 'The Development of Sanjiangyuan National Ecosystem Protection Experimental Region.' http://www.qh.gov.cn/system/2012/01/14/010619802.shtml, accessed 22 May 2013.

MGGT (*rMa rgyal gangs thigs*). [s.a.] 'Bod kyi skye khams lta ba dang 'brel ba'i bklag deb.' *rMa rgyal gangs thigs*. [s.l.]: rMa yul skyes khams khor yug srung skyob tsogs pa, p. 11.

Miller, Daniel. 2000. 'Tough Times for Tibetan Nomads in Western China. Snowstorms, Settling down, Fences, and the Demise of Traditional Nomadic Pastoralism.' *Nomadic Peoples*. 4:1, pp. 83-109.

Miller, Daniel. 2001. *Poverty among Tibetan Nomads in Western China. Profiles of Poverty and Strategies for Poverty Reduction.* Paper prepared for the Tibet Development Symposium, Brandeis University, May 4-6, 2001, http://www.cwru.edu/affil/tibet/booksAndPapers/papers/miller2.htm, accessed 15 April 2014.

Mintz, Sidney W. 2004. 'The Employment of Capital by Market Women in Haiti.' In *Capital, Saving and Credit in Peasant Societies. Studies from Asia, Oceania, the Caribbean and Middle America.* Raymond Firth and B.S. Yamey (eds.), London: Routledge, pp. 256-286.

Morcom, Anna. 2008. 'Getting Heard in Tibet. Music, Media and Markets.' In *Consumption, Markets and Culture*. 11:4, pp. 259-285.

Morcom, Anna. 2011. 'History, Traditions, Identities and Nationalism. Drawing and Redrawing the Musical Cultural Map of Tibet.' In *Tibetan Studies. An Anthology. Proceedings of the 11th Seminar of International Association for Tibetan Studies.* Saadet Arslan and Peter Schwieger (eds.), Andiast: International Institute for Tibetan and Buddhist Studies GmbH, pp. 385-417.

Mu, Jingliang, and Huiying Zhu. 2005. *Yushu dongchong xiacao caiwa jumian shikong de chengyin fenxi.* (*Analysis of the Causes Contributing to the Situation of Unearthing Chinese Caterpillar Fungus in Yushu being out of Control*). http://wenku.baidu.com/view/00185871o1f69e3143329425, accessed 25 March 2014.

Mu, Wang, Kampot Tashi, Zhuoga, and Dan Ba. 2011. 'Contribution of *Cordyceps sinensis* to Tibetan Pastoralists Income and Problems in its Sustainable Use.' In *Pastoralists and Rangeland Management on the Tibetan Plateau in the Context of Climate and Global Change.* Hermann Kreutzmann, Yang Yong and Jürgen Richter (eds.), Bonn: Deutsche Gesellschaft für Internationale Zusammenarbeit, pp. 165-176.

Næss, Marius Warg. 2004. *Living with Risk and Uncertainty. The Case of the Nomadic Pastoralists in the Aru Basin, Tibet.* Cand. Polit. thesis, Department of Social Anthropology, Faculty of Social Science, University of Tromsø, Norway.

Namgyel, Phuntsho. 2003. '*Walk the Extra Mile and Make the Difference.* Household Income, Property Rights and Sustainable Use of NTFP in Subsistence. Mountain Economy. The Case of *Cordyceps* and *Matsutake* in Bhutan Himalayas.' Paper presented at the Regional CBNRM Workshop, 04-07.11. 2003.

Negi, Chandra. S., Prithvi Raj Koranga, and Hira Singh Ghinga. 2006. '*Yartsa Gumba* (*Cordyceps sinensis*). A Call for its Sustainable Exploitation.' *International Journal of Sustainable Development and World Ecology.* 13:3, pp. 1-8.

Norbu, Namkhai. 1997. *Journey among the Tibetan Nomads. An Account of a Remote Civilization.* Dharamsala: Library of Tibetan Works and Archives.

Oi, Jean C. 1991. *State and Peasant in Contemporary China. The Political Economy of Village Government.* Berkeley: University of California Press.

Pad ma Tshe ring. 2004. 'rTa bo gong zhol dang dme tshang sgur ru tsho drug sogs sbra nag sde ba'i lo rgyus.' ('History of the Wranakh *dewa* Dawu Gong, [Dawu] Zhu, Metsang, Guru Tsodruk and Others.') In *mGo log rig gnas lo rgyus* (*A Cultural History of Golok*). Vol. 4. Ziling: Srid gros mgo log khul u slob sbyong lo rgyus dang tshan slob rig 'phrod u yon lhan khang.

Panda, Ashok Kumar, and Kailash Chandra Swain. 2011. 'Traditional Uses and Medicinal Potential of *Cordyceps sinensis* of Sikkim.' *Journal of Ayurveda and Integrative Medicine*. 2:1, pp. 9-13.

Paterson, R. Russell M. 2008. '*Cordyceps* – a Traditional Chinese Medicine and another Fungal Therapeutic Biofactory?' *Phytochemistry*. 69, pp. 1469-1495.

Peebles, Gustav. 2012. 'Filth and Lucre: The Dirty Money Complex as a Taxation Regime.' *Anthropological Quarterly*. 85:4, pp. 1229-1255.

Petech, Luciano. 1990. *Central Tibet and the Mongols. The Yüan-Sa-skya Period of Tibetan History*. Rome: Istituto Italiano per il Medio ed Estremo Oriente.

'Phrin las. 2008. *Mgo log khul gyi dgon sde khag gi lo rgyus snying bsdus*. (*A Compendium of History of Monasteries of Golok Prefecture*). Xining: mTsho sngon mi rigs dpe skrun khang.

Pirie, Fernanda. 2005. *Tribe and State in Eastern Tibet. Feuding, Mediation and the Negotiation of Authority among the Amdo Nomads*. Working Paper No. 72. Halle: Max Planck Institute for Social Anthropology.

Pirie, Fernanda. 2008. 'Violence and Opposition among the Nomads of Amdo. Expectations of Leadership and Religious Authority.' In *Conflict and Social Order in Tibet and Inner Asia*. Fernanda Pirie and Toni Huber (eds.), Leiden: Brill, pp. 217-240.

Pirie, Fernanda. 2012. 'Legal Dramas on the Amdo Grasslands. Abolition, Transformation or Survival?' In *Revisiting Rituals in a Changing Tibetan World*. Katia Buffetrille (ed.), Leiden: Brill, pp. 83-107.

Pirie, Fernanda. 2013. 'The Limits of the State. Coercion and Consent in Chinese Tibet.' *Journal of Asian Studies*. 72, pp. 69-89.

Plattner, Stuart. 1985. 'Equilibrating Market Relationships.' In *Markets and Marketing*. Stuart Plattner (ed.), Lanham: University Press of America, pp. 133-152.

Podgorecki, Adam (Podgórecki, Adam). 1962. *Socjologia prawa* (*Sociology of Law*). Warsaw: Wiedza Powszechna.

Polanyi, Karl. 2001. *The Great Transformation. The Political and Economic Origins of Our Time*. Boston: Beacon Press.

Pohle, Perdita. 1990. *Useful Plants of Manang District*. Stuttgart: Franz Steiner Verlag.

Potter, Sulamith H., and Jack M. Potter. 1991. *China's Peasants. The Anthropology of a Revolution*. Cambridge: Cambridge University Press.

Procacci, Giovanna. 1991. 'The Social Economy of Poverty.' In *The Foucault Effect. Studies in Governmentality*. Graham Burchell, Colin Gordon and Peter Miller (eds.), Chicago: University of Chicago Press, pp. 151-158.

Przhevalsky, Nikolay M. [Prejevalsky, Nikolai M.] 1876. *Mongolia, the Tangut Country, and the Solitudes of Northern Tibet being a Narrative of Three Year's Travel in Eastern High Asia*. London: Sampson Low, Marston, Searle and Rivington.

Ptackova, Jarmila (Ptáčková, Jarmila). 2011. 'Sedentarization of Tibetan Nomads in China: Implementation of the Nomadic Settlement Project in the Tibetan Amdo Area; Qinghai and Sichuan Provinces.' *Pastoralism: Research, Policy and Practice*. 1:4. http://www.pastoralism-journal.com/content/pdf/2041-7136-1-4.pdf, accessed 30 May 2011.

Ptackova, Jarmila (Ptáčková, Jarmila). 2013. *The Great Opening of the West Development Strategy and its Impact on the Life and Livelihood of Tibetan Pastoralists. Sedentarization of Tibetan Pastoralists in Zeku County as a Result of Implementation of Socioeconomic and Environmental Development Projects in Qinghai Province, P.R. China*. PhD dissertation, Humboldt University in Berlin. http://edoc.hu-berlin.de/dissertationen/ptackova-jarmila-2013-06-19/PDF/ptackova.pdf, accessed 12 February 2014.

QPSY (Qinghai Province Statistical Yearbook). 2009. *Qinghai Statistical Yearbook*. Xining: Qinghai Province Statistics Bureau. CD-ROM.

Rahnema, Majid, 1992. 'Poverty.' In *The Development Dictionary. A Guide to Knowledge as Power.* Wolfgang Sachs (ed.), London: Zed Books, pp. 158-176.

Rahnema, Majid, and Victoria Bawtree (eds.) 1997. *The Post-Development Reader.* London: Zed Books.

Ramble, Charles. 1993. 'Whither, Indeed, the *Tsampa* Eaters?' *Himal.* 6:5, pp. 21-25.

Rao, Aparna. 1987. 'The Concept of Peripatetics. An Introduction.' In *The Other Nomads. Peripatetic Minorities in Cross-Cultural Perspective.* Aparna Rao (ed.), Cologne: Böhlau Verlag, pp. 1-32.

RFA (Radio Free Asia). 2012a. 'Tibetan Town Tense after Protests.' http://www.rfa.org/english/news/tibet/tibetans-03232009160253.html, accessed 30 December 2012.

RFA (Radio Free Asia). 2012b. 'Tibetans Clash with Muslims in China's Qinghai.' http://www.rfa.org/english/tibet/tibetan_muslims-20070809.html, accessed 27 September 2012.

Ribot, Jesse C. 1998. 'Theorizing Access. Forest Profits along Senegal's Charcoal Commodity Chain.' *Development and Change.* 29, pp. 307-341.

Rist, Gilbert. 2003. *The History of Development. From Western Origins to Global Faith.* London: Zed Books.

Roborovskii, Vsevolod I. 1900. *Trudy ekspedicii Imperatorskago Obshchestva po Tsentral'noi Azii sovershennoi v 1893-1895 pod nachal'stvom V.I. Roborovskago.* (*The Works of the Expedition of the Imperial Society to Central Asia Carried out in 1893-1895 under the Command of V.I. Roborovskii*). Vol. 1, St. Petersburg: M.M. Stasjulevich.

Rock, Joseph F. 1956. *The Amnye Ma-chhen Range and Adjacent Regions.* Rome: Istituto Italiano per il Medio ed Estremo Oriente.

Rockhill, William W. 1891. *The Land of the Lamas. Notes of a Journey through China, Mongolia and Tibet.* London: Longmans, Green and Co.

Rockhill, William W. 1894. *Diary of a Journey through Mongolia and Tibet in 1891 and 1892.* Washington: Smithsonian Institute.

Roerich, George. 1933. 'Les Goloks et leur caractère ethnique.' In *XVe Congrès International d'Anthropologie et d'Archéologie Préhistorique. Paris 20-27 Septembre 1931.* Paris: Libraire E. Nourry, pp. 736-741.

Roerich, George. 1958. *Le parler de l'Amdo. Étude d'un dialecte archaïque du Tibet.* Rome: Istituto Italiano per il Medio ed Estremo Oriente.

Roerich, George [Rerikh, Iurii N.]. 1961. 'Kochevye plemena Tibeta.' ('Nomadic Tribes of Tibet.') In *Strany i Narody Vostoka* (*Countries and Peoples of the East*). 2, pp. 7-12.

Rong, Ma. 2011. *Population and Society in Contemporary Tibet.* Hong Kong: Hong Kong University Press.

RPRCPWP (Regulations of the People's Republic of China on Wild Plants Protection). 2001. Regulations of the People's Republic of China on Wild Plants Protection. In *Laws and Regulations of the People's Republic of China.* Beijing: China Legal Publishing House, pp. 98-102.

Ruf, Gregory. 1998. *Cadres and Kin: Making a Socialist Village in West China, 1921-91.* Stanford: Stanford University Press.

Sachs, Wolfgang (ed.). 1992. *The Development Dictionary. A Guide to Knowledge as Power.* London: Zed Books.

Samuel, Geoffrey. 1993. *Civilized Shamans. Buddhism in Tibetan Societies.* Kathmandu: Mandala.

Schneider, Nicola. 2013. *Le renoncement au féminin. Couvents et nonnes dans le bouddhisme tibétain.* Paris: Presses Universitaires de Paris Ouest.

Schrempf, Mona. 2010-2011. 'Re-Production at Stake. Experiences of Family Planning and Fertility among Amdo Tibetan Women.' *Asian Medicine*. 6, pp. 321-347.

Scott, James. 1985. *Weapons of the Weak. Everyday Forms of Peasant Resistance*. New Haven: Yale University Press.

SFAMA (State Forestry Administration and Ministry of Agriculture). 1999. 'State Forestry Administration and Ministry of Agriculture, Order No. 4. 1999. The List of the Wild Plants under the State Emphasized Protection.' http://www.gov.cn/gongbao/content/2000/content_60072.htm, accessed 24 March 2014.

Shakabpa, Tsepon Wangchuk Deden. 2010. *One Hundred Thousand Moons. An Advanced Political History of Tibet*. Derek F. Maher (ed.), 2 vols, Leiden: Brill, vol. 1.

Shakya, Tsering W. 1993. 'Whither the *Tsampa* Eaters?' *Himal*. 6:5, pp. 8-11.

Shakya, Tsering W. 1999. *The Dragon in the Land of Snows. A History of Modern Tibet since 1947*. London: Pimlico.

Sharma, Subrat. 2004. 'Trade of *Cordyceps sinensis* from High Altitudes of the Indian Himalaya. Conservation and Biotechnological Priorities.' *Current Science*. 86:12, pp. 1614-1619.

Shelton, Albert. 1921. *Pioneering in Tibet. A Personal Record of Life and Experience in Mission Fields*. New York: Fleming H. Revell Company.

Shen, Nanyin. 2010. 'Clarification of the Names of Chinese Caterpillar Fungus.' Paper presented at the 2010 International Conference on *Ophiocordyceps sinensis*. Xining, China, 08-10.06.2010.

Shipton, Parker M. 1989. *Bitter Money. Cultural Economy and Some African Meanings of Forbidden Commodities*. Washington: American Anthropological Association.

Simmel, Georg. 2004. *The Philosophy of Money*. London: Routledge.

Sindzingre, Alice. 2006. 'The Relevance of the Concepts of Formality and Informality. A Theoretical Appraisal.' In *Linking the Formal and Informal Economy. Concepts and Policies*. Basudeb Guha-Khasnobis, Ravi Kanbur and Elinor Ostrom (eds.), Oxford: Oxford University Press, pp. 58-74.

Singh, Narayan, Rakshita Pathak, Arjun Singh Kathait, Deepak Rautela, and Anoop Dubey. 2010. 'Collection of *Cordyceps sinensis* (Berk.) Sacc. in the Interior Villages of Chamoli District in Garhwal Himalaya (Uttarakhand) and its Social Impacts.' *Journal of American Science*. 6:6, pp. 5-9.

Smelser, Neil J. 1973. 'Introduction.' In *Karl Marx on Society and Social Change*. Neil J. Smelser (ed.), Chicago: University of Chicago Press, pp. vii-xxxix.

Sodnamkyid and Emilia Roza Sulek (Emilia Róża Sułek). 2017. ''Everything costs Money'. Livelihood and Economics in the 'New Resettled Village' of Sogrima, Golok (Qinghai Province).' *Nomadic Peoples*. 21:1, pp. 136-151.

Solinger, Dorothy. 1995. 'The Floating Population in the Cities: Chances for Assimilation?' In *Urban Spaces in Contemporary China. The Potential for Autonomy and Community in Post-Mao China*. Deborah S. Davies, Richard Kraus, Barry Naughton and Elisabeth J. Perry (eds.), Cambridge: Woodrow Wilson Center Press and Cambridge University Press, pp. 113-139.

Sonam Lindub. 2008. *The Social Exclusion of Tibetan Nomads in the Caterpillar Fungus Trade. A Case Study from Zaduo County, Yushu Prefecture, Qinghai Province, PR China*. MA thesis, Miriam College, Quezon City, Philippines.

SQSY (*Sanjiangyuan Qinghai Sanjiangyuan Yaoye*). 2009. 'Zhaxi Caiji. Zhanfang zai Qingzang gaoyuan de Xuelianhua.' ('Zhaxi Caiji. The Blooming Snow Lotus on the Qinghai-Tibetan Plateau.') *Sanjiangyuan Qinghai Sanjiangyuan Yaoye*. May 2009, pp. 15-18.

Stammler, Florian, 2005. *Reindeer Nomads Meet the Market. Culture, Property and Globalization at the End of the Land*. Münster: LIT Verlag.

Steinkraus, Donald, and James B. Whitfield. 1994. 'Chinese Caterpillar Fungus and World Record Runners.' ***American Entomologist***. 40:4, pp. 235-239.

Stewart, Michelle O. 2009. 'The 'Himalayan Gold' Rush: Prospectors' Practices and Implications for Management.' In *Contemporary Visions in Tibetan Studies: Proceedings of the First International Seminar of Young Tibetologists*. Brandon Dotson, Kalsang Norbu Gurung, Georgios Halkias and Tim Myatt (eds.), Chicago: Serindia Publications, pp. 67-93.

Stewart, Michelle O. 2014. 'Constructing and Deconstructing the Commons: Caterpillar Fungus Governance in Developing Yunnan.' In *Mapping Shangrila: Nature, Personhood and Sovereignty in the Sino-Tibetan Borderlands*. Chris Coggins and Emily T. Yeh (eds.), Seattle: University of Washington Press, pp. 175-197.

Stubel, Hans. 1958. *The Mewu Fantzu. A Tibetan Tribe of Kansu*. New Haven: HRAF Press.

Sulek, Antoni (Sułek, Antoni). 1986. 'Malinowski in Mexico.' (Review of Bronislaw Malinowski, Julio de la Fuente, *The Economics of a Mexican Market System*. London: Routledge and Kegan Paul, 1982). *The Polish Sociological Bulletin*. 3-4, pp. 67-69.

Sulek, Emilia Roza (Sułek, Emilia Róża). 2006. 'Imagining Tibet in Poland. A Contribution to Anthropology of Imagined Countries.' *Tibet Journal*. 2, pp. 49-68.

Sulek, Emilia Roza (Sułek, Emilia Róża). 2008. 'Tseren's Last Gold Rush. Tales of *Yartsa* Hunting in Tibet.' *International Institute for Asian Studies Newsletter*. 46, pp. 20-21.

Sulek, Emilia Roza (Sułek, Emilia Róża). 2009. 'In the Land of Checkpoints. *Yartsa Gunbu* Trade in Golok 2007, a Preliminary Report from the Field.' In *Contemporary Visions in Tibetan Studies*. Brandon Dotson, Kalsang Norbu Gurung, Georgios Halkias and Tim Myatt (eds.), Chicago: Serindia, pp. 15-44.

Sulek, Emilia Roza (Sułek, Emilia Róża). 2010a. 'Disappearing Sheep. The Unexpected Consequences of the Emergence of the Caterpillar Fungus Economy in Golok, Qinghai, China.' *Himalaya*. 30:1-2, pp. 9-22.

Sulek, Emilia Roza (Sułek, Emilia Róża). 2010b. 'The Wranahs. A Story of Discovering and Forgetting the Tibetan Tribes of Amnye Machen.' *Rocznik Orientalistyczny* (*Altaica et Tibetica. A Volume in Honour of Prof. S. Godziński on his Seventieth Birthday*). Agata Bareja-Starzyńska, Filip Majkowski, Jan Rogala and Jerzy Tulisow (eds.), 63:1, pp. 237-252.

Sulek, Emilia Roza (Sułek, Emilia Róża). 2012. ''Everybody likes Houses. Even Birds are Coming!' Housing Tibetan Pastoralists in Golok. Policies and Everyday Realities.' In *Pastoral Practices in High Asia. Agency of 'Development' Effected by Modernisation, Resettlement and Transformation*. Hermann Kreutzmann (ed.), Dordrecht: Springer Verlag, pp. 235-255.

Sulek, Emilia Roza (Sułek, Emilia Róża). 2014. 'Invisible Mongols. Observations from Fieldwork in Tibet.' In *A Window on Others. Anniversary Volume with Essays on Mongolian, Manchu and Turkic Studies Dedicated to Jerzy Tulisow on His Seventieth Birthday*. Agata Bareja-Starzyńska (ed.), Warsaw: Elipsa, pp. 239-257.

Sulek, Emilia Roza (Sułek, Emilia Róża). 2016. 'Caterpillar Fungus and the Economy of Sinning. On Entangled Relations between 'Religious' and 'Economic' in a Tibetan Pastoral Region of Golog, Qinghai, China.' *Études Mongoles et Sibériennes, Centrasiatiques et Tibétaines*. 47, http://emscat.revues.org/2769, accessed 15 October 2017.

Sulek, Emilia Roza (Sułek, Emilia Róża). 2017. 'I Don't Sing About Love.' ***Tethys. Central Asia** Everyday*, http://www.tethys.caoss.org/i-dont-sing-about-love, accessed 15 October 2017.

Sulek, Emilia Roza (Sułek, Emilia Róża). In Press. 'Tibetan Nomadic Architecture.' In *Encyclopedia of Vernacular Architecture of the World*. Marcel Vellinga (ed.), New York: Bloomsbury Press.

Sulek, Emilia Roza (Sułek, Emilia Róża) and Jarmila Ptackova (Jarmila Ptáčková), 2017. 'Mapping Amdo. People and Places in an Ongoing Transition.' In *Mapping Amdo. Dynamics of Change*. Jarmila Ptackova and Adrian Zenz (eds.), Prague: Oriental Institute CAS, pp. 9-21.

Sun, Faping. 2008. 'Qinghai dongchong xiacao ziyuan baohu yu kaifa.' ('Conservation and Development of Caterpillar Fungus Resources in Qinghai.') *Qinghai Jingji Yanju* (*Qinghai Economics Study*). 2:146, pp. 4-12.

Tan, Gillian. 2013. 'An Ethnography of Life and Changes among Tibetan Nomads of Minyag Dora Karmo, Ganzi Tibetan Autonomous Prefecture, Sichuan Province.' *Études Mongoles et Sibériennes, Centrasiatiques et Tibétaines.* 43-44, pp. 1-18.

Tan, Gillian. 2016. '„Life" and „freeing life" (*tshe thar*) among Pastoralists of Kham: Intersecting Religion and Environment.' *Études Mongoles et Sibériennes, Centrasiatiques et Tibétaines.* 47, pp. 1-15.

Tan, Gillian. 2017. *In the Circle of White Stones. Moving through Seasons with Nomads of Eastern Tibet.* Seattle: University of Washington Press.

Tan, Gillian. 2018. *Pastures of Change. Contemporary Adaptations and Transformations among Nomadic Pastoralists of Eastern Tibet.* Berlin: Springer.

Taussig, Michael. 1980. *The Devil and Commodity Fetishism in South America.* Chapel Hill: The University of North Carolina Press.

Taylor, Charles. 2001. 'Two Theories of Modernity.' In *Alternative Modernities.* Dilip Parameshwar Gaonkar (ed.), Durham: Duke University Press, pp. 172-196.

TCHRD (Tibetan Center for Human Rights and Democracy). 2007. 'Commotion between Two Communities Lead to Arrest of 30 Tibetans.' July 2007, p. 4.

Terrone, Antonio. 2008. 'Tibetan Buddhism Beyond the Monastery: Revelation and Identity in *rNying ma* Communities of Present-day Kham.' In *Images of Tibet in the 19th and 20th Centuries.* Monica Esposito (ed.), *Études thématiques.* 22:2, Paris, École Française d'Extrême-Orient, pp. 747-779.

Thargyal, Rinzin. 2007. *Nomads of Eastern Tibet. Social Organization and Economy of a Pastoral Estate in the Kingdom of Dege.* Toni Huber (ed.), Leiden: Brill.

Thomas, Frederick W. 1948. *Nam. An Ancient Language of the Sino-Tibetan Borderland. Text with Introduction and Linguistic Studies.* London: Oxford University Press.

Tugs chen skyabs. 1991. 'A skyong dpon mo tshang gi lo rgyus.' ('A History of the Archung Huonmo Tribe.') In *Mgo log rig gnas lo rgyus* (*A Cultural History of Golok*). Vol. I, Xining, pp. 218-220.

Tsai, Kellee. 2002. *Back-Alley Banking. Private Entrepreneurs in China.* Ithaca: Cornell University Press.

Tsing, Anna L. 2015. *The Mushroom at the End of the World: On the Possibility of Life in Capitalist Ruins.* Princeton: Princeton University Press.

van Schendel, Willem. 2005. 'Spaces of Engagement. How Borderlands, Illicit Flows, and Territorial States Interlock.' In *Illicit Flows and Criminal Things. States, Borders, and the Other Side of Globalization.* Itty Abraham and Willem van Schendel (eds.), Bloomington: Indiana University Press, pp. 38-68.

Veblen, Thorstein. 1987[1899]. *The Theory of the Leisure Class.* New York: Penguin Books.

Veldwisch GJA. 2008. 'Authoritarianism, Validity, and Security. Researching Water Distribution in Khorezm, Uzbekistan.' In *Fieldwork in Difficult Environments. Methodology as Boundary Work in Development Research.* Caleb R.L. Wall and Peter P. Molinga (eds.), Berlin: LIT Verlag, pp. 161-181.

Walravens, Hartmut (ed.). 2003. *Joseph Franz Rock. Expedition zum Amnye Machhen in Südwest-China im Jahre 1926 im Spiegel von Tagebüchern und Briefen.* Wiesbaden: Harrassowitz.

Wang, Feiling. 2005. *Organizing through Division and Exclusion. China's* Hukou *System.* Stanford: Stanford University Press.

Wang, Huailin. 2000. *Xanzhao Kangba.* (*Searching for Kham*). Chengdu: Sichuan chubanshe.

Willerslev, Rane. 2007. *Soul Hunters. Hunting, Animism and Personhood among the Siberian Yukaghirs.* Berkeley: University of California Press.

Williams, Dee Mack. 1997. 'The Desert Discourse of Modern China.' *Modern China.* 23:3, pp. 328-355.

Winkler, Daniel. 2005. '*Yartsa Gunbu – Cordyceps sinensis*. Economy, Ecology and Ethno-Mycology of a Fungus Endemic to the Tibetan Plateau.' In *Wildlife and Plants in Traditional and Modern Tibet. Conceptions, Exploitation and Conservation.* Alessandro Boesi and Francesca Cardi (eds.), *Memorie della Società Italiana di Scienze Naturali e del Museo Civico di Storia Naturale di Milano.* 33:1, pp. 69-85.

Winkler, Daniel. 2008a. '*Yartsa Gunbu* (*Cordyceps sinensis*) and the Fungal Commodification of Tibet's Rural Economy.' *Economic Botany.* 62:3, pp. 291-305.

Winkler, Daniel. 2008b. 'The Mushrooming Fungi Market in Tibet Exemplified by *Cordyceps sinensis* and *Tricholoma matsutake*.' *Journal of the International Association of Tibetan Studies.* 4, http://www.thlib.org/collections/texts/jiats/#jiats=/04/winkler/, accessed 27 September 2010.

Winkler, Daniel. 2008c. "It Reaches into Every Aspect of Rural Tibetan Life.' *Yartsa Gunbu*: Chances, Challenges and Prospects. An Interview with Daniel Winkler.' http://www.tibetinfonet.net/content/update/127, accessed 24 April 2014.

Winkler, Daniel. 2008d. '*Yartsa Gunbu*. Tibet's Underground Cash Cow.' http://www.tibetinfonet.net/content/update/98, accessed 24 April 2011.

Winkler, Daniel. 2009. 'Caterpillar Fungus (*Ophiocordyceps sinensis*). Production and Sustainability on the Tibetan Plateau and in the Himalayas.' *Asian Medicine.* 5, pp. 291-316.

Winkler, Daniel. 2010. 'Morels in Eastern Tibet.' *Fungi.* 3:2, pp. 22-24.

Winkler, Daniel. 2013. 'Steps towards Sustainable Harvest of *Yartsa Gunbu* (Caterpillar Fungus, *Ophiocordyceps sinensis*).' In *Proceedings of the 7th International Medicinal Mushroom Conference, Beijing, China, August 26-29, 2013.* Beijing: [s.e.], pp. 635-644, http://mushroaming.com/sites/default/files/blog_images/Winkler%20Caterpillar%20fungus%20sustainability%20IMMC7%202013.pdf, accessed 6 April 2014.

Woodhouse, Emily, Martin Mills, Philip McGowan, and Eleanor Milner-Gulland. 2015. 'Religious Relationships with the Environment in a Tibetan Rural Community: Interactions and Contrasts with Popular Notions of Indigenous Environmentalism.' *Human Ecology.* 43:2, pp. 295-307, https://www.ncbi.nlm.nih.gov/pmc/articles/PMC4422863/, accessed 27 September 2017.

WTNN (World Tibetan Network News). 2007. 'Gun Battle between Rival Tibetan Groups in Kardze Leaves 6 Dead.' http://www.tibet.ca/en/wtnarchive/2007/7/20_6.html, accessed 29 August 2007.

Wu, Ning, Zhaoli Yan, and Tao Lu. 2012. 'Enclosure and Resettlement in the Eastern Tibetan Plateau. Dilemma of Pastoral Development during the Last Three Decades.' In *Pastoral Practices in High Asia. Agency of 'Development' Effected by Modernisation, Resettlement and Transformation.* Hermann Kreutzmann (ed.), Dordrecht: Springer Verlag, pp. 291-306.

Xing, Haining. 1994. *Guoluo zangzu shehui.* (*Golok Tibetan Society*). Beijing: Zhongguo zangxue chubanshe.

Xun, Lili, and Zhiming Bao. 2008. 'Government, Market and Households in the Ecological Relocation Process. A Sociological Analysis of Ecological Relocation in S Banner.' *Social Sciences in China.* 29:1, pp. 113-128.

Yan, Zhaoli, Guangrong Luo, and Ning Wu. 2002. 'A Probe into the Pastoral Production System in Hongyuan, Eastern Qinghai-Tibetan Plateau.' In *Yak Production in Central Asian Highlands. Proceedings of the Third International Congress on Yak Held in Lhasa, PRC, 4-9 September 2000.* Han Jianlin, Camille Richard, Olivier Hanotte, C. McVeigh and J.E.O. Rege (eds.), Nairobi: International Livestock Research Institute, pp. 120-127.

Yang, Darong. 2010. 'The Distribution Pattern of *Cordyceps sinensis* in China and the Response to Environmental Changes.' Paper presented at the 2010 International Conference on *Ophiocordyceps sinensis.* Xining, China, 08-10.06.2010.

Yang, Darong, Yanqiong Peng, Jiyue Chen, Yongqiang Cao, and Pei Yang. 2010. 'Zhongguo dongchong xiacao fenbu geju yu huanjing bianhua dui qi fenbu de yingxiang.' ('The Distribution Pattern of *Cordyceps sinensis* in China and the Response to Environmental Changes.') *Chinese Journal of Grassland.* 32 (Suppl.), pp. 23-27.

Yang, Enhong, 2006. 'Guoluo nüwang buluo koushushi. Nüwang buluo houdai duoji zhuoma fangtanlu.' ('Oral History of the Golok Queen Tribe. An Interview with Duojizhuoma, Descendant of the Golok Queen Tribe.') In Enhong Yang, *Zangzu funü koushushi* (*Oral Histories of Tibetan Women*), Beijing: Zhongguo zangzuxue chubanshe, pp. 127-140.

Yao, Yijian. 2010. 'Integrative Biological Study on *Ophiocordyceps sinensis.*' Paper presented at the 2010 International Conference on *Ophiocordyceps sinensis.* Xining, China, 08-10.06.2010.

Yeh, Emily. 2000. 'Forest Claims, Conflicts and Commodification: The Political Ecology of Tibetan Mushroom-Harvesting Villages in Yunnan Province, China.' *The China Quarterly.* 161, pp. 264-278.

Yeh, Emily T. 2003. 'Tibetan Range Wars. Spatial Politics and Authority on the Grasslands of Amdo.' *Development and Change.* 34:3, pp. 1-25.

Yeh, Emily T. 2005. ''An Open Lhasa Welcomes You'. Disciplining the Researcher in Tibet.' In *Doing Fieldwork in China.* Stig Thøgersen and Maria Heimer (eds.), Honolulu: University of Hawai'i Press, pp. 96-109.

Yeh, Emily T. 2013a. *Taming Tibet. Landscape Transformation and the Gift of Chinese Development.* Ithaca: Cornell University Press.

Yeh, Emily T. 2013b. 'Blazing Pelts and Burning Passions: Nationalism, Cultural Politics, and Spectacular Decommodification in Tibet.' *The Journal of Asian Studies.* 72:2, pp. 319-344.

Yeh, Emily T. and Kunga T. Lama. 2013. 'Following the Caterpillar Fungus. Nature, Commodity Chains and the Place of Tibet in China's Uneven Geographies.' *Social and Cultural Geography.* 14:3, pp. 318-340.

Zelizer, Viviana A. 1997. *The Social Meaning of Money. Pin Money, Paychecks, Poor Relief, and Other Currencies.* Princeton: Princeton University Press.

Zhang, Jiangcai. 2003. 'Guoluo zhou chongcao ziyuan xianzhuang ji guanli.' ('The Situation of Caterpillar Fungus Resource and Managing in Golok Prefecture.') *Qinghai Prataculture.* 12, pp. 33-37.

Zhou, Huakun, Xiaohui Wang, Jun Wen, Jinfu Zhu, Xin Ye, Wenying Wang, and Chen Zhe. 2012. 'Guoluo zhou maqin xian caoyuan maochong chonghai fasheng yu qihou yinzi de xianghu guanxi.' ('The Relationship between Damage of Grassland Caterpillar and Climate Factors in the Machen County of Golok Prefecture.') *Pratacultural Science.* 29:1, pp. 128-134.

Zhou, Mingxiu. 2004. 'Guoluo zhou dongchong xiacao ziyuan fenbu ji qi liyong.' ('Distribution and Utilize of *Cordyceps sinensis* in Golok Prefecture.') *Chinese Qinghai Journal of Animal and Veterinary Science.* 34:3, pp. 37-38.

Zhu, Jiashi, Georges M. Halpern, and Kenneth Jones. 1998. 'The Scientific Rediscovery of an Ancient Chinese Herbal Medicine: *Cordyceps sinensis.* Part 1.' *The Journal of Alternative and Complementary* Medicine. 4:3, pp. 289-303.

Znoj, Heinzpeter. 1998. 'Hot Money and War Debts: Transactional Regimes in Southwest Sumatra.' *Comparative Study of Society and History.* 40:2, pp. 193-222.

Zukosky, Michael L. 2007. 'Making Pastoral Settlement Visible in China.' *Nomadic Peoples.* 11:2, pp. 107-133.

Index

For Product Safety Concerns and Information please contact our EU representative GPSR@taylorandfrancis.com Taylor & Francis Verlag GmbH, Kaufingerstraße 24, 80331 München, Germany

Batch number: 10722100

Printed by Printforce, the Netherlands